T0214308

Lecture Notes in Computer Science 11936

More information about this series at http://www.springer.com/series/7412

Zhen Cui · Jinshan Pan · Shanshan Zhang ·
Liang Xiao · Jian Yang (Eds.)

Intelligence Science and Big Data Engineering

Big Data and Machine Learning

9th International Conference, IScIDE 2019
Nanjing, China, October 17–20, 2019
Proceedings, Part II

 Springer

Editors
Zhen Cui
Nanjing University of Science
and Technology
Nanjing, China

Jinshan Pan
Nanjing University of Science
and Technology
Nanjing, China

Shanshan Zhang
Nanjing University of Science
and Technology
Nanjing, China

Liang Xiao
Nanjing University of Science
and Technology
Nanjing, China

Jian Yang
Nanjing University of Science
and Technology
Nanjing, China

ISSN 0302-9743 ISSN 1611-3349 (electronic)
Lecture Notes in Computer Science
ISBN 978-3-030-36203-4 ISBN 978-3-030-36204-1 (eBook)
https://doi.org/10.1007/978-3-030-36204-1

LNCS Sublibrary: SL6 – Image Processing, Computer Vision, Pattern Recognition, and Graphics

This Springer imprint is published by the registered company Springer Nature Switzerland AG
The registered company address is: Gewerbestrasse 11, 6330 Cham, Switzerland

Preface

The International Conference on Intelligence Science and Big Data Engineering (IScIDE 2019), took place in Nanjing, China, during October 17–20, 2019. As one of the annual events organized by the Chinese Golden Triangle ISIS (Information Science and Intelligence Science) Forum, this meeting was scheduled as the 9th in a series of annual meetings promoting the academic exchange of research on various areas of intelligence science and big data engineering in China and abroad.

We received a total of 225 submissions, each of which was reviewed by at least 3 reviewers. Finally, 84 papers were accepted for presentation at the conference, with an acceptance rate of 37.33%. Among the accepted papers, 14 were selected for oral presentations, 35 for spotlight presentations, and 35 for poster presentations. We would like to thank all the reviewers for spending their precious time on reviewing the papers and for providing valuable comments that helped significantly in the paper selection process. We also included an invited paper in the proceedings entitled "Deep IA-BI and Five Actions in Circling" by Prof. Lei Xu.

We are grateful to the conference general co-chairs, Lei Xu, Xinbo Gao and Jian Yang, for their leadership, advice, and help on crucial matters concerning the conference. We would like to thank all members of the Steering Committee, Program Committee, Organizing Committee, and Publication Committee for their hard work. We give special thanks to Prof. Xu Zhang, Prof. Steve S. Chen, Prof. Lei Xu, Prof. Ming-Hsuan Yang, Prof. Masashi Sugiyama, Prof. Jingyi Yu, Prof. Dong Xu, and Prof. Kun Zhang for delivering the keynote speeches. We would also like to thank Prof. Lei Xu for contributing a high-quality invited paper. Finally, we greatly appreciate all the authors' contributions to the high quality of this conference. We count on your continued support of the ISIS community in the future.

October 2019

Zhen Cui
Jinshan Pan
Shanshan Zhang
Liang Xiao
Jian Yang

Organization

General Chairs

Lei Xu — Shanghai Jiao Tong University, China
Xinbo Gao — Xidian University, China
Jian Yang — Nanjing University of Science and Technology, China

Program Chairs

Huafu Chen — University of Electronic Science and Technology of China, China
Zhouchen Lin — Peking University, China
Kun Zhang — Carnegie Mellon University, USA
Zhen Cui — Nanjing University of Science and Technology, China

Organization Chairs

Liang Xiao — Nanjing University of Science and Technology, China
Chen Gong — Nanjing University of Science and Technology, China

Special Issue Chairs

Mingming Cheng — Nankai University, China
Jinshan Pan — Nanjing University of Science and Technology, China

Publication Chairs

Shanshan Zhang — Nanjing University of Science and Technology, China
Wankou Yang — Southeast University, China

Program Committee

Mingming Gong — The University of Melbourne, Australia
Joseph Ramsey — Carnegie Mellon University, USA
Biwei Huang — Carnegie Mellon University, USA
Daniel Malinsky — Johns Hopkins University, USA
Ruben Sanchez-Romero — Rutgers University, USA
Shohei Shimizu — RIKEN, Japan
Ruichu Cai — Guangdong University of Technology, China
Shuigeng Zhou — Fudan University, China
Changdong Wang — Sun Yat-sen University, China
Tao Lei — Shaanxi University of Science & Technology, China

Xianye Ben	Shandong University, China
Jorma Rissanen	Emeritus of Tampere University of Technology, Finland
Alan L. Yuille	Johns Hopkins University, USA
Andrey S. Krylov	Lomonosov Moscow State University, Russia
Jinbo Xu	Toyota Technological Institute at Chicago, University of Chicago, USA
Nathan Srebro	Toyota Technological Institute at Chicago, University of Chicago, USA
Raquel Urtasun	Uber ATG Toronto, Canada
Hava T. Siegelmann	University of Massachusetts Amherst, USA
Jurgen Schmidhuber	European Academy of Sciences and Arts, Austria
Sayan Mukherjee	Duke University, USA
Vincent Tseng	National Cheng Kung University, Taiwan
Alessandro Giua	University of Cagliari, Italy
Shu-Heng Chen	National Chengchi University, Taiwan
Seungjin Choi	Pohang University of Science and Tech, South Korea
Kenji Fukumizu	The Institute of Statistical Mathematics, Japan
Kalviainen Heikki	Lappeenranta University of Technology, Finland
Akira Hirose	The University of Tokyo, Japan
Tu Bao Ho	JAIST, Japan
Derek Hoiem	University of Illinois at Urbana-Champaign, USA
Ikeda Kazushi	NARA Institute of Science and Technology, Japan
Seiichi Ozawa	Kobe University, Japan
Yishi Wang	UNC Wilmington, USA
Cuixian Chen	UNC Wilmington, USA
Karl Ricanek	UNC Wilmington, USA
Hichem Sahli	Vrije Universiteit Brussel, Belgium
Fiori Simone	Universita Politecnica delle Marche, Italy
Cox Stephen	The Australian National University, Australia
Vincent Tseng	Cheng Kung University, Taiwan
Qiang Yang	Hong Kong University of Science and Technology, Hong Kong, China
Chengjun Liu	New Jersey Institute of Technology, USA
Shuicheng Yan	National University of Singapore, Singapore
Jieping Ye	University of Michigan, USA
Wai-Kiang Yeap	Auckland University of Technology, New Zealand
Hujun Yin	The University of Manchester, UK
Lei Zhang	Hong Kong Polytechnic University, Hong Kong, China
Qinfeng Shi	University of Adelaide, Australia
Wanli Ouyang	The University of Sydney, Australia
Yida Xu	University of Technology, Sydney, Australia
Hongyan Wang	Dalian University of Technology, USA
Yazhou Yao	University of Technology, Sydney, Australia
Xiaoning Song	Jiangnan University, China
Yong Xia	Northwestern Polytechnical University, China

Lei Zhang   Chongqing University, China
Tao Wang   Nanjing University of Science and Technology, China
Changxing Ding   South China University of Technology, China
Xin Liu   Huaqiao University, China
Yang Liu   Dalian University of Technology, USA
Ying Tai   YouTu Lab, Tencent, China
Minqiang Yang   Lanzhou University, China
Guangwei Gao   Nanjing University of Posts and Telecommunications,
                         China
Shuzhe Wu   Chinese Academy of Sciences, China
Youyong Kong   Southeast University, China
Qiguang Miao   Xidian University, China
Chang Xu   The University of Sydney, Australia
Tengfei Song   Southeast University, China
Xingpeng Jiang   Central China Normal University, China
Wei-Shi Zheng   Sun Yat-sen University, China
Yu Chen   Motovis Inc., Australia
Zebin Wu   Nanjing University of Science and Technology, China
Wei Luo   South China Agricultural University, China
Minxian Li   Nanjing University of Science and Technology, China
Ruiping Wang   Chinese Academy of Sciences, China
Jia Liu   Xidian University, China
Yang He   Max Planck Institute for Informatics, Germany
Xiaobo Chen   Jiangsu University, China
Xiangbo Shu   Nanjing University of Science and Technology, China
Yun Gu   Shanghai Jiao Tong University, China
Xin Geng   Southeast University, China
Zheng Wang   National Institute of Informatics, Japan
Lefei Zhang   Wuhan University, China
Liping Xie   Southeast University, China
Xiangyuan Lan   Hong Kong Baptist University, Hong Kong, China
Xi Peng   Agency for Science, Technology and Research
                         (A*STAR), Singapore
Yuxin Peng   Peking University, China
Cheng Deng   Xidian University, China
Dong Gong   The University of Adelaide, Australia
Meina Kan   Chinese Academy of Sciences, China
Hualong Yu   Jiangsu University of Science and Technology, China
Kazushi Ikeda   NARA Institute of Science and Technology, Japan
Meng Yang   Sun Yat-Sen University, China
Ping Du   Shandong Normal University, China
Jufeng Yang   Nankai University, China
Andrey Krylov   Lomonosov Moscow State University, Russia
Shun Zhang   Northwestern Polytechnical University, China
Di Huang   Beihang University, China
Shuaiqi Liu   Tianjin Normal University, China

Chun-Guang Li	Beijing University of Posts and Telecommunications, China
Huimin Ma	Tsinghua University, China
Longyu Jiang	Southeast University, China
Shikui Tu	Shanghai Jiao Tong University, China
Lijun Wang	Dalian University of Technology, USA
Xiao-Yuan Jing	Wuhan University, China
Shiliang Sun	East China Normal University, China
Zhenzhen Hu	HeFei University of Technology, China
Ningzhong Liu	NUAA, China
Hiroyuki Iida	JAIST, Japan
Jinxia Zhang	Southeast University, China
Ying Fu	Beijing Institute of Technology, China
Tongliang Liu	The University of Sydney, Australia
Weihong Deng	Beijing University of Posts and Telecommunications, China
Wen Zhang	Wuhan University, China
Dong Wang	Dalian University of Technology, USA
Hang Dong	Xi'an Jiaotong University, China
Dongwei Ren	Tianjin University, China
Xiaohe Wu	Harbin Institute of Technology, China
Qianru Sun	National University of Singapore, Singapore
Yunchao Wei	University of Illinois at Urbana-Champaign, USA
Wenqi Ren	Chinese Academy of Sciences, China
Wenda Zhao	Dalian University of Technology, USA
Jiwen Lu	Tsinghua University, China
Yukai Shi	Sun Yat-sen University, China
Enmei Tu	Shanghai Jiao Tong University, China
Yufeng Li	Nanjing University, China
Qilong Wang	Tianjin University, China
Baoyao Yang	Hong Kong Baptist University, Hong Kong, China
Qiuhong Ke	Max Planck Institute for Informatics, Germany
Guanyu Yang	Southeast University, China
Jiale Cao	Tianjin University, China
Zhuo Su	Sun Yat-sen University, China
Zhao Zhang	HeFei University of Technology, China
Hong Pan	Southeast University, China
Hu Han	Chinese Academy of Sciences, China
Hanjiang Lai	Sun Yat-Sen University, China
Xin Li	Harbin Institute of Technology, Shenzhen, China
Dingwen Zhang	Northwestern Polytechnical University, China
Guo-Sen Xie	Inception Institute of Artificial Intelligence, UAE
Xibei Yang	Jiangsu University of Science and Technology, China
Wang Haixian	Southeast University, China
Wangmeng Zuo	Harbin Institute of Technology, China
Weiwei Liu	University of Technology, Sydney, Australia

Shuhang Gu	ETH Zurich, Switzerland
Hanli Wang	Tongji University, China
Zequn Jie	Tencent, China
Xiaobin Zhu	University of Science and Technology Beijing, China
Gou Jin	Huaqiao University, China
Junchi Yan	Shanghai Jiao Tong University, China
Bineng Zhong	Huaqiao University, China
Nannan Wang	Xidian University, China
Bo Han	RIKEN, Japan
Xiaopeng Hong	Xi'an Jiaotong University, China
Yuchao Dai	Northwestern Polytechnical University, China
Wenming Zheng	Southeast University, China
Lixin Duan	University of Science and Technology of China, China
Hu Zhu	Nanjing University of Posts and Telecommunications, China
Xiaojun Chang	Carnegie Mellon University, USA

Contents – Part II

Contents – Part I

Analysis of WLAN's Receiving Signal Strength Indication for Indoor Positioning

Minmin Lin[1,2], Zhisen Wei[1,2], Baoxing Chen[1,2], Wenjie Zhang[1,2],
and Jingmin Yang[1,2(✉)]

[1] School of Computer Science, Minnan Normal University, Zhangzhou 363000, China
yjm4201@163.com
[2] Key Laboratory of Data Science and Intelligence Application,
Fujian Province University, Zhangzhou 363000, Fujian, China

Abstract. The location method based on Received Signal Strength Indication (RSSI) ranging has the advantages of low development cost and simple implementation mechanism, and is the mainstream indoor location method nowadays. However, at present, there is a lack of systematic research on the characteristics of Access Point (AP) received signal strength, which cannot meet the signal strength characteristics required by specific indoor positioning environment. Detailedly, in this study, the indoor distribution characteristics of 2.4 GHz RSSI were analyzed experimentally from three influencing factors including antenna orientation of the receiver, type of wireless network card of the receiver and height difference of the transmitting and receiving antenna. The experimental results show that: (1) The RSSI value measured is the strongest When the antenna of the receiver is vertically oriented to the antenna of the transmitter, while the antenna is vertically backward to the transmitter, it is the opposite. The difference between the strongest signal and the weakest signal is 20%–25% at the same test point. (2) The network card with a large measurement range should be selected for data collection on the premise that the quantization step size is 1 and the smoothness is small. Avoiding using the network card for positioning with the maximum limit of measurement range is proposed also. (3) The path loss index of the ranging model is affected by the height difference of the antenna, resulting in a deviation of 0.1–0.2.

Keywords: Characteristic · Data analysis · Indoor positioning · RSSI ranging

---

This work was supported by the Natural Science Funds of China under Grant 61701213, the Cooperative Education Project of Ministry of Education under Grant 201702098015 and Grant 201702057020, the Special Research Fund for Higher Education of Fujian under Grant JK2017031, Zhangzhou Municipal Natural Science Foundation under Grant ZZ2018J21.

Z. Cui et al. (Eds.): IScIDE 2019, LNCS 11936, pp. 1–14, 2019.
https://doi.org/10.1007/978-3-030-36204-1_1

1   Introduction

The positioning method based on RSSI ranging has the characteristics of low development cost and easy realization. It is the mainstream indoor positioning method nowadays [1]. Compared with the existing technologies based on Time of Arrival (TOA) [2], Time Difference of Arrival (TDOA) [3], Angle of Arrival (AOA) [4] and Channel State Information (CSI) [5], RSSI-based location methods are based on wireless network interface network card [6]. At the receiving terminal, there is no need for dedicated hardware, and the existing widely deployed WLAN infrastructure can be directly used to locate the location. Compared with the location technology based on location fingerprint, this method does not need to consume a lot of manpower and material resources to construct the location fingerprint database in the offline stage.

Because Wi-Fi signal is easily disturbed by environment or human factors when it propagates indoors, it produces multi-path effects such as reflection, diffraction, scattering and refraction, and the accuracy of ranging is easily affected [7]. RSSI mean, standard deviation, range, spatial variation and dependence on hardware characteristics of WLAN card and antenna orientation of receiver are important for understanding and modeling RSSI ranging model for indoor positioning. However, in the existing research results of indoor positioning based on RSSI ranging, there are some deficiencies in the analysis of WLAN received signal strength based on positioning angle. This paper chooses different RSSI signal influencing factors, through a large number of statistical data, to measure and analyze the distribution characteristics of RSSI in a specific environment more comprehensively and completely, laying the foundation for the next positioning in this environment.

The positioning method based on RSSI ranging is based on the characteristic of the location correlation of RSSI, i.e. the signal strength changes with the propagation distance, to achieve positioning, which is theoretically analyzed in the third section. It illustrates the feasibility of using RSSI-based ranging technology in indoor three-dimensional space positioning system and the instability of RSSI signal, which firmly shows that our research is meaningful.

The organizational structure of this paper is as follows. Firstly, the related work investigating the characteristics of RSSI for indoor positioning systems are briefly reviewed in Sect. 2. Next, Sect. 3 theoretically analyses the position correlation characteristics of RSSI and the data preprocessing methods. In Sect. 4, the measurement devices and experimental environment settings are described. Then, Sect. 5 discusses three influencing factors including antenna orientation of the receiver, type of wireless network card of the receiver and height difference of the transmitting and receiving antenna. Finally, the full text is summarized in Sect. 6.

2   Related Works

The existing literature on indoor positioning system mainly focuses on the improvement of positioning accuracy performance and location estimation algorithm, and only a few studies on RSSI random vectors. Zhang Heng et al. analyzed the relationship between the angle of the transmitting antenna and the horizontal space and the path loss, as well as the relationship between the height of the transmitting antenna and the path loss for 2.4 GHz wireless signal in indoor corridor environment [8]. However, the research was based on the same horizontal height of antenna at the transmitter and receiver, and did not consider the influence of height difference between them on signal. Hu Guoqiang studied the correlation between signal propagation characteristics and factors such as communication distance, AP height, propagation path, etc. [9]. However, this study focused on 5.8 GHz radio frequency signal rather than 2.4 GHz. Literature [10] studies the effects of user presence and user orientation on RSSI. It is emphasized that the influence of human body should be included in RSSI collection in human-related indoor positioning. The authors in [11] made an observation that user's orientation could lead to a variation in RSSI level, and then suggested that orientation should be included in computing the users location. However, these studies do not propose the criterion of how to select the optimal user orientation for positioning purposes and does not consider the influence of antenna orientation of receiver on RSSI without human body effect. The authors in document [12] described that hardware variance can significantly degrade the positional accuracy of RSSI-based Wi-Fi localization systems. However, no analysis of the RSSI data was provided. Meanwhile, in reference [13], the influence of hardware factors on RSSI are studied and analyzed. It is considered that the best WLAN card should have the two characteristics of the widest range or the lowest fluctuation of RSSI in order to distinguish the location. However, the research in [13] is based on the indoor positioning of position fingerprint technology, not on the analysis of relevant factors of RSSI ranging technology. The research in document [14] shows that the height difference of the antenna will bring significant changes to the signal intensity received by the user terminal. However, this study aims at the outdoor to indoor integrated environment, rather than the indoor environment to measure and analyze the influencing factors. In reference [15], aiming at the AP deployment height of airport terminal is the key factor affecting positioning, a three-dimensional positioning technology based on propagation model is proposed. However, no analysis of RSSI data is provided.

To summarize, there is a lack of systematic research on the distribution characteristics of AP received signal strength, which can not meet the needs of specific indoor positioning environment. Based on the review of relevant work, this paper makes a more comprehensive and complete data measurement and analysis of the characteristics of 2.4 GHz radio frequency signal in view of its shortcomings.

3 Theoretical Analysis

The eigenvalue of RSSI signal can describe the approximation degree of the environment of two locations to a certain extent [16]. For a specific combination of AP and receiving terminal, assuming that $p(d)$ and $p(d_0)$ respectively represent the received signal strength at any distance d and reference distance d_0 from AP, there are:

$$p(d) = p(d_0) - 10nlg(\frac{d}{d_0}) + X \tag{1}$$

$$= 10lg(\frac{P_{AP}G_{AP}G_M\lambda^2}{16\pi^2 d_0^2 f}) - 10nlog(\frac{d}{d_0}) + X \tag{2}$$

Among them, P_{AP} is the transmitting power of AP, G_{AP} is the antenna gain of AP, G_M is the antenna gain of receiving terminal, f is the system loss factor, λ is the wavelength of wireless signal, n is the path loss factor, X is a normal random variable, $X \frown (0, \sigma^2)$.

When the signal intensity changes with time or space, small-scale fading can describe the random variance of its change. When neglecting small-scale fading, i.e. $\sigma = 0$, the signal propagation model describes the deterministic relationship between RSSI and distance. At present, all kinds of indoor propagation models are based on logarithmic model, so no matter which model is used, RSSI and distance are logarithmic. It can be seen that at different locations, the difference of each RSSI reflects different distances. The position correlation characteristics of RSSI are correlated with the position, and these characteristics are used to infer the position, which is the characteristic of the propagation model that can be used by the positioning system.

In practice, the receiver can receive n RSSI values in a certain period of time. Because of the influence of objective environment, these RSSI values have abnormal phenomena, so it is necessary to remove the abnormal values from the collected data to get a smoother and accurate value before practical application.

Most studies show that RSSI random variables obey or approximately obey normal distribution in most cases, so this paper uses the 3σ criterion to discriminate gross errors and eliminate outliers.

3σ criterion is also called Laida criterion. Specifically, it assumes that a set of detection data contains only random errors, calculates and processes the original data to obtain standard deviation, and then determines an interval according to a certain probability, and considers that the error exceeding this interval belongs to abnormal value.

Assuming that RSSI data in the environment obey the Gauss distribution with mean u and variance σ^2, the probability density function of RSSI data is:

$$f(RSSI) = \frac{1}{\sigma\sqrt{2\pi}}e^{\frac{-(RSSI-u)^2}{2\sigma^2}} \tag{3}$$

In the formula, $u = \frac{1}{n}\sum_{i=1}^{n} RSSI(i)$, $\sigma = \sqrt{\frac{1}{n-1}\sum_{i=1}^{n}(RSSI(i) - u)^2}$.

In our study, we select the range of 2σ to eliminate the outliers of the data. The final RSSI value can be obtained by taking out all RSSI values in $[-2\sigma + u, 2\sigma + u]$ range and calculating the average value.

4 Measurement Devices and Experimental Environment Settings

In Sect. 5, we use WirelessMon Version 4.0 Signal Collection Software to measure the Received Signal Intensity. Each study was measured by sampling RSSI data at a rate of one sample every 8 seconds. We study the characteristics of a single RSSI set based on 2.4 GHz radio frequency. In order to reduce co-frequency and adjacent-frequency interference, the wireless channel used by AP in the experiment is Channel 3, which is not used by other AP in our test environment. The AP model used in this study is TL-WR742N. Its omni-directional antenna is vertical to the ground. Its transmitting power is 100 mW. It adopts 802.11n standard and its maximum transmission rate is 150 Mbps.

The experiments were set in the corridor of the second floor of the LiZhi building and the 221 wireless network training room of Minnan Normal University. As shown in Fig. 1, the triangle represents the known node AP. Set the AP point as origin coordinate (0,0). In the corridor environment, set a test point every 1 m from left to right along the middle, and set a total of six test points, respectively, L1–L6. Meanwhile, in the wireless network training room, AP was taken as the origin, and a test point was set every 0.6 meters from left to right along a straight line, with a total of 10 test points (the circle in the figure represents the test points), which were M1–M10 respectively.

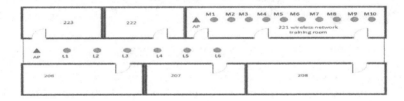

Fig. 1. The schematic diagram of experimental site and point distribution.

5 Characteristic Analysis of Received Signal Strength

5.1 Effect of Antenna Orientation of the Receiver on RSSI

In indoor positioning system based on Wireless Local Area Network (WLAN), the receiving terminals equipped with wireless network card are needed to receive the wireless signal. In the absence of human blocking, when the receiving antenna is in different directions, the transmitted signal received has a different path, resulting in a change in the distance between the transceiver antenna, so that the received signal strength is also different, which can affect the accuracy of the ranging. In order to obtain the best orientation of the terminal antenna, a comparative experiment is carried out.

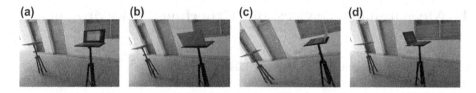

Fig. 2. (a)Test site picture in corridor environment when the receiver's antenna is oriented to the east. (b)Test site picture in corridor environment when the receiver's antenna is oriented to the west. (c)Test site picture in corridor environment when the receiver's antenna is oriented to the south. (d)Test site picture in corridor environment when the receiver's antenna is oriented to the north.

In order to study the effect of antenna orientation on RSSI, it is necessary to measure and analyze the change of received signal intensity caused by the change of the antenna's orientation. In order to be general, this section uses two test environments in Fig. 1 to collect signals. The test scenarios are shown in Figs. 2 and 3.

Fig. 3. (a)Test site picture in wireless network training room when the receiver's antenna is oriented to the east. (b)Test site picture in wireless network training room when the receiver's antenna is oriented to the west. (c)Test site picture in wireless network training room when the receiver's antenna is oriented to the south. (d)Test site picture in wireless network training room when the receiver's antenna is oriented to the north.

The measurement is carried out in four directions (facing the north, west, south and east of the building). Under the condition of visual distance, the transmitter and receiver are fixed at a height of 1.1 m (which is higher than the height of the desktop in the training room, thus avoiding the influence of obstacles such as tables on signal transmission). Maintain the fixed position of the transmitter, let the transmitter be located in the eastern part of the building in the corridor environment, and let the transmitter be located in the northern part of the building in the wireless network training room environment. In the two test environments of Fig. 1, we collect four sets of data at each test point, which are the signal strength from the same AP received when the antenna of the receiver faces four different directions, east, west, north and south. In this group of experiments, 100 RSSI data were collected from AP node in each direction. During the measurements, no people moved and no users occluded, and the door of the training room remained closed until the data acquisition of all test points was completed.

Table 1. RSSI (dBm) received by test point L1–L6 in corridor environment when the antenna of receiver is in different directions.

Category	Direction	L1	L2	L3	L4	L5	L6
Distance/m		1	2	3	4	5	6
Mean	North	−43.63	−42.90	−43.33	−47.27	−52.27	−55.32
	South	−41.91	−47.04	−47.81	−44.24	−49.83	−55.09
	West	−49.81	−53.64	−52.78	−54.96	−57.37	−57.12
	East	−35.29	−40.04	−42.94	−39.65	−51.63	−52.72
Standard deviation	North	2.47	1.42	1.22	1.16	1.83	1.31
	South	2.26	1.00	0.96	0.83	1.06	1.12
	West	0.82	1.27	1.06	1.44	1.70	1.19
	East	2.26	0.85	1.53	1.29	1.06	1.16
Range	North	11	7	5	5	7	9
	South	5	3	5	5	9	7
	West	5	7	7	9	9	7
	East	9	5	7	5	5	7

Statistical results of the mean, standard deviation and range of RSSI after eliminating outliers by 3σ criterion in two scenarios are shown in Tables 1 and 2. The calculation method of RSSI range is as follows:

$$RSSI_{range} = RSSI_{max} - RSSI_{min} + 1 \tag{4}$$

In Formula (4), $RSSI_{max}$ and $RSSI_{min}$ represent the maximum and minimum values in the RSSI data sets, respectively.

From Tables 1 and 2, it can be seen that in the case of no human body occlusion, at a fixed test point, the received signal intensity varies greatly when the antenna of the receiver is in different directions. Although the values of standard deviation and range produced in each direction differ slightly, the mean values of RSSI are quite different. For example, at the second test point in Table 1, when the receiving antenna faces eastward, the RSSI sample average value is −40.04 dBm, while it is −53.64 dBm when facing westward, and RSSI is attenuated by 13.60 dB. Generally, most of the research work of indoor positioning system is to calculate the average value of RSSI and apply it to ranging calculation, that is, different mean values will produce different ranging results. This shows that the direction of the receiving antenna is very important and should be included in the factors considered in indoor positioning. At the same time, it can be observed that the mean of RSSI collected at each test point is generally the strongest when the receiver's antenna is facing AP. This is because when the terminal's antenna

Table 2. RSSI (dBm) received by test point M1–M10 when the antenna of receiver is in different directions in wireless network training room environment.

Category	Direction	M1	M2	M3	M4	M5	M6	M7	M8	M9	M10
Distance/m		0.6	1.2	1.8	2.4	3.0	3.6	4.2	4.8	5.4	6.0
Mean	North	−28.56	−34.70	−40.24	−42.00	−43.24	−40.15	−39.07	−44.00	−42.00	−42.65
	South	−39.98	−41.48	−51.47	−52.59	−52.90	−52.42	−49.37	−50.67	−50.00	−50.00
	West	−29.55	−38.55	−43.80	−44.00	−45.50	−46.94	−50.89	−46.00	−48.61	−44.92
	East	−35.27	−43.41	−45.03	−43.54	−44.75	−48.53	−42.59	−44.96	−47.46	−42.64
Standard deviation	North	0.90	0.96	1.18	0	0.97	0.53	1.00	0	0	0.94
	South	1.17	0.88	0.89	0.92	1.00	0.82	0.93	0.95	0	0
	West	0.84	0.90	2.12	0	0.87	1.00	1.00	0	0.92	1.00
	East	0.97	0.92	1.00	0.85	0.97	0.89	0.92	1.00	0.89	0.94
Range	North	3	3	5	1	3	3	3	1	1	3
	South	5	3	3	3	3	3	3	3	1	1
	West	3	3	9	1	3	3	3	1	3	3
	East	3	3	3	3	3	3	3	3	3	3

is facing AP directly, the distance between AP and antenna is the shortest, the effect on signal transmission is small, the multi-path utility is relatively weak, and the signal attenuation is small. On the contrary, we find that RSSI is the weakest when the antenna of the receiver is back to AP, because the distance between the receiver and the transmitter is the longest, and the signal is easily affected and attenuated.

The conclusion of this section is that the orientation of the receiver antenna has a certain effect on RSSI. When the antenna of the receiver is vertically oriented to the antenna of the transmitter, the RSSI value measured is the strongest, while when the antenna is vertically backward to the transmitter, it is the opposite. In order to ensure the accuracy of positioning, the consistency of antenna direction should be maintained in the process of data collection.

5.2   Effect of Type of Wireless Network Card of the Receiver on RSSI

Because the actual RSSI value mapping may vary from hardware vendor to hardware, different types of network cards cause hardware-related parameters to change, such as antenna gain, etc., the choice of WLAN card will affect the performance of the indoor positioning system.

In order to study the influences of different types of network cards on RSSI, we collect the data of three types of wireless network cards at the third test point (M3) in the wireless network training room environment of Fig. 1, namely, D-Link DWL-G122 Wireless G USB Adapter (rev.F1), 802.11n USB Wireless LAN Card and D-Link DWA-160 Xtreme N Dual Band USB Adapter (rev.B2). For convenience, in the following, they are called network card 1, network card 2 and network card 3 respectively. During the measurement period, there is no movement and user blocking. The whole environment is in a relatively stable state. The height of the tested equipment is also maintained at 1.1 m, and the antennas of the receiver are oriented eastward. In this group of experiments, each network card collected 200 RSSI data for node AP. The results of RSSI statistics under three network cards are compared as shown in Table 3. In order to reflect the mapping range of RSSI values for different types of network cards more truly, we do not process the data in this experiment.

Table 3. Comparison of RSSI (dBm) means and standard deviations collected by different types of wireless network cards in wireless network training room.

Types of Network Cards	Mean of RSSI	Standard deviation of RSSI
Network Card 1	−51.77	10.34
Network Card 2	−20.60	4.15
Network Card 3	−44.87	1.01

As can be seen from Table 3, the signal intensity received by different types of network cards at a fixed test point varies greatly. For example, the RSSI sample average of network card 1 is lower than the maximum RSSI of −20.60 dBm of the network card 2, which is −51.77 dBm, and RSSI attenuation is 31.17 dB. This indicates that the type of network card is very important and should be included in the factors considered in indoor positioning. At the same time, we find that the standard deviations of data collected by different types of wireless network cards are also quite different. For communication purposes, WLAN cards with high average received signal level at a fixed location are considered to be better cards. However, a high level of received signal strength is not necessarily important for location. We consider that in indoor positioning based on RSSI ranging model, the most important attributes affecting positioning are the quantization step of RSSI value (i.e. the difference between the two nearest values in a set of RSSI data), measurement range and smoothness.

Firstly, different types of wireless network cards convert real signal energy into RSSI value through different conversion mechanisms and provide it to test software. Large quantization step can easily cause real signal energy in a certain range to map to the same RSSI value, thus reducing the discrimination of signal energy. Large measurement range and small quantization step mean that the conversion of real signal energy is more meticulous, so that the measured data can better reflect the real signal strength.

To observe the RSSI range and quantization step size of the three types of network cards measured, we show the RSSI distribution in Fig. 4. From the observation of RSSI distribution of network card 3 in Fig. 4, it is shown that the quantization step of network card 3 is 2 and the RSSI measurement range is the shortest. This shows that different real signals can easily be mapped to the same RSSI value, that is, some RSSI values in dBm will never be reported, which makes the measured RSSI value not reflect the real signal well. On the other hand, this also explains the reason why the standard deviation of network card 3 shown in Table 3 is the smallest. Similarly, by observing the RSSI distribution of Network

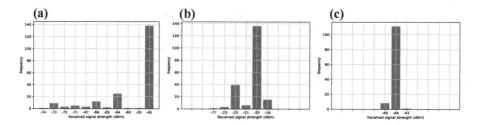

Fig. 4. (a)RSSI distribution of Network Card 1 in wireless network training room. (b)RSSI distribution of Network Card 2 in wireless network training room. (c)RSSI distribution of Network Card 3 in wireless network training room.

Card 1 in Fig. 4, we find that although the quantization step of Network Card 1 is equal to 1 as that of Network Card 2 and the RSSI measurement range is the widest. However, the data distribution of network card 1 presents the phenomenon that the maximum value is the mode number. Therefore, we guess that the measurement of Network Card 1 has the maximum upper limit for data, which is -45 dBm. That is to say, all the real signal values measured by Network Card 1 above -45 dBm will be mapped to -45 dBm.

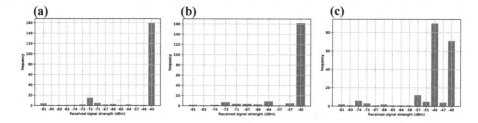

Fig. 5. (a)RSSI distribution of Network Card 1 at the first test point in corridor environment. (b)RSSI distribution of Network Card 1 at the second test point in corridor environment. (c)RSSI distribution of Network Card 1 at the third test point in corridor environment.

To prove this conjecture, we collected data at the first, second and third test points in the corridor environment using network card 1 and made the antenna of the receiver face east. Figure 5 shows the bar chart of RSSI frequency distribution at these three test points.

Based on the observation, we find that the maximum value of RSSI is -45 dBm at these three test points, although the distribution of RSSI is quite different, which verifies our conjecture. Obviously, the network card with this characteristic greatly reduces the accuracy of the measured data and is not suitable for indoor positioning.

Secondly, although signal fluctuations usually occur in any wireless communication due to environmental changes, we believe that the better the quality of

RSSI, which is useful for improving spatial resolution. The main reason is that it is unlikely to display the same RSSI values in different locations. In order to quantify the impact of different types of network cards on positioning performance, we use the curve smoothness index represented by S as a measure. According to [17], the curve smoothness index is defined as:

$$S = \sum_{i=2}^{N-1} \sqrt{(RSSI_i - \frac{RSSI_{i-1} + RSSI_i + RSSI_{i+1}}{3})^2} \tag{5}$$

The smaller the S value, the smoother the signal curve and the better the quality of RSSI. Among them, N is the number of sampling points (i.e. the number of location points) on the curve, and here $RSSI_i$ is the average value of the $RSSI$ values collected at the ith location point. Using the RSSI data collected by Scene 2 Wireless Network Training Room, 10 locations ($N = 10$) were selected along the straight line.

The smoothness index of the original RSSI observations of these three types of network cards are 2.6100, 1.7667 and 2.4150 respectively. As can be seen that different types of wireless network cards produce different RSSI smoothness indices under the same conditions, and the smoothness index of network card 2 is the smallest, so in summary, network card 2 is most suitable for indoor positioning among the three kinds of network cards.

The conclusion of this section is that, for positioning purposes, the network card with large measurement range should be selected for data collection on the premise that the quantization step is 1 and the smoothness is small, and the network card for positioning with the maximum limit of measurement range should be avoided.

5.3 Effect of Height Difference of the Transmitting and Receiving Antenna on RSSI

At present, the RSSI ranging models commonly used for positioning only involve statistical parameters such as frequency, antenna horizontal distance, walls and floor obstructions. They do not describe the height information of user's terminals in real positioning scenarios, resulting in the failure to reflect the signal changes when the receiving terminals is at different heights.

In order to verify the influence of antenna's height difference on received signal strength, at the third sample point of the wireless network training room in Fig. 1, this experiment simulated three different antenna height differences generated by the user was in three different states in the real positioning scene, and collected the data. They were the height at which the user put the terminal into his pocket while sitting, the height at which he put the terminal into his pocket while standing and the height at which he held the terminal while standing. The corresponding heights are 0.4 m, 0.9 m and 1.5 m, respectively. The transmitter AP's height is modulated to 2 m and the location of AP remains unchanged. At each height, we used the same network card and set the receiving terminal in the same orientation to collect RSSI samples. In this group of experiments, 250 RSSI data were collected from AP node at each height. The mean values of RSSI at three different heights are shown in Table 4. Meanwhile, we use MatLab software to conduct regression analysis on the values of n by using the measured RSSI values and Formula 1. The fitting values are also shown in Table 4.

Table 4. Fitting value of path loss index n and RSSI mean value at different heights of receiver

Receiver's heights (m)	Sample mean of RSSI (dBm)	n
0.4	−48.83	2.2356
0.9	−42.30	2.1001
1.5	−47.52	2.3002

As can be seen from Table 4, at a fixed test point, the received signal strength varies greatly when the receiving terminal is at different heights. For example, the average value of RSSI samples at 0.4 m height is lower than the maximum RSSI of −42.30 dBm at 0.9 m height, which is −48.83 dBm, and RSSI attenuates more than 6.53 dB. This shows that the antenna's height differences is very important and should be included in the factors considered in indoor positioning. At the same time, Table 4 shows that when AP is at different heights, the value of n has a certain deviation. The conclusion of this section is that the path loss index of the ranging model is affected by the height difference of the antenna, resulting in a deviation of 0.1–0.2, which makes the RSSI ranging model used for positioning different.

6   Conclusion

The characteristics of RSSI signal are analyzed from three aspects: antenna orientation of the receiver, type of wireless network card of the receiver and height difference of the transmitting and receiving antenna. We point out that the antenna orientation of the receiver has a great influence on the RSSI value without human influence. In order to ensure the accuracy of positioning, the consistency of antenna direction at the receiving terminal should be maintained during data collection. We also analyze the influence of network card type on RSSI, and consider that the quantization step, measurement range and smoothness of RSSI value are the most important attributes affecting positioning. A possible way to improve positioning accuracy is to select a network card with large measurement range to collect data on the premise that the quantization step is 1 and the smoothness is small. Finally, we analyze the influence of antenna height difference on the received signal strength, and use linear regression to fit the data. The results show that the path loss index of the ranging model is biased by the influence of antenna height difference, which leads to the difference of RSSI ranging model used for positioning.

The results of this paper can provide a deep understanding of the mechanism behind the indoor positioning system based on RSSI ranging model. The future work is to build a more accurate RSSI ranging model for indoor positioning based on the results of statistical analysis and the viewpoints put forward in this paper.

References

1. Hossain, A.K.M.M., Soh, W.S.: A survey of calibration-free indoor positioning systems. Comput. Commun. **66**, 1–13 (2015)
2. Al-Jazzar, S.O., Caffery, J., You, H.R.: A scattering model based approach to NLOS mitigation in TOA location systems. In: IEEE Vehicular Technology Conference (2002)
3. Keunecke, K., Scholl, G.: IEEE 802.11 n-based TDOA performance evaluation in an indoor multipath environment. In: European Conference on Antennas & Propagation. IEEE (2014)
4. Xie, Y., Wang, Y., Zhu, P., You, X.: Grid-search-based hybrid TOA/AOA location techniques for NLOS environments. IEEE Commun. Lett. **13**(4), 254–256 (2009)
5. Halperin, D., Hu, W., Sheth, A., Wetherall, D.: Predictable 802.11 packet delivery from wireless channel measurements. Acm Sigcomm Comput. Commun. Rev. **40**(4), 159–170 (2010)
6. Chen, Y.: Research on Wi-Fi indoor positioning technology based on rssi ranging. Master, Southwest University of Science and Technology (2015) (in Chinese)
7. Zeng, C., Liu, H.L., Xu, K., Han, W.J., Zhang, H.T.: RSSI characteristic analysis of indoor three-dimensional space positioning system. Comput. Eng. Appl. **50**(11), 70–74 (2014). (in Chinese)
8. Zhang, H., Wang, X., Qu, X.D., Yang, J.: Analysis of propagation characteristics of wireless signals in indoor corridors. Comput. Eng. **44**(10), 154–159+174 (2018). (in Chinese)
9. Hu, G.Q.: Study on propagation characteristics of indoor 5.8 GHz wireless signal. Comput. Digit. Eng. **45**(8), 1611–1614 (2017). (in Chinese)
10. Kaemarungsi, K., Krishnamurthy, P.: Properties of indoor received signal strength for WLAN location fingerprinting. In: International Conference on Mobile & Ubiquitous Systems: Networking & Services, pp. 14–23. IEEE (2004)
11. Bahl, P., Padmanabhan, V.N.: RADAR: An in-building RF-based user location and tracking system. In: Nineteenth Annual Joint Conference of the IEEE Computer and Communications Societies, INFOCOM 2000, pp. 775C784. IEEE (2000)
12. Tsui, A.W., Chuang, Y.H., Chu, H.H.: Unsupervised learning for solving RSS hardware variance problem in wifi localization. Mobile Netw. Appl. **14**(5), 677–691 (2009)
13. Kaemarungsi, K., Krishnamurthy, P.: Analysis of WLANs received signal strength indication for indoor location fingerprinting. Pervasive Mobile Comput. **8**(2), 292–316 (2012)
14. Li, C.: Measurement and modeling of the influence of transceiver antenna height on wireless channel propagation characteristics. Master, Beijing University of Posts and Telecommunications (2015) (in Chinese)
15. Wang, Z.M.: Research on indoor positioning technology based on WLAN in airport waiting hall. Master, University of Electronic Science and Technology of China (2014) (in Chinese)

16. Chen, Y.L., Zhu, H.S., Sun, L.M.: A visual distance fingerprint location algorithm resistant to multipath and shadows. Comput. Res. Dev. **50**(03), 524–531 (2013). (in Chinese)
17. Borel, C.C.: Surface emissivity and temperature retrieval for a hyperspectral sensor. In: IEEE International Geoscience & Remote Sensing Symposium. IEEE (1998)

Computational Decomposition of Style for Controllable and Enhanced Style Transfer

Minchao Li, Shikui Tu$^{(\boxtimes)}$, and Lei Xu$^{(\boxtimes)}$

Department of Computer Science and Engineering,
Center for Cognitive Machines and Computational Health (CMaCH),
Shanghai Jiao Tong University, Shanghai, China
{marshal_lmc,tushikui,leixu}@sjtu.edu.cn

Abstract. Neural style transfer has been demonstrated to be powerful in creating artistic images with help of Convolutional Neural Networks (CNN), but continuously controllable transfer is still a challenging task. This paper provides a computational decomposition of the style into basic factors, which aim to be factorized, interpretable representations of the artistic styles. We propose to decompose the style by not only spectrum based methods including Fast Fourier Transform and Discrete Cosine Transform, but also latent variable models such as Principal Component Analysis, Independent Component Analysis, and so on. Such decomposition induces various ways of controlling the style factors to generate enhanced, diversified styled images. We mix or intervene the style basis from more than one styles so that compound style or new style could be generated to produce styled images. To implement our method, we derive a simple, effective computational module, which can be embedded into state-of-the-art style transfer algorithms. Experiments demonstrate the effectiveness of our method on not only painting style transfer but also other possible applications such as picture-to-sketch problems.

Keywords: Style transfer · Representation learning · Deep neural network

1 Introduction

Painting art has attracted people for many years and is one of the most popular art forms for creative expression of the conceptual intention of the practitioner. Since 1990's, researches have been made by computer scientists on the artistic work, in order to understand art from the perspective of computer or to turn a camera photo into an artistic image automatically. One early attempt is Non-photorealistic rendering (NPR) [17], an area of computer graphics, which focuses on enabling artistic styles such as oil painting and drawing for digital images. However, NPR is limited to images with simple profiles and is hard to generalize to produce styled images for arbitrary artistic styles.

One significant advancement was made by [8], called neural style transfer, which could separate the representations of the image content and style learned

© Springer Nature Switzerland AG 2019
Z. Cui et al. (Eds.): IScIDE 2019, LNCS 11936, pp. 15–39, 2019.
https://doi.org/10.1007/978-3-030-36204-1_2

by deep Convolutional Neural Networks (CNN) and then recombine the image content from one and the image style from another to obtain styled images. During this neural style transfer process, fantastic stylized images were produced with the appearance similar to a given real artistic work, such as Vincent Van Gogh's "The Starry Night". The success of the style transfer indicates that artistic styles are computable and are able to be migrated from one image to another. Thus, we could learn to draw like some artists apparently without being trained for years.

Following [8], a lot of efforts have been made to improve or extend the neural style transfer algorithm. The content-aware style-transfer configuration was considered in [23]. In [14], CNN was discriminatively trained and then combined with the classical Markov Random Field based texture synthesis for better mesostructure preservation in synthesized images. Semantic annotations were introduced in [1] to achieve semantic transfer. To improve efficiency, a fast neural style transfer method was introduced in [13,21], which is a feed-forward network to deal with a large set of images per training. Results were further improved by an adversarial training network [15]. For a systematic review on neural style transfer, please refer to [12].

The recent progress on style transfer relied on the separable representation learned by deep CNN, in which the layers of convolutional filters automatically learns low-level or abstract representations in a more expressive feature space than the raw pixel-based images. However, it is still challenging to use CNN representations for style transfer due to their uncontrollable behavior as a black-box, and thus it is still difficult to select appropriate composition of styles (e.g. textures, colors, strokes) from images due to the risk of incorporation of unpredictable or incorrect patterns. In this paper, we propose computational analysis of the artistic styles and decompose them into basis elements that are easy to be selected and combined to obtain enhanced and controllable style transfer. Specifically, we propose two types of decomposition methods, i.e., spectrum based methods featured by Fast Fourier Transform (FFT), Discrete Cosine Transform (DCT), and latent variable models such as Principal Component Analysis (PCA), Independent Component Analysis (ICA). Then, we suggest methods of combination of styles by intervention and mixing. The computational decomposition of styles could be embedded as a module to state-of-the-art neural transfer algorithms. Experiments demonstrate the effectiveness of style decomposition in style transfer. We also demonstrate that controlling the style bases enables us to transfer the Chinese landscape paintings well and to transfer the sketch style similar to picture-to-sketch [2,19].

2   Related Work

Style transfer generates a styled image having similar semantic content as the content image and similar style as the style image. Conventional style transfer is realized by patch-based texture synthesis methods [5,22] where style is approximated as texture. Given a texture image, patch-based texture synthesis methods can automatically generate new image with the same texture. However, texture images are quite different from arbitrary style images [22] in duplicated patterns, which limits the functional ability of patch-based texture synthesis method in style transfer. Moreover, control of the texture by varying the patch size (shown in Fig. 2 of [5]) is limited due to the duplicated patterns in the texture image.

The neural style transfer algorithm proposed in [8] is a further development from the work in [7] which pioneers to take advantage of pre-trained CNN on ImageNet [3]. Rather than using texture synthesis methods which are implemented directly on the pixels of raw images, the feature maps of images are used which preserves better semantic information of the image. The algorithm starts with a noise image and finally converges to the styled image by iterative learning. The loss function $\mathcal{L}$ is composed of the content loss $\mathcal{L}_{content}$ and the style loss $\mathcal{L}_{style}$:

$$\mathcal{L} = \alpha\mathcal{L}_{content} + \beta\mathcal{L}_{style}, \tag{1}$$

$$\mathcal{L}_{content} = \frac{1}{2}(\mathcal{F}_l^{pred} - \mathcal{F}_l^{content})^2, \tag{2}$$

$$\mathcal{L}_{style} = \sum_l e_l \frac{1}{4h_l^2 w_l^2 c_l^2}(G_l^{pred} - G_l^{style})^2, \tag{3}$$

where $\mathcal{F}_l^{pred}$, $\mathcal{F}_l^{content}$ and $\mathcal{F}_l^{style}$ denote the feature maps of the synthesized styled image, content image and style image separately, $\mathcal{F}_l$ is treated as 2-dimensional data ($\mathcal{F}_l \in \mathcal{R}^{(h_l w_l) \times c_l}$), G_l is the Gram matrix of layer l by $G_l = \mathcal{F}_l^T \times \mathcal{F}_l, G_l \in \mathcal{R}^{c_l \times c_l}$, and h_l, w_l, c_l denote the height, width and the channel number of the feature map. Notice that the content loss is measured at the level of feature map, while the style loss is measured at the level of Gram matrix.

Methods were proposed in [9] for spatial control, color control and scale control for neural style transfer. Spatial control is to transfer style of specific regions of the style image via guided feature maps. Given the binary spatial guidance channels for both content and style image, one way to generate the guided feature map is to multiply the guidance channel with the feature map in an element-wise manner while another way is to concatenate the guidance channel with the feature map. Color control is realized by YUV color space and histogram matching methods [10] as post-processing methods. Although both ways are feasible, it is not able to control color to a specified degree, i.e., the control is binary, either transferring all colors of style image or preserving all colors of content image. Moreover, scale control [9] depends on different layers in CNN which represents different abstract levels. Since the number of layers in CNN is finite (19 layers at most for VGG19), the scale of style can only be controlled in finite degrees.

The limitation of control over neural style transfer proposed by pre-processing and post-processing methods in [9] derives from the lack of computational analysis of the artistic style which is the foundation of continuous control for neural style transfer. Inspired by spatial control in [9] that operations on the feature map could affect the style transferred, we analyze the feature map and decompose the style via projecting feature map into a latent space that is expanded by style basis, such as color, stroke, and so on. Since every point in the latent space can be regarded as a representation of a style and can be decoded back to feature map, style control become continuous and precise. Meanwhile, our work facilitates the mixing or intervention of the style basis from more than one styles so that compound style or new style could be generated, enhancing the diversity of styled images.

3   Methods

We propose to decompose the feature map of the style image into style basis in a latent space, in which it becomes easy to mix or intervene style bases of different styles to generate compound styles or new styles which are then projected back to the feature map space. Such decomposition process enables us to continuously control the composition of the style basis and enhance the diversity of the synthesized styled image. An overview of our method is demonstrated in Fig. 1.

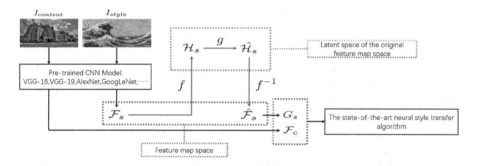

Fig. 1. An overview of our method which is indicated by red part. The red rectangle represents the latent space expanded by the style basis, f denotes computational decomposition of the style feature map $\mathcal{F}_s$, g denotes mixing or intervention within the latent space. The red part works as a computational module embedded in Gatys' or other neural style transfer algorithms. (Color figure online)

Given the content image $I_{content}$ and style image I_{style}, we decompose the style by function f from the feature map $\mathcal{F}_s$ of the style image to $\mathcal{H}_s$ in the latent space which is expanded by style basis $\{S_i\}$. We can mix or intervene the style basis via function g which is operated on style basis to generate the desired style coded by $\hat{\mathcal{H}}_s$. Using the inverse function f^{-1}, $\hat{\mathcal{H}}_s$ is projected back to the feature map space to get $\hat{F}_s$, which replace the original $\mathcal{F}_s$ for style transfer. Our method can serve as embedded module for the state-of-the-art neural style transfer algorithms, as shown in Fig. 1 by red.

It can be noted that the module can be regarded as a general transformation from original style feature map $\mathcal{F}_s$ to new style feature map $\hat{\mathcal{F}}_s$. If we let $\hat{\mathcal{F}}_s = \mathcal{F}_s$, our method degenerates back to traditional neural style transfer [8].

Since the transformation of the feature map is only done on the feature map of the style image, we simply notate $\mathcal{F}_s$ as $\mathcal{F}$ the denote the feature map of the style image and $\mathcal{H}_s$ as $\mathcal{H}$ in the rest of the paper. We notate h and w as the height and width of each channel in the feature map. Next, we introduce two types of decomposition function f and also suggest some control functions g.

3.1 Decomposed by Spectrum Transforms

We adopt 2-dimensional Fast Fourier Transform (FFT) and 2-dimensional Discrete Cosine Transform (DCT) as the decomposition function with details given in Table 1. Both methods are implemented in channel level of $\mathcal{F}$ where each channel is treated as 2-dimensional gray image.

Through the transform by 2-d FFT and 2-d DCT, the style feature map was decomposed as frequencies in the spectrum space where the style is coded by frequency that forms style bases. We will see that some style bases, such as stroke and color, actually correspond to different level of frequencies. With help of decomposition, similar styles are quantified to be close to each other as a cluster in the spectrum space, and it is easy to combine the existing styles to generate compound styles or new styles $\hat{\mathcal{H}}$ by appropriately varying the style codes. $\hat{\mathcal{H}}$ is then projected back to the feature map space via the inverse function of 2-d FFT and 2-d DCT shown in Table 1.

3.2 Decomposed by Latent Variable Models

We consider another type of decomposition by latent variable models, such as Principal Component Analysis (PCA) or Independent Component Analysis (ICA), which decompose the input signal into uncorrelated or independent components. Details are referred to Table 1, where each channel of the feature map $\mathcal{F}$ is vectorized as one input vector.

- **Principal Component Analysis (PCA)**
  We implement PCA from the perspective of matrix factorization. The eigenvectors are computed via Singular Value Decomposition (SVD). Then, the style is coded as linear combination of orthogonal eigenvectors, which could be regarded as style bases. By varying the combination of eigenvectors, compound styles or new styles are generated and then projected back to feature map space via the inverse of the matrix of the eigenvectors.
- **Independent Component Analysis (ICA)**
  We implement ICA via the fastICA algorithm [11], so that we decompose the style feature map into statistically independent components, which could be regarded as the style bases. Similar to PCA, we could control the combination of independent components to obtain compound styles or new styles, and then project them back to the feature map space.

3.3   Control Function g

The control function g in Fig. 1 defines style operations in the latent space expanded by the decomposed style basis. Instead of operating directly on the feature map space, such operations within the latent space have several advantages. First, after decomposition, style bases are of least redundancy or independent to each other, operations on them are easier to control; Second, the latent space could be made as a low dimensional manifold against noise, by focusing on several key frequencies for the spectrum or principal components in terms of maximum data variation; Third, continuous operations, such as linear mixing, intervention, and interpolation, are possible, and thus the diversity of the output style is enhanced, and even new styles could be sampled from the latent space. Fourth, multiple styles are able to be better mixed and transferred simultaneously.

Table 1. The mathematical details of f and f^{-1}

Method	Decomposition function f	Projection back
2d FFT	$\mathcal{H}(u,v) = \frac{1}{hw}\sum_{x=0}^{h-1}\sum_{y=0}^{w-1}\mathcal{F}(x,y)e^{-2(\frac{ux}{h}+\frac{vy}{w})\pi i}, \mathcal{F} \in \mathcal{R}^{h\times w\times c}$	Inverse 2d FFT
2d DCT	$\mathcal{H}(u,v)$ $= c(u)c(v)\sum_{x=0}^{h-1}\sum_{y=0}^{w-1}\mathcal{F}(x,y)cos[\frac{(x+0.5)\pi}{h}u]cos[\frac{(y+0.5)\pi}{w}v]$ $\mathcal{F}\in\mathcal{R}^{h\times w\times c}, c(u)=\sqrt{\frac{1}{N}}, u=0 \text{ and } c(u)=\sqrt{\frac{2}{N}}, u\neq 0$	Inverse 2d DCT
PCA	$\mathcal{F}=UDV^T, U=[v_1,\ldots,v_{hw}], \mathcal{H}=U\times\mathcal{F}$ $\mathcal{F}\in\mathcal{R}^{(hw)\times c}, U\in\mathcal{R}^{(hw)\times(hw)}$	$\hat{\mathcal{F}}=U^T\times\hat{\mathcal{H}}$
ICA	$[S,A]=fastICA(\mathcal{F}), \mathcal{H}=S, \mathcal{F}\in\mathcal{R}^{(hw)\times c}$ $S\in\mathcal{R}^{c\times(hw)}, A\in\mathcal{R}^{c\times c}$	$\hat{\mathcal{F}}=(A\times\hat{\mathcal{H}})^T$

Let $S_i^{(n)}, i \in \mathbb{Z}$ denote the i-th style basis of n-th style image. Notate $\{S_i^{(n)}|i \in I\}, I \subset \mathbb{Z}$ as $S_I^{(n)}$.

- **Single style basis:** Project the latent space on one of the style basis S_j. That is $S_i = 0$ if $i \neq j$
- **Intervention:** Reduce or amplify the effect of one style basis S_j by multiplying I while keeping other style bases unchanged. That is $S_i = I * S_i$ if $i = j$
- **Mixing:** Combine the style bases of n styles. That is $S = \{S_I^{(1)}, S_J^{(2)}, \ldots, S_K^{(n)}\}$

4   Experiments

We demonstrate the performance of our method using the fast neural style transfer algorithm [6,13,21]. We take the feature map 'relu4_1' from pre-trained VGG-19 model [18] as input to our style decomposition method because we try every single activation layer in VGG-19 and find that 'relu4_1' is suitable for style transfer.

4.1 Inferiority of Feature Map

Here, we demonstrate that it is not suitable for the style control function g to be applied on the feature map space directly because feature map space is possibly formed by a complicated mixture of style bases. To check whether the basis of feature map $\mathcal{F}$ can form the style bases, we experimented on the channels of $\mathcal{F}$ and the pixels of $\mathcal{F}$.

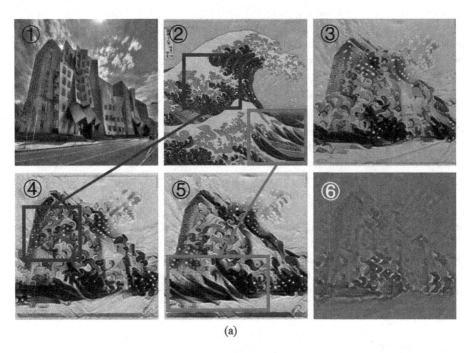

(a)

Fig. 2. (a)① content image (Stata Center); ② style image ("The Great Wave off Kanagawa" by Katsushika Hokusai); ③ styled image by traditional neural style transfer; ④⑤⑥ are the results of implementing control function directly on the feature map $\mathcal{F}$. Specifically, we amplify some pixels of $\mathcal{F}$ which generate ④⑤ and preserve a subset of channels of $\mathcal{F}$ which generate ⑥. (Color figure online)

Channels of $\mathcal{F}$. Assume styles are encoded in space $\mathcal{H}$ which is expanded by style basis $\{S_1, S_2, \ldots, S_n\}$. A good formulation of $\mathcal{H}$ may imply that if two styles are intuitively similar from certain aspects, there should be at least one style basis S_i such that the projections of the two styles onto S_i are close in Euclidean distance. Based on the above assumption, we generate the subset C of channels of $\mathcal{F}$ that could possibly represent color basis with semi-supervised method using style images in Fig. 13(a–c). It can be noticed that Chinese paintings and pen sketches (Fig. 13(a, c)) share the same color style while oil painting

(Fig. 13(b)) has an exclusive one. We iteratively find the largest channel set C_{max} (384 channels included) whose clustering result out of K-means [16] conforms to the following **clustering standard for color basis:**

- No cluster contains both oil painting and Chinese painting or pen sketch.
- One cluster contains only one or two points, since K-means is not adaptive to the cluster number and the cluster number is set as 3.

However, if we only use C_{max} to transfer style, the styled image (Fig. 2(a)⑥) isn't well stylized and doesn't indicate any color style of the style image (Fig. 2②), which probably indicates that the channels of $\mathcal{F}$ are not suitable to form independent style basis. However, by proper decomposition functions f introduced in Table 1, a good formulation of latent space $\mathcal{H}$ is feasible to reach the above clustering standard and to generate reasonably styled images.

Pixels of $\mathcal{F}$. We give intervention $I = 2.0$ to certain region of each channel of $\mathcal{F}$ to see if any intuitive style basis is amplified. The styled images are shown in Fig. 2(a)④⑤. Rectangles in style image (Fig. 2(a)②) are the intervened regions correspondingly. Compared to the styled image using [8] (Fig. 2(a)③), when small waves in style image is intervened, the effect of small blue circles in the styled image are amplified (green rectangle) and when large waves in style image is intervened, the effect of long blue curves in the styled image are amplified (red rectangle). Actually, implementing control function g on the pixels of the channels of $\mathcal{F}$ is quite similar to the methods proposed for spatial control of neural style transfer [9] which controls style transfer via a spatially guided feature map defined by a binary or real-valued mask on a region of the feature map. Yet it fails to computationally decompose the style basis.

4.2   Transfer by Single Style Basis

To check whether $\mathcal{H}$ is composed of style bases, we transfer style with single style basis preserved. We conduct experiments on $\mathcal{H}$ formulated by different decomposition functions, including FFT, DCT, PCA as well as ICA, with details mentioned in Sect. 3. The results are shown in Fig. 3. As is shown in Fig. 3(c), (d), The DC component only represents the color of style while the rest frequency components represent the wave-like stroke, which indicates that FFT is feasible for style decomposition. The result of DCT is quite similar to that of FFT, with DC component representing color and the rest representing stroke.

Besides, we analyze the spectrum space via Isomap [20], which can analytically demonstrate the effectiveness and robustness of spectrum based methods. Color (low frequency) and stroke (high frequency) forms X-axis and Y-axis of the 2-dimensional plane respectively where every style is encoded as a point. We experiment on 3 artistic styles, including Chinese painting, oil painting and pen sketch, and each style contains 10 pictures which is shown in Fig. 13(a–c). Chinese paintings and pen sketches share similar color style which is sharply

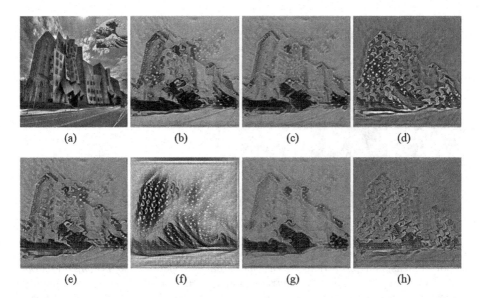

Fig. 3. (a) the original content and style images; (b) styled image by traditional neural style transfer; (c–h) results of preserving one style basis by different methods. Specifically, (c–d) FFT; (e–f) PCA; (g–h) ICA where (c, e, g) aim to transfer the color of style image while (d, f, h) aim to transfer the stroke of style image.

distinguished with oil paintings' while the stroke of three artistic styles are quite different from each other, which conforms to the result shown as Fig. 13(d).

Via PCA, the first principal component (Fig. 3(e)) fails to separate color and stroke, while the rest components (Fig. 3(f)) fail to represent any style basis, which indicates PCA is not suitable for style decomposition (Fig. 4).

The results of ICA (Fig. 3(g), (h)) are as good as the results of FFT but show significant differences. The color basis (Fig. 3(g)) is more murky than Fig. 3(c) while the stroke basis (Fig. 3(h)) retains the profile of curves with less stroke color preserved compared to Fig. 3(d). The stroke basis consists of $S_{arg_i}, i \in [0, n-1] \cup [c-n, c-1]$ while the rest forms the color basis. $arg \in \mathcal{R}^c$ is the ascent order of $A^{sum} \in \mathcal{R}^c$ where A_i^{sum} is the sum of ith column of mixing matrix A.

4.3   Transfer by Intervention

We give intervention to the stroke basis via control function g to demonstrate the controllable diversified styles and distinguish the difference in stroke basis between spectrum based methods and ICA.

As is shown in Fig. 5, the strokes of 'wave' are curves with light and dark blue while the strokes of 'aMuse' are black bold lines and coarse powder-like dots.

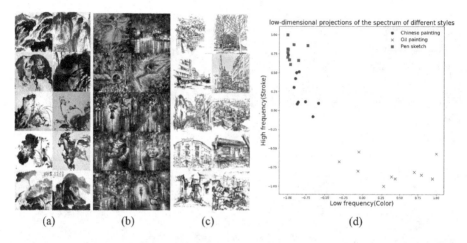

Fig. 4. (a) Chinese paintings; (b) Oil paintings (by Leonid Afremov); (c) Pen sketches; (d) low-dimensional projections of the spectrum of style(a-c) via Isomap; (e) low-dimensional projections of the spectrum of large scale of style images via Isomap. The size of each image shown above does not indicate any other information, but is set to prevent the overlap of the images only.

Intervention using spectrum based methods affects both color and outline of the strokes while intervention using ICA only influences stroke outline but not its color.

4.4 Transfer by Mixing

Current style mixing method, interpolation, cannot mix the style bases of different styles because styles are integrally mixed however interpolation weights are modified (Fig. 6(g–i)), which limits the diversity of style mixing. Based on the success spectrum based methods and ICA in style decomposition, we experiment to mix the stroke of 'wave' with the color of 'aMuse' to check whether such newly compound artistic style can be transferred to the styled image.

Specifically, for ICA, we not only need to replace the color basis of 'wave' with that of 'aMuse' but also should replace the rows of mixing matrix A corresponding to the exchanged signals. Both spectrum based methods (Fig. 6(d–f)) and ICA (Fig. 6(j–l)) work well in mixing style bases of different styles. Moreover, we can intervene the style basis when mixing, which further enhances the diversity of style mixing.

4.5 Sketch Style Transfer

The picture-to-sketch problem challenges whether computer can understand and represent the concept of objects both abstractly and semantically. The proposed controllable neural style transfer tackles the obstacle in current State-of-the-art

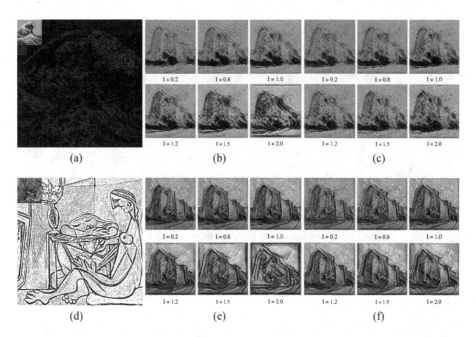

Fig. 5. (a, d) the stroke of style image 'wave' and 'aMuse'; (b, c, e, f) the results with different intervention I to the stroke basis. Specifically, (b, e) use spectrum based method and (c, f) use ICA. (Color figure online)

methods [2,19] which is caused by inconsistency in sketch style because diversified styles can in turn increase the style diversity of output images. Moreover, as is shown in Fig. 7, our method can control the abstract level by reserving major semantic details and minor ones automatically. Our method does not require vector sketch dataset. As a result, we cannot generate sketches in a stroke-drawing way [2,19].

4.6   Chinese Painting Style Transfer

Chinese painting is an exclusive artistic style having much less color than Western painting and represents the artistic conception by strokes. As is shown in Fig. 8, with effective controls over stroke via our methods, the Chinese painting styled image can be either misty-like known as freehand-brush or meticulous representation known as fine-brush.

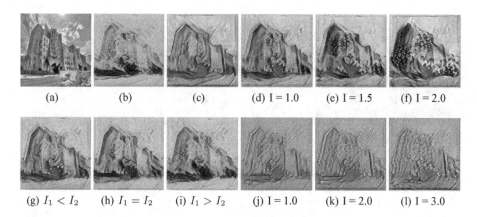

(a)   (b)   (c)   (d) I = 1.0   (e) I = 1.5   (f) I = 2.0

(g) $I_1 < I_2$ (h) $I_1 = I_2$ (i) $I_1 > I_2$ (j) I = 1.0   (k) I = 2.0   (l) I = 3.0

Fig. 6. (b), (c) styled image of single style; (g–i) interpolation mixing where I_1 and I_2 are the weights of 'wave' and 'aMuse'; (d–f, j–l) results of mixing the color of 'aMuse' and the stroke of 'wave' where I is the intervention to the stroke of 'wave'. Specifically, (d–f) use FFT; (j–l) use ICA.

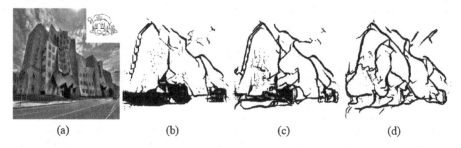

(a)   (b)   (c)   (d)

Fig. 7. Picture-to-sketch using style transfer and binarization. (a) content image and style image; (b–d) styled images. From (b) to (d), the number of stroke increases as more details of the content image are restored.

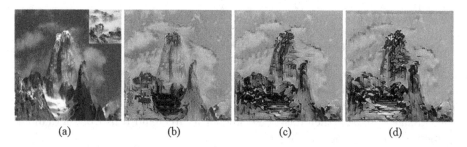

(a)   (b)   (c)   (d)

Fig. 8. Neural style transfer of Chinese painting with stroke controlled. (a) content image and style image (by Zaixin Miao); (b–d) styled images. From (b) to (d), the strokes are getting more detailed which gradually turns freehand style into finebrush style.

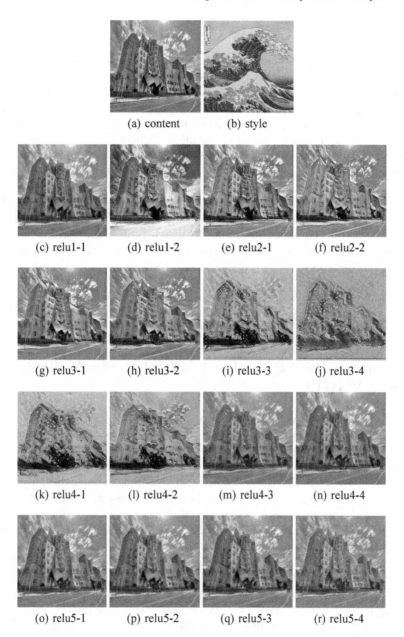

(a) content   (b) style

(c) relu1-1   (d) relu1-2   (e) relu2-1   (f) relu2-2

(g) relu3-1   (h) relu3-2   (i) relu3-3   (j) relu3-4

(k) relu4-1   (l) relu4-2   (m) relu4-3   (n) relu4-4

(o) relu5-1   (p) relu5-2   (q) relu5-3   (r) relu5-4

Fig. 9. Styled images using [8] with same epochs using every single activation layer from the pre-trained VGG19.

5   Conclusion

Artistic styles are made of basic elements, each with distinct characteristics and functionality. Developing such a style decomposition method facilitate the quantitative control of the styles in one or more images to be transferred to another natural image, while still keeping the basic content of natural image. In this paper, we proposed a novel computational decomposition method, and demonstrated its strengths via extensive experiments. The method could serve as a computational module embedded in those neural style transfer algorithms. We implemented the decomposition function by spectrum transform or latent variable models, and thus it enabled us to computationally and continuously control the styles by linear mixing or intervention. Experiments showed that our method enhanced the flexibility of style mixing and the diversity of stylization. Moreover, our method could be applied in picture-to-sketch problems by transferring the sketch style, and it captures the key feature and facilitates the stylization of the Chinese painting style.

Acknowledgement. This work was supported by the Zhi-Yuan Chair Professorship Start-up Grant (WF220103010), and Startup Fund (WF220403029) for Youngman Research, from Shanghai Jiao Tong University.

Appendix 1: The Stylization Effect of Every Activation Layer in VGG19

Since different layers in VGG19 [18] represent different abstract levels, we experiment the stylization effect of every activation layer on a couple of different style images as are shown in Figs. 9 and 10. From our experiments, it can be noticed that not every layer is effective in style transfer and among those that work, shallow layers only transfer the coarse scale of style (color) while deep layers can transfer both the coarse scale (color) and detailed scale (stroke) of style, which conforms to the result of scale control in [9]. Since 'relu4_1' performs the best in style transfer after same learning epochs, we determine to study the feature map of 'relu4_1' in our research.

We further visualize each channel of the feature map of the style image using t-SNE [4], as are shown in Figs. 11 and 12 where the similarity of both the results is quite interesting. However, we could not explain those specific patterns shown in the visualization results yet. What's more, the relationship between the similarity of the visualization results with the similarity of the stylization effects of different VGG layers remains further study as well.

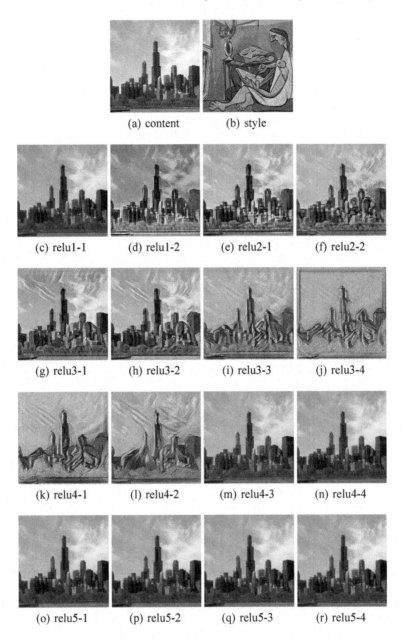

Fig. 10. Styled images using [8] with same epochs using every single activation layer from the pre-trained VGG19.

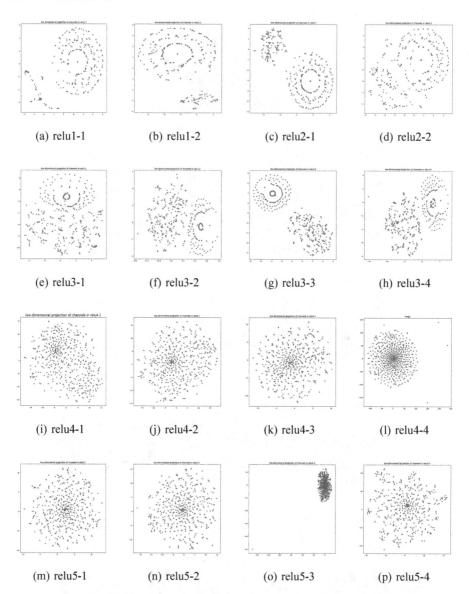

(a) relu1-1   (b) relu1-2   (c) relu2-1   (d) relu2-2

(e) relu3-1   (f) relu3-2   (g) relu3-3   (h) relu3-4

(i) relu4-1   (j) relu4-2   (k) relu4-3   (l) relu4-4

(m) relu5-1   (n) relu5-2   (o) relu5-3   (p) relu5-4

Fig. 11. Low dimensional projection of all channels of every single layer of the style image (Fig. 9(b)) via t-SNE [4].

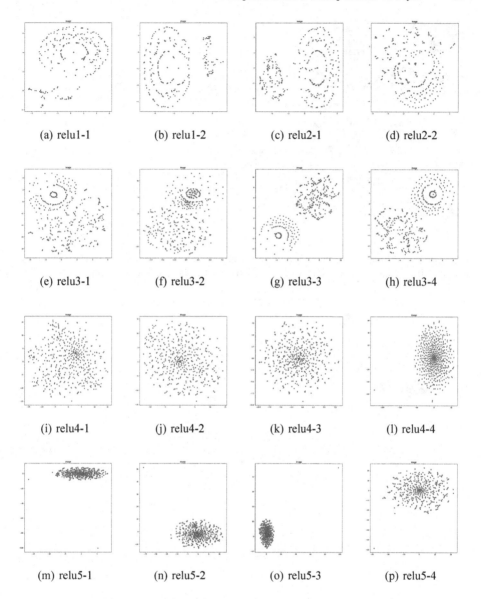

(a) relu1-1   (b) relu1-2   (c) relu2-1   (d) relu2-2

(e) relu3-1   (f) relu3-2   (g) relu3-3   (h) relu3-4

(i) relu4-1   (j) relu4-2   (k) relu4-3   (l) relu4-4

(m) relu5-1   (n) relu5-2   (o) relu5-3   (p) relu5-4

Fig. 12. Low dimensional projection of all channels of every single layer of style image (Fig. 10(b)) via t-SNE [4].

Appendix 2: The Manifold of Spectrum Based Methods

We analyze the spectrum space by projecting the style bases via Isomap [20] into low dimensional space where the X-axis represents the color basis and the Y-axis represents the stroke basis, which can analytically demonstrate the effectiveness and robustness of spectrum based methods. Three artistic styles are experimented (shown in Fig. 13(a–c)). Chinese paintings and pen sketches share

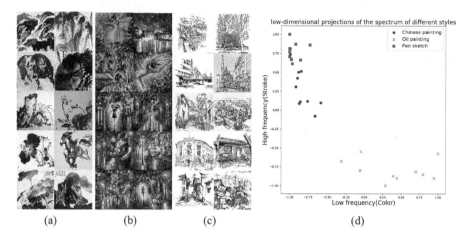

(a)   (b)   (c)   (d)

Fig. 13. (a) Chinese paintings; (b) Oil paintings (by Leonid Afremov); (c) Pen sketches; (d) low-dimensional projections of the spectrum of style (a–c) via Isomap [20].

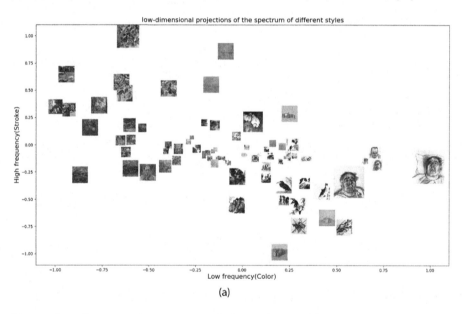

(a)

Fig. 14. Low-dimensional projections of the spectrum of large scale of style images via Isomap [20]. The size of each image shown above does not indicate any other information, but is set to prevent the overlap of the images only.

similar color style which is sharply distinguished with oil paintings' while the stroke of three artistic styles are quite different from each other. Thus, as in shown in Fig. 13(d), Chinese paintings and pen sketches are close to each other and both stay away from oil paintings in X-axis which represents color while three styles are respectively separable in Y-axis which represents stroke, which completely satisfies our analysis of the three artistic styles.

When we apply the same method to large scale of style images (Fig. 14), X-axis clearly represents the linear transition from dull-colored to rich-colored. However, we fail to conclude any notable linear transition for Y-axis from the 2-dimensional visualization probably because it is hard to describe the style of stroke (boldface, length, curvity, etc.) using only one dimension.

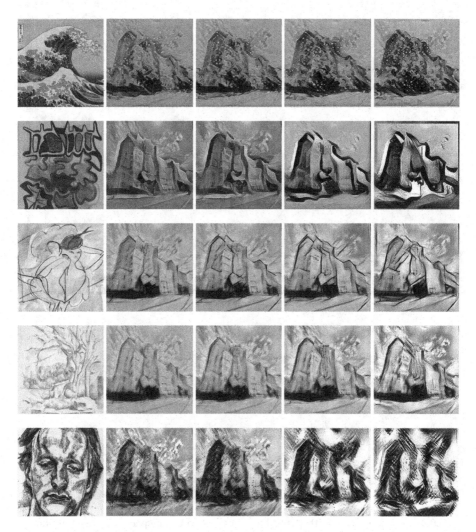

Fig. 15. The styled image with stroke basis intervened using spectrum based methods. The left most row shows the style images (From top to bottom: The Great Wave off Kanagawa - Katsushika Hokusai; Composition - Alberto Magnelli; Dancer - Ernst Ludwig Kirchner; Pistachio Tree in the Courtyard of the Chateau Noir - Paul Cezanne; Potrait - Lucian Freud). From left to right of each row, the effect of stroke is increasingly amplified.

Appendix 3: Stroke Intervention

We demonstrate more styled images with stroke basis intervened using spectrum based method (Figs. 15 and 16) and ICA (Figs. 17 and 18) respectively.

Fig. 16. The styled image with stroke basis intervened using spectrum based methods. The left most row shows the style images (From top to bottom: A Muse (La Muse) - Pablo Picasso; Number 4 (Gray and Red) - Jackson Pollock; Shipwreck - J.M.W. Turner; Natura Morta - Giorgio Morandi; The Scream - Edvard Munch). From left to right of each row, the effect of stroke is increasingly amplified. (Color figure online)

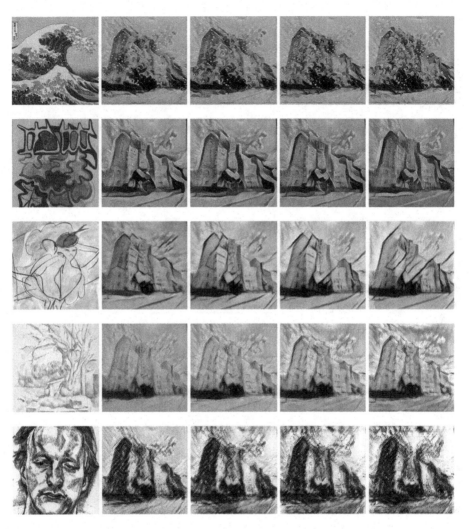

Fig. 17. The styled image with stroke basis intervened using ICA. The left most row shows the style images (From top to bottom: The Great Wave off Kanagawa - Katsushika Hokusai; Composition - Alberto Magnelli; Dancer - Ernst Ludwig Kirchner; Pistachio Tree in the Courtyard of the Chateau Noir - Paul Cezanne; Potrait - Lucian Freud). From left to right of each row, the effect of stroke is increasingly amplified.

Appendix 4: Style Mixing

We demonstrate more styled images transferred with compound style generated by mixing the color basis and stroke basis of two different styles. The results of both spectrum based method and ICA method are shown in Fig. 19 with comparison with traditional mixing method - interpolation.

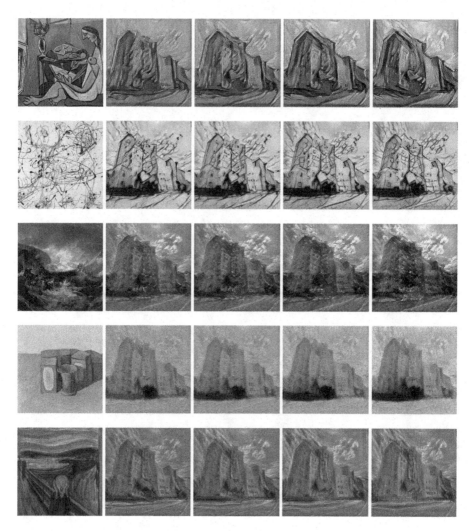

Fig. 18. The styled image with stroke basis intervened using ICA. The left most row shows the style images (From top to bottom: A Muse (La Muse) - Pablo Picasso; Number 4 (Gray and Red) - Jackson Pollock; Shipwreck - J.M.W. Turner; Natura Morta - Giorgio Morandi; The Scream - Edvard Munch). From left to right of each row, the effect of stroke is increasingly amplified. (Color figure online)

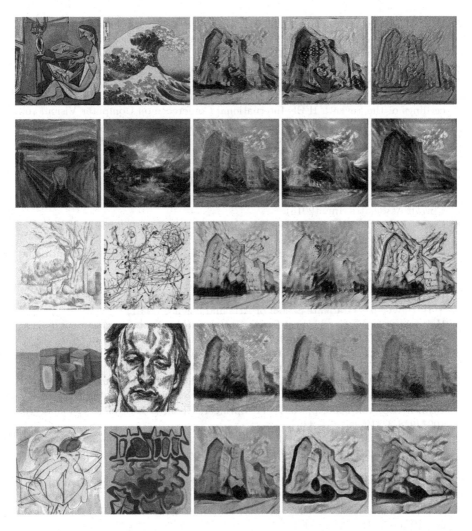

Fig. 19. The left two columns are the style images used for mixing. Specifically, we mix the color of the most left one with the stroke of the second left one. The third left column shows the styled images with traditional interpolation method. The second right column shows the styled images using spectrum mixing method. The most right column shows the styled images using ICA mixing method.

References

1. Champandard, A.J.: Semantic style transfer and turning two-bit doodles into fine artworks. arXiv preprint arXiv:1603.01768 (2016)
2. Chen, Y., Tu, S., Yi, Y., Xu, L.: Sketch-pix2seq: a model to generate sketches of multiple categories. CoRR abs/1709.04121 (2017)

3. Deng, J., Dong, W., Socher, R., Li, L.J., Li, K., Fei-Fei, L.: ImageNet: a large-scale hierarchical image database. In: 2009 IEEE Conference on Computer Vision and Pattern Recognition, pp. 248–255. IEEE (2009)
4. Der Maaten, L.V., Hinton, G.E.: Visualizing data using t-SNE. J. Mach. Learn. Res. **9**, 2579–2605 (2008)
5. Efros, A.A., Leung, T.K.: Texture synthesis by non-parametric sampling. In: Proceedings of the Seventh IEEE International Conference on Computer Vision, vol. 2, pp. 1033–1038. IEEE (1999)
6. Engstrom, L.: Fast style transfer (2016). https://github.com/lengstrom/fast-style-transfer/
7. Gatys, L., Ecker, A.S., Bethge, M.: Texture synthesis using convolutional neural networks. In: Cortes, C., Lawrence, N.D., Lee, D.D., Sugiyama, M., Garnett, R. (eds.) Advances in Neural Information Processing Systems, vol. 28, pp. 262–270. Curran Associates, Inc. (2015)
8. Gatys, L.A., Ecker, A.S., Bethge, M.: A neural algorithm of artistic style. arXiv preprint arXiv:1508.06576 (2015)
9. Gatys, L.A., Ecker, A.S., Bethge, M., Hertzmann, A., Shechtman, E.: Controlling perceptual factors in neural style transfer. In: Proceedings of the IEEE Conference on Computer Vision and Pattern Recognition, pp. 3985–3993 (2017)
10. Hertzmann, A.P.: Algorithms for rendering in artistic styles. Ph.D. thesis, New York University, Graduate School of Arts and Science (2001)
11. Hyvarinen, A.: Fast and robust fixed-point algorithms for independent component analysis. IEEE Trans. Neural Netw. **10**(3), 626–634 (1999)
12. Jing, Y., Yang, Y., Feng, Z., Ye, J., Yu, Y., Song, M.: Neural style transfer: a review. IEEE Trans. Vis. Comput. Graph. (2019)
13. Johnson, J., Alahi, A., Fei-Fei, L.: Perceptual losses for real-time style transfer and super-resolution. In: Leibe, B., Matas, J., Sebe, N., Welling, M. (eds.) ECCV 2016. LNCS, vol. 9906, pp. 694–711. Springer, Cham (2016). https://doi.org/10.1007/978-3-319-46475-6_43
14. Li, C., Wand, M.: Combining Markov random fields and convolutional neural networks for image synthesis. In: Proceedings of the IEEE Conference on Computer Vision and Pattern Recognition, pp. 2479–2486 (2016)
15. Li, C., Wand, M.: Precomputed real-time texture synthesis with Markovian generative adversarial networks. In: Leibe, B., Matas, J., Sebe, N., Welling, M. (eds.) ECCV 2016. LNCS, vol. 9907, pp. 702–716. Springer, Cham (2016). https://doi.org/10.1007/978-3-319-46487-9_43
16. Lloyd, S.P.: Least squares quantization in PCM. IEEE Trans. Inf. Theory **28**(2), 129–137 (1982)
17. Rosin, P., Collomosse, J.: Image and Video-Based Artistic Stylisation, vol. 42. Springer, London (2012). https://doi.org/10.1007/978-1-4471-4519-6
18. Simonyan, K., Zisserman, A.: Very deep convolutional networks for large-scale image recognition. arXiv preprint arXiv:1409.1556 (2014)
19. Song, J., Pang, K., Song, Y.Z., Xiang, T., Hospedales, T.M.: Learning to sketch with shortcut cycle consistency. In: Proceedings of the IEEE Conference on Computer Vision and Pattern Recognition, pp. 801–810 (2018)
20. Tenenbaum, J.B., De Silva, V., Langford, J.: A global geometric framework for nonlinear dimensionality reduction. Science **290**(5500), 2319–2323 (2000)
21. Ulyanov, D., Lebedev, V., Vedaldi, A., Lempitsky, V.: Texture networks: feed-forward synthesis of textures and stylized images. In: Proceedings of the 33rd International Conference on International Conference on Machine Learning, ICML 2016, vol. 48, pp. 1349–1357 (2016)

22. Wei, L.Y., Levoy, M.: Fast texture synthesis using tree-structured vector quantization. In: Proceedings of the 27th Annual Conference on Computer Graphics and Interactive Techniques, SIGGRAPH 2000, pp. 479–488. ACM Press/Addison-Wesley Publishing Co., New York (2000)
23. Yin, R.: Content aware neural style transfer. CoRR abs/1601.04568 (2016). http://arxiv.org/abs/1601.04568

Laplacian Welsch Regularization for Robust Semi-supervised Dictionary Learning

Jingchen Ke, Chen Gong$^{(\boxtimes)}$, and Lin Zhao

PCA Lab, Key Lab of Intelligent Perception and Systems for High-Dimensional Information of Ministry of Education, and Jiangsu Key Lab of Image and Video Understanding for Social Security, School of Computer Science and Engineering, Nanjing University of Science and Technology, Nanjing, China
{csjke,chen.gong,linzhao}@njust.edu.cn

Abstract. Semi-supervised dictionary learning aims to find a suitable dictionary by utilizing limited labeled examples and massive unlabeled examples, so that any input can be sparsely reconstructed by the atoms in a proper way. However, existing algorithms will suffer from large reconstruction error due to the presence of outliers. To enhance the robustness of existing methods, this paper introduces an upper-bounded, smooth and nonconvex Welsch loss which is able to constrain the adverse effect brought by outliers. Besides, we adopt the Laplacian regularizer to enforce similar examples to share similar reconstruction coefficients. By combining Laplacian regularizer and Welsch loss into a unified framework, we propose a novel semi-supervised dictionary learning algorithm termed "Laplacian Welsch Regularization" (LWR). To handle the model non-convexity caused by the Welsch loss, we adopt Half-Quadratic (HQ) optimization algorithm to solve the model efficiently. Experimental results on various real-world datasets show that LWR performs robustly to outliers and achieves the top-level results when compared with the existing algorithms.

Keywords: Semi-supervised dictionary learning · Welsch loss · Half-Quadratic optimization

1 Introduction

Dictionary Learning (DL) has gained intensive attention in recent years, and has been applied to various computer vision and image processing problems including image classification [10], image restoration [6], image alignment [12], and so on. The target of DL is to find a suitable dictionary $\mathbf{D}$ such that any input signal $\mathbf{x}$ can be sparsely reconstructed by its atoms, namely $\mathbf{x} \approx \mathbf{D}\mathbf{a}$ where $\mathbf{a}$ is a sparse coding vector. By incorporating the supervision information to the traditional DL setting, supervised dictionary learning has been proposed which aims to obtain a discriminative dictionary by exploiting the labels of examples.

© Springer Nature Switzerland AG 2019
Z. Cui et al. (Eds.): IScIDE 2019, LNCS 11936, pp. 40–52, 2019.
https://doi.org/10.1007/978-3-030-36204-1_3

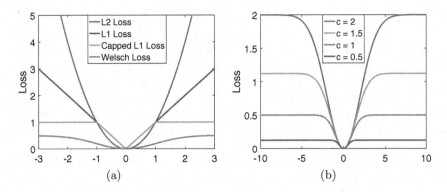

Fig. 1. (a) The motivation of our method: the comparison of our adopted Welsch loss with the existing ℓ_2 loss, ℓ_1 loss, and capped ℓ_1 loss. (b) The Welsch loss with different selections of parameter c.

Supervised dictionary learning algorithms can achieve satisfactory performance when labeled training examples are sufficient.

However, in many practical cases, it is very expensive to label massive data due to both time and labor cost, so traditional supervised dictionary learning models cannot be used due to the rarity of labeled examples. On the other hand, abundant unlabeled examples can often be easily and cheaply collected. Therefore, semi-supervised dictionary learning has been proposed which can effectively learn a good dictionary with limited labeled examples. Semi-supervised dictionary learning algorithms are able to take advantages of the supervision information carried by labeled examples and the distribution information revealed by the unlabeled examples. For example, Shrivastava *et al.* [7] utilize the connection between labeled examples and unlabeled examples to estimate the confidence matrix of unlabeled examples. Furthermore, they define the within-class scatter matrix and the between-class scatter matrix to obtain discriminative dictionaries. Recently, Wang *et al.* [9] suggest that not all the coding vectors of labeled examples are helpful to the classifier. They claim that only some coding vectors of labeled examples which named as *active point* should be employed to enhance the dictionary discriminative capability. Wang *et al.* [8] prefer to use $\ell_{p,p}$ norm than the conventional ℓ_1 norm for learning the dictionary since $\ell_{p,p}$ norm inherits the advantages of both ℓ_1 norm and ℓ_2 norm. Moreover, they enforce the same class examples to share the same atoms by minimizing the error which represents the sparse relationship between labeled and unlabeled examples. Yang *et al.* [10] adopt the entropy regularization term to estimate the class probability of unlabeled examples which helps to find the discriminative representations of training examples.

Although several semi-supervised dictionary learning algorithms have been proposed so far, they share the same drawback that they are not robust to the outliers in the feature space. Most of the existing algorithms utilize ℓ_2 norm or Frobenius norm to characterize the reconstruction error for the input data.

Consequently, the reconstruction error will be greatly increased when there are outliers in the dataset. Due to the penalization of such huge error, semi-supervised dictionary learning algorithms will produce biased dictionary which leads to poor classifications. To overcome the shortcoming, in this paper, we focus on the loss function and resort to the Welsch loss to solve the robustness problem. Unlike the traditional ℓ_2 norm, the reconstruction error computed by Welsch loss will not be arbitrarily large in the presence of outliers. To show this point, we plot ℓ_2 loss, ℓ_1 loss, capped ℓ_1 loss [3], and the adopted Welsch loss in Fig. 1(a). It is obvious that normally used ℓ_2 loss and ℓ_1 loss do not have an upper bound which leads to the unlimited reconstruction and classification error. In contrast, our employed Welsch loss is upper bounded, so the outliers will not have a large impact on the learning of the dictionary. Although the recently proposed capped ℓ_1 loss can also suppress the huge error, the loss function is nonsmooth which brings about the non-differentiable objective function. In contrast, only our Welsch loss is bounded and smooth, therefore it is more robust and easier to implement than the normally used loss functions. Moreover, we introduce the Laplacian regularizer to make use of the data distribution information revealed by the scarce labeled examples and massive unlabeled examples, such that the similar examples in the feature space will also obtain similar reconstruction coefficients. By combining Welsch loss and Laplacian regularization, our method is thus dubbed "**L**aplacian **W**elsch **R**egularization for robust Semi-Supervised Dictionary Learning" ("LWR" for short). Moreover, to handle the non-convexity brought by the Welsch loss, we adopt the Half-Quadratic (HQ) optimization technique [2] so that the nonconvex problem is transformed into several convex sub-problems. In the experiments, we compare our algorithm on real-world datasets with two state-of-the-art supervised dictionary learning algorithms and two semi-supervised dictionary learning algorithms, and the results confirm the effectiveness of our algorithm in the presence of outliers. Also, the convergence of our designed iterative optimization process is also verified.

The organization of the paper is as follows. In Sect. 2, we describe the LWR model and its optimization algorithm. Experimental results are shown in Sect. 3. Finally, this paper is summarized in Sect. 4.

2   Model Description

In this section, we first introduce some basic background of semi-supervised dictionary learning, and then present our proposed model with its solution. At last, we show the classification strategy of LWR for a text example.

2.1   Preliminaries for Semi-supervised Dictionary Learning

The semi-supervised dictionary learning problem is mathematically defined as follows. Given a training set $\mathbf{X} = [\mathbf{X}_l, \mathbf{X}_u] \in \mathbb{R}^{d \times n}$ where d is the feature dimension and $n = l + u$ is the size of dataset. Let $\mathbf{X}_l = [\mathbf{x}_1, \dots, \mathbf{x}_l] \in \mathbb{R}^{d \times l}$ be the set of labeled data with the corresponding label matrix $\mathbf{Y}_l \in \mathbb{R}^{r \times l}$ where r is the number of classes, and $\mathbf{X}_u = [\mathbf{x}_{l+1}, \ldots, \mathbf{x}_{l+u}] \in \mathbb{R}^{d \times u}$ be the set of unlabeled data with the all-zero matrix $\mathbf{Y}_u \in \mathbb{R}^{r \times u}$. Let $\mathbf{D} \in \mathbb{R}^{d \times k}$ be the dictionary with k atoms and $\mathbf{A} = [\mathbf{A}_l, \mathbf{A}_u] \in \mathbb{R}^{k \times n}$ be the matrix of the sparse coding, where $\mathbf{A}_l = [\mathbf{a}_1, \ldots, \mathbf{a}_l]_{k \times l}$ and $\mathbf{A}_u = [\mathbf{a}_{l+1}, \ldots, \mathbf{a}_{l+u}]_{k \times u}$ show the matrices of the sparse coding of the labeled and unlabeled data, respectively. Here we utilize a linear classifier $f(\mathbf{A}, \mathbf{W}) = \mathbf{W}\mathbf{A}$ for classification problem, where $\mathbf{W} \in \mathbb{R}^{r \times k}$ is the parameter matrix.

2.2 Laplacian Welsch Regularization for Semi-supervised Dictionary Learning

As mentioned in the Introduction, the traditional loss function adopted by existing semi-supervised dictionary learning algorithms is unbounded, which is easy to cause huge reconstruction and classification error when outliers appear. Due to the adverse impact caused by the outliers, these algorithms cannot gain discriminative dictionaries which lead to an unexpected classification result. Moreover, since semi-supervised dictionary learning only has limited labeled training examples, the negative impact will be apparently magnified on account of the non-robust loss.

To handle the influence caused by the outliers, we introduce Welsch loss to the framework of semi-supervised dictionary learning. Welsch loss is a bounded, smooth and nonconvex loss which is robust to the outliers. It is defined as

$$V(z) = \frac{c^2}{2} \left[1 - \exp\left(-\frac{z^2}{2c^2} \right) \right], \tag{1}$$

where c is a tuning parameter controlling the upper bound of Welsch loss. Figure 1(b) plots the Welsch loss $V(z)$ under different values of c which changes from 0.5 to 2. We obtain that the upper bound of Welsch loss increases and converges slowly when the parameter gradually increases. By driving z to infinity, it can be easily found that the upper bound of Welsch loss is $c^2/2$ from (1) which means that the unexpected impact can be bounded. Due to the excellent robustness of Welsch loss, we introduce the loss (1) to semi-supervised dictionary learning to estimate the reconstruction error and the classification error as follows

$$\mathcal{W}(\mathbf{X}, \mathbf{D}, \mathbf{Y}, \mathbf{W}, \mathbf{A})$$

$$= \sum_{i=1}^{n} \frac{c^2}{2} \left[1 - \exp\left(-\frac{\|\mathbf{x}_i - \mathbf{D}\mathbf{a}_i\|_2^2}{2c^2} \right) \right] + \lambda \sum_{i=1}^{n} \frac{c^2}{2} \left[1 - \exp\left(-\frac{\|\mathbf{y}_i - \mathbf{W}\mathbf{a}_i \mathbf{J}_i\|_2^2}{2c^2} \right) \right]$$

$$s.t. \quad \|\mathbf{d}_i\|_2^2 \le q, \forall i, \tag{2}$$

where q is a constant; λ is an nonnegative parameter; and $\mathbf{J}_i$ is the i-th diagonal element of the $n \times n$ diagonal matrix $\mathbf{J}$ which is given by $\mathbf{J} = diag(1, \ldots, 1, 0, \ldots, 0)$ with the first l diagonal entries being 1 and the rest u elements being 0. The first term of (2) computes the reconstruction error on both labeled examples and unlabeled examples. According to the diagonal matrix $\mathbf{J}$, the second term of (2) only calculates the classification error of labeled examples.

Based on the assumption that similar examples share similar reconstruction coefficients, we try to explore the information revealed by unlabeled examples by introducing Laplacian regularizer to semi-supervised dictionary learning. $\mathbf{P}$ is the adjacency matrix of the whole dataset $\mathbf{X}$ which can be defined as $p_{ij} = \exp\left(-\|\mathbf{x}_i - \mathbf{x}_j\|^2 / \left(2\sigma^2\right)\right)$, where p_{ij} denotes the similarity between examples $\mathbf{x}_i$ and $\mathbf{x}_j$; the variance σ is a free parameter that should be manually tuned. Accordingly, the graph Laplacian constraint based on $\mathbf{P}$ can be written as

$$\text{tr}\left(\mathbf{ALA}^\top\right) = \frac{1}{2}\sum_{i=1}^{n}\sum_{j=1}^{n} p_{ij}\|\mathbf{x}_i - \mathbf{x}_j\|_2^2, \tag{3}$$

where $\mathbf{L} = \mathbf{M} - \mathbf{P}$ is the Laplacian matrix, and $\mathbf{M} \in \mathbb{R}^{n \times n}$ is a diagonal matrix with its i-th diagonal element being equal to the sum of the i-th row of $\mathbf{P}$.

Above all, in this paper, we incorporate Welsch loss and Laplacian regularizer to the general framework of semi-supervised dictionary learning, and thus the proposed method is termed "Laplacian Welsch Regularization for robust semi-supervised dictionary learning" (LWR). Consequently, our LWR model is formally represented as

$$\min_{\mathbf{D},\mathbf{W},\mathbf{A}} \mathcal{W}\left(\mathbf{X},\mathbf{D},\mathbf{Y},\mathbf{W},\mathbf{A}\right) + \frac{\beta}{2}\|\mathbf{A}\|_1 + \frac{\mu}{2}\text{tr}\left(\mathbf{ALA}^\top\right) \quad s.t. \ \|\mathbf{d}_i\|_2^2 \le q, \forall i, \tag{4}$$

where $\mathcal{W}$ has been shown in (2), and β and μ are nonnegative regularization parameters. Our goal is to find a discriminative dictionary $\mathbf{D}$, a well representation sparse matrix $\mathbf{A}$, and a suitable parameter matrix $\mathbf{W}$ by minimizing (4). However, the optimization of (4) is nonconvex due to the adopted Welsch loss. To handle this problem, we introduce Half-Quadratic (HQ) method [2] here to obtain the suitable solution.

2.3   HQ Optimization of LWR

Before we utilize the HQ optimization algorithm to optimize LWR, we can rewrite (4) as

$$\max_{\mathbf{D},\mathbf{W},\mathbf{A}} G_1\left(\mathbf{D},\mathbf{W},\mathbf{A}\right) \quad s.t. \ \|\mathbf{d}_i\|_2^2 \le q, \forall i, \tag{5}$$

where

$$G_1\left(\mathbf{D},\mathbf{W},\mathbf{A}\right) = \sum_{i=1}^{n}\exp\left(-\frac{\|\mathbf{x}_i - \mathbf{Da}_i\|_2^2}{2c^2}\right) + \lambda\sum_{i=1}^{n}\exp\left(-\frac{\|\mathbf{y}_i - \mathbf{Wa}_i\mathbf{J}_i\|_2^2}{2c^2}\right)$$
$$- \frac{\beta}{c^2}\|\mathbf{A}\|_1 - \frac{\mu}{c^2}\text{tr}\left(\mathbf{ALA}^\top\right). \tag{6}$$

Based on the conjugate function theory [1], we define a convex function $g(v) = -v \log(-v) + v$ where $v < 0$ and then we can obtain

$$\exp\left(-\frac{(y - f(b))^2}{2c^2}\right) = \sup_{v<0}\left\{v\frac{(y - f(b))^2}{2c^2} - g(v)\right\}, \tag{7}$$

of which the upper bound is achieved at

$$v = -\exp\left(-\frac{(y - f(b))^2}{2c^2}\right). \tag{8}$$

By using (7), we further derive (5) as

$$\max_{\mathbf{D},\mathbf{W},\mathbf{A},\mathbf{v}_1<0,\mathbf{v}_2<0} G_2(\mathbf{D}, \mathbf{W}, \mathbf{A}, \mathbf{v}_1, \mathbf{v}_2) \quad s.t. \quad \|\mathbf{d}_i\|_2^2 \leq q, \forall i, \tag{9}$$

where

$$G_2(\mathbf{D}, \mathbf{W}, \mathbf{A}, \mathbf{v}_1, \mathbf{v}_2) = \sum_{i=1}^{n}\left[v_{1i}\frac{\|\mathbf{x}_i - \mathbf{D}\mathbf{a}_i\|_2^2}{2c^2} - g(v_{1i})\right] + \lambda\sum_{i=1}^{n}\left[v_{2i}\frac{\|\mathbf{y}_i - \mathbf{W}\mathbf{a}_i\mathbf{J}_i\|_2^2}{2c^2} - g(v_{2i})\right]$$
$$- \frac{\beta}{c^2}\|\mathbf{A}\|_1 - \frac{\mu}{c^2}\operatorname{tr}\left(\mathbf{A}\mathbf{L}\mathbf{A}^\top\right). \tag{10}$$

Now an alternating optimization procedure can be adopted to solve (9). Note that there are five variables to be optimized in (9), we may first optimize $\mathbf{v}_1$ and $\mathbf{v}_2$ while keeping $\{\mathbf{D}, \mathbf{W}, \mathbf{A}\}$ unchanged, and then optimize the rest variables alternatively.

Update $\{\mathbf{v}_1, \mathbf{v}_2\}$. With $\{\mathbf{D}, \mathbf{W}, \mathbf{A}\}$ fixed, we can optimize $\mathbf{v}_1$ and $\mathbf{v}_2$ in parallel due to that they are irrelevant to each other. Then the objectives associated with $\mathbf{v}_1$ and $\mathbf{v}_2$ become

$$\max_{\mathbf{v}_1<0}\sum_{i=1}^{n}\left[v_{1i}\frac{\|\mathbf{x}_i - \mathbf{D}\mathbf{a}_i\|_2^2}{2c^2} - g(v_{1i})\right] \text{ and } \max_{\mathbf{v}_2<0}\lambda\sum_{i=1}^{n}\left[v_{2i}\frac{\|\mathbf{y}_i - \mathbf{W}\mathbf{a}_i\mathbf{J}_i\|_2^2}{2c^2} - g(v_{2i})\right]. \tag{11}$$

According to (7) and (8), the analytical solutions to (11) are

$$v_{1i} = -\exp\left(-\frac{\|\mathbf{x}_i - \mathbf{D}\mathbf{a}_i\|_2^2}{2c^2}\right) \quad \text{and} \quad v_{2i} = -\exp\left(-\frac{\|\mathbf{y}_i - \mathbf{W}\mathbf{a}_i\mathbf{J}_i\|_2^2}{2c^2}\right). \tag{12}$$

Update $\mathbf{W}$. With $\{\mathbf{D}, \mathbf{A}, \mathbf{v}_1, \mathbf{v}_2\}$ fixed, we solve the subproblem of $\mathbf{W}$ through optimizing

$$\min_{\mathbf{W}} \mathcal{Z}_1(\mathbf{W}) = \frac{\lambda}{2c^2}\operatorname{tr}((\mathbf{Y} - \mathbf{W}\mathbf{A}\mathbf{J})\mathbf{\Omega}_2(\mathbf{Y} - \mathbf{W}\mathbf{A}\mathbf{J})^\top), \tag{13}$$

where $\mathbf{\Omega}_2$ is $n \times n$ diagonal matrices given by $\mathbf{\Omega}_2 = diag(-\mathbf{v}_2)$.

By computing the derivative of (13) to $\mathbf{W}$ and setting it zero, we obtain the following closed-form solution

$$\mathbf{W} = \mathbf{Y}\boldsymbol{\Omega}_2 \left(\mathbf{AJ}\right)^\top \left[\mathbf{AJ}\boldsymbol{\Omega}_2 \left(\mathbf{AJ}\right)^\top\right]^{-1}. \tag{14}$$

Update A. When $\{\mathbf{D}, \mathbf{W}, \mathbf{v}_1, \mathbf{v}_2\}$ are fixed, we solve $\mathbf{A}$ by minimizing the following problem

$$\begin{aligned}
\min_{\mathbf{A}} \mathcal{Z}_2\left(\mathbf{A}\right) = {} & \frac{1}{2c^2}\operatorname{tr}((\mathbf{X} - \mathbf{DA})\boldsymbol{\Omega}_1 (\mathbf{X} - \mathbf{DA})^\top) \\
& + \frac{\lambda}{2c^2}\operatorname{tr}((\mathbf{Y} - \mathbf{WAJ})\boldsymbol{\Omega}_2 (\mathbf{Y} - \mathbf{WAJ})^\top) \\
& + \frac{\beta}{c^2}\|\mathbf{A}\|_1 + \frac{\mu}{c^2}\operatorname{tr}(\mathbf{ALA}^\top),
\end{aligned} \tag{15}$$

where $\boldsymbol{\Omega}_1$ is $n \times n$ diagonal matrices given by $\boldsymbol{\Omega}_1 = diag\left(-\mathbf{v}_1\right)$.

Due to the ℓ_1 norm adopted in (15), we cannot obtain a closed-form solution. Therefore, the gradient descent algorithm is used to solve (15). The sub-gradient of $\mathbf{A}$ in (15) can be computed as

$$\frac{\partial \mathcal{Z}_2\left(\mathbf{A}\right)}{\partial \mathbf{A}} = \frac{1}{c^2}\mathbf{D}^\top\left(\mathbf{DA} - \mathbf{X}\right)\boldsymbol{\Omega}_1 + \frac{\lambda}{c^2}\mathbf{W}^\top\left(\mathbf{WAJ} - \mathbf{Y}\right)\boldsymbol{\Omega}_2\mathbf{J} + \frac{\beta}{c^2}sign\left(\mathbf{A}\right) + \frac{2\mu}{c^2}\mathbf{AL}. \tag{16}$$

Then we use the gradient decent method to update $\mathbf{A}$ at step t, namely $\mathbf{A}^{t+1} = \mathbf{A}^t - \eta\partial \mathcal{Z}_2\left(\mathbf{A}\right)/\partial \mathbf{A}$, where η is the learning rate.

Update D. With the updated $\{\mathbf{W}, \mathbf{A}, \mathbf{v}_1, \mathbf{v}_2\}$, we solve the subproblem of $\mathbf{D}$ by optimizing

$$\min_{\mathbf{D}} \mathcal{Z}_3\left(\mathbf{D}\right) = \frac{1}{2c^2}\operatorname{tr}((\mathbf{X} - \mathbf{DA})\boldsymbol{\Omega}_1 (\mathbf{X} - \mathbf{DA})^\top) \quad s.t. \quad \|\mathbf{d}_i\|_2^2 \leq q, \forall i. \tag{17}$$

Due to the additional constraints, the closed-form solution is unavailable. In this case, we rewrite (17) as

$$\min_{\mathbf{D}} \mathcal{Z}_3\left(\mathbf{D}\right) = \|\hat{\mathbf{X}} - \mathbf{D}\hat{\mathbf{A}}\|_F^2 \quad s.t. \quad \|\mathbf{d}_i\|_2^2 \leq q, \forall i, \tag{18}$$

where $\hat{\mathbf{X}} = \left[\sqrt{-v_{11}/c^2}\mathbf{x}_1, \ldots, \sqrt{-v_{1n}/c^2}\mathbf{x}_n\right]$ and $\hat{\mathbf{A}} = \left[\sqrt{-v_{11}/c^2}\mathbf{a}_1, \ldots, \sqrt{-v_{1n}/c^2}\mathbf{a}_n\right]$. Then (18) can be solved effectively by the Lagrange dual method [5].

For clarity, the entire algorithm of solving the problem (4) is summarized in Algorithm 1, in which $\mathbf{v}_1$ and $\mathbf{v}_2$ are initialized to -1. We set the initial value of $\{\mathbf{D}, \mathbf{W}, \mathbf{A}\}$ to be random number between 0 and 1. Also we set the convergence condition as $|G_1(\mathbf{D}^s, \mathbf{W}^s, \mathbf{A}^s) - G_1(\mathbf{D}^{s-1}, \mathbf{W}^{s-1}, \mathbf{A}^{s-1})| < \psi$ where $\psi = 10^{-3}$.

Moreover, we analyse the computation cost of our algorithm as follows. In Line 4, the complexity for computing $\{\mathbf{v}_1, \mathbf{v}_2\}$ is $\mathcal{O}\left(dk\right)$. Due to the pseudo-inverse of $\mathbf{AJ}$, the complexity of finding $\mathbf{W}$ is $\mathcal{O}\left(max\left(rn^2, kn^2\right)\right)$. Since we use the gradient decent method to update $\mathbf{A}$, the complexity of Line 6 is upper

---

Algorithm 1. The HQ optimization algorithm for the objective function (4).

Input: The data matrix $\mathbf{X}$; the label matrix $\mathbf{Y}$; the free parameters c, λ, β, μ, and q; and the maximum iteration number S.

1: Set $s = 0$, $\psi = 10^{-3}$;
2: Initialize $\mathbf{D}$, $\mathbf{W}$, $\mathbf{A}$, $\mathbf{v}_1$ and $\mathbf{v}_2$;
3: **while** not converged & $s < S$ **do**
4:    Construct $\mathbf{\Omega}_1 = diag\left(-\mathbf{v}_1\right)$ and $\mathbf{\Omega}_2 = diag\left(-\mathbf{v}_2\right)$, where $\mathbf{v}_1$ and $\mathbf{v}_2$ are updated via (12);
5:    Update $\mathbf{W}$ by solving (14);
6:    Update $\mathbf{A}$ by solving (16) using gradient descent algorithm;
7:    Update $\mathbf{D}$ by solving (18) using Lagrange dual method;
8:    Set $s = s + 1$;
9: **end while**
Output: The dictionary matrix $\mathbf{D}$; the sparse coding matrix $\mathbf{A}$; and the parameter matrix $\mathbf{W}$.

---

bounded by $\mathcal{O}\left(dtn\right)$ where t is the number of iterations of gradient descent. Besides, we can get the complexity for computing $\mathbf{D}$ as $\mathcal{O}\left(dkn\right)$. Therefore, the overall complexity of our procedure is $\mathcal{O}\left(s\left(dk + max\left(rn^2, kn^2\right) + dtn + dkn\right)\right)$ in which s represents the number of iterations of Algorithm 1.

2.4 Classification Strategy

Once the structured dictionary $\mathbf{D}$ and the parameter matrix $\mathbf{W}$ have been learned, the classification procedure of our model can be described as follows. For the new test data $\tilde{\mathbf{X}}$, we first compute the sparse coding matrix $\tilde{\mathbf{A}}$ over the learned $\mathbf{D}$ by solving the following problem

$$\tilde{\mathbf{A}} = \|\tilde{\mathbf{X}} - \mathbf{D}\tilde{\mathbf{A}}\|_{\mathrm{F}}^2 + \frac{\beta}{2}\|\tilde{\mathbf{A}}\|_1. \tag{19}$$

We apply the linear classifier $f(\cdot)$ on $\tilde{\mathbf{A}}$ by $\tilde{\mathbf{Y}} = \mathbf{W}\tilde{\mathbf{A}}$. Then we find the maximum value of each column in $\tilde{\mathbf{Y}}$ which represents the predicted label for test example.

3 Experiment

In this section, we compare the proposed LWR with two well-known supervised dictionary learning algorithms including Discriminative K-SVD (DKSVD) [11] and Label Consistent K-SVD (LCKSVD1 and LCKSVD2) [4], as well as two state-of-the-art semi-supervised dictionary learning methods including Unified Semi-Supervised Dictionary Learning (USSDL) [9] and Discriminative Semi-Supervised Dictionary Learning (DSSDL) [10] on three real-world collections which contain two object recognition datasets and one digit classification dataset. Furthermore, we investigate the convergence property of our method. In the experiments, we select 70% examples from the entire dataset to form the training data, and the rest 30% examples are treated as test data. Moreover, we

Table 1. The test accuracies of compared methods on *COIL20* dataset. ●/○ indicates that LWR is significantly better/worse than the corresponding method (paired *t*-test with 95% confidence level).

	10%	15%	20%
DKSVD [11]	48.75 ± 5.28 ●	49.61 ± 6.13 ●	52.52 ± 5.11 ●
LCKSVD1 [4]	70.67 ± 2.03 ●	71.63 ± 2.34 ●	74.94 ± 1.44 ●
LCKSVD2 [4]	71.83 ± 1.38 ●	73.75 ± 2.10 ●	75.77 ± 2.17 ●
USSDL [9]	70.88 ± 3.50 ●	74.26 ± 2.98 ●	77.88 ± 3.64 ●
DSSDL [10]	57.71 ± 3.35 ●	59.81 ± 3.36 ●	60.95 ± 3.22 ●
LWR (Ours)	75.21 ± 1.26	79.89 ± 1.86	83.64 ± 1.13

consider three different cases in which 10%, 15% and 20% of training examples are labeled. We repeat above process 10 times for each dataset and then calculate the average test accuracies of compared algorithms with the standard deviation to measure their performances. Because the parameter c determines the upper bound of the loss function in our LWR, we fix $c = 1$ to get small losses for the outliers in all experiments. Moreover, we set $S = 10$ to determine the maximum iteration of the algorithm and use the paired t-test with 95% confidence level to examine whether the accuracies of LWR are significantly higher than other algorithms.

3.1   Object Recognition

Object recognition has been widely studied as a traditional research area of computer vision because of the extensive practical demands. We apply the proposed LWR on *COIL20*[1] and *Textures*[2] for object recognition.

COIL20 is one of the most well-known object recognition datasets, which contains 1440 object images belonging to 20 classes. The size of each image is 32×32 with 256 grey levels per pixel. Therefore, each image is represented by a 1024-dimensional feature vector. We build a 7-NN graph with $\lambda = 10$, and set $\beta = 0.01$, $\mu = 100$ and $q = 120$ for LWR in *COIL20*. For a fair comparison, we set the same dictionary size $k = 50$ for all the algorithms. As for the parameters of baselines, we adopt the default values which are provided by the authors. Table 1 shows the mean accuracies with the standard deviation of ten independent runs of all methods on *COIL20*, in which the best result under each labeling rate is marked in red. From Table 1, we observe that LWR achieves the best performance on all three ratios of labeling examples when compared with baselines. In particular, approximately 80% accuracy can be acquired when only 15% of training examples are labeled, which is better than other algorithms when 20% labeled examples are available. With 20% labeled examples, LWR can achieve

---

[1] http://www.cs.columbia.edu/CAVE/databases.
[2] http://www.robots.ox.ac.uk/~vgg/data/dtd/.

Table 2. The test accuracies of compared methods on *Textures* dataset. •/○ indicates that LWR is significantly better/worse than the corresponding method (paired t-test with 95% confidence level).

	10%	15%	20%
DKSVD [11]	17.30 ± 2.70 •	18.61 ± 2.59 •	19.17 ± 1.78 •
LCKSVD1 [4]	49.14 ± 3.65 •	49.59 ± 3.94 •	50.77 ± 1.69 •
LCKSVD2 [4]	48.59 ± 3.67 •	49.98 ± 1.74 •	51.27 ± 2.21 •
USSDL [9]	48.50 ± 2.39 •	52.33 ± 4.73 •	57.50 ± 3.80
DSSDL [10]	45.33 ± 3.71 •	46.67 ± 2.97 •	47.42 ± 3.17 •
LWR (Ours)	53.92 ± 2.86	58.92 ± 2.54	61.17 ± 2.64

Table 3. The test accuracies of compared methods on *USPS* dataset. •/○ indicates that LWR is significantly better/worse than the corresponding method (paired t-test with 95% confidence level).

	10%	15%	20%
DKSVD [11]	66.55 ± 3.96 •	67.75 ± 2.71 •	68.82 ± 3.35 •
LCKSVD1 [4]	81.27 ± 1.74 •	82.05 ± 0.97 •	82.51 ± 0.84 •
LCKSVD2 [4]	79.07 ± 2.13 •	80.16 ± 1.29 •	80.75 ± 1.15 •
USSDL [9]	68.90 ± 3.83 •	72.63 ± 3.62 •	78.10 ± 3.20 •
DSSDL [10]	66.23 ± 3.95 •	70.83 ± 3.55 •	72.62 ± 1.98 •
LWR (Ours)	83.65 ± 0.92	84.48 ± 0.75	85.72 ± 0.60

83% accuracy which is 6% higher than the best baseline model. Moreover, we use the paired t-test with 95% confidence level to examine whether the accuracies obtained by LWR are significantly higher than the baselines. As we can see, LWR consistently generates better performance than all the baselines on this dataset.

Textures is an evolving collection of textural images in the wild which contains 5640 images, organized according to a list of 47 categories inspired from human perception. Here we choose 120 images from each of 6 classes as the dataset including "Banded", "Chequered", "Cobwebbed", "Interlaced", "Knitted" and "Waffled". We use the low-level features including PHOG, GIST and LBP. We build a 10-NN graph for model comparison, and the key parameters in LWR are $\lambda = 2$, $\beta = 0.0001$, $\mu = 110$ and $q = 15$. As for the dictionary size, we set $k = 35$ for all methods. The results of all methodologies are illustrated in Table 2, in which the best result under each labeling rate is marked in red. We observe that DKSVD generates very low accuracy on this dataset, of which the accuracy is around 20%. In contrast, LCKSVD1 and LCKSVD2 are slightly better than DSSDL. Among the baseline methods, USSDL achieves very high accuracy. However, LWR is able to achieve better performance than USSDL. The

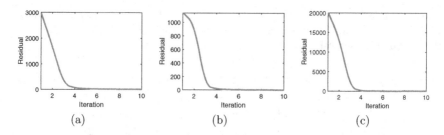

Fig. 2. Convergence behaviors of LWR on the three practical datasets. Subfigures (a) represents the convergence curves on *COIL20* dataset, (b) shows the curves on *Textures* dataset, and (c) is about *USPS* dataset.

paired t-test also statistically confirms the superiority of LWR to the compared baselines.

3.2 Digit Classification

Besides object recognition, digit classification is also an important task which is worth exploring. We evaluate the performance on the handwritten text recognition dataset *USPS*[3] which contains 9298 images consisting of 10 classes. Each image is represented by a 256-dimensional vector. In this experiment, we build a 10-NN graph with $\lambda = 2$, $\beta = 0.001$, $\mu = 100$ and $q = 140$ for LWR and set the size of dictionary $k = 20$ for all algorithms. The results are presented in Table 3 where the best performance has been marked as red. From Table 3 we observe that LCKSVD1 and LCKSVD2 are generally the best methods among the baselines. In contrast, our LWR can still obtain better results than them with fewer labeled examples. Furthermore, we conduct the t-test on the *USPS* which shows that LWR is significantly better than all the baselines.

3.3 Convergence Study

Due to that the adopted Welsch loss is nonconvex, we propose to use HQ optimization algorithm to solve the problem (4) in Algorithm 1. In this section, we empirically study the convergence property of Algorithm 1.

In Fig. 2, we plot the convergence curves of our algorithm on *COIL20*, *Textures* and *USPS* datasets with 15% training data labeled, where the y-axis represents the residual of object function and x-axis is the iteration times s. All the parameters remain the same as mentioned above. From Fig. 2, we can observe that the differences between successive iterations of object function gradually go down and converge within ten iterations on all datasets. This observation justifies that LWR can converge to a stable point rapidly and explains the reason that we set the maximum iteration number to 10 in all the above experiments.

---

[3] https://www.csie.ntu.edu.tw/~cjlin/libsvmtools/datasets/multiclass.html.

4 Conclusion

This paper proposed a novel semi-supervised dictionary learning algorithm called LWR to handle the outliers. Different from the previous algorithms, LWR takes advantages of the Welsch loss to restrict the reconstruction error and the classification error caused by the outliers. Moreover, we try to explore the data distribution information revealed by the unlabeled examples by introducing the Laplacian regularizer. Due to the non-convexity caused by Welsch loss, we use Half-Quadratic (HQ) optimization algorithm to solve the model. The experimental results on several real-world datasets show that LWR has superior performances than the existing state-of-art algorithms due to the robust Welsch loss.

Acknowledgments. This research is supported by NSF of China (No: 61602246, No: 61802189), NSF of Jiangsu Province (No: BK20171430, No: BK20180464), the Fundamental Research Funds for the Central Universities (No: 30918011319, No: 30918014107), the "Summit of the Six Top Talents Program" (No: DZXX-027), the "Innovative and Entrepreneurial Doctor Program" of Jiangsu Province, the "Young Elite Scientists Sponsorship Program" by Jiangsu Province, and the "Young Elite Scientists Sponsorship Program" by CAST (No: 2018QNRC001).

References

1. Boyd, S.P., Vandenberghe, L.: Convex Optimization. Cambridge University Press, Cambridge (2004)
2. He, R., Hu, B., Yuan, X., Wang, L.: M-estimators and half-quadratic minimization. In: Robust Recognition via Information Theoretic Learning. SCS, pp. 3–11. Springer, Cham (2014). https://doi.org/10.1007/978-3-319-07416-0_2
3. Jiang, W., Nie, F., Huang, H.: Robust dictionary learning with capped l1-norm. In: Proceedings of the International Joint Conference on Artificial Intelligence (2015)
4. Jiang, Z., Lin, Z., Davis, L.S.: Label consistent K-SVD: learning a discriminative dictionary for recognition. IEEE Trans. Pattern Anal. Mach. Intell. **11**, 2651–2664 (2013)
5. Lee, H., Battle, A., Raina, R., Ng, A.Y.: Efficient sparse coding algorithms. In: Proceedings of the Advances in Neural Information Processing Systems, pp. 801–808 (2007)
6. Mairal, J., Bach, F.R., Ponce, J., Sapiro, G., Zisserman, A.: Non-local sparse models for image restoration. In: Proceedings of the International Conference on Computer Vision, pp. 54–62 (2009)
7. Shrivastava, A., Pillai, J.K., Patel, V.M., Chellappa, R.: Learning discriminative dictionaries with partially labeled data. In: Proceedings of the International Conference on Image Processing, pp. 3113–3116 (2012)
8. Wang, D., Zhang, X., Fan, M., Ye, X.: Semi-supervised dictionary learning via structural sparse preserving. In: Proceedings of the AAAI Conference on Artificial Intelligence (2016)
9. Wang, X., Guo, X., Li, S.Z.: Adaptively unified semi-supervised dictionary learning with active points. In: Proceedings of the International Conference on Computer Vision, pp. 1787–1795 (2015)

10. Yang, M., Chen, L.: Discriminative semi-supervised dictionary learning with entropy regularization for pattern classification. In: Proceedings AAAI Conference on Artificial Intelligence (2017)
11. Zhang, Q., Li, B.: Discriminative K-SVD for dictionary learning in face recognition. In: Proceedings of the Computer Society Conference on Computer Vision and Pattern Recognition, pp. 2691–2698 (2010)
12. Zhang, X., Wang, D., Zhou, Z., Ma, Y.: Simultaneous rectification and alignment via robust recovery of low-rank tensors. In: Proceedings of the Advances in Neural Information Processing Systems, pp. 1637–1645 (2013)

Non-local MMDenseNet
with Cross-Band Features
for Audio Source Separation

Yi Huang[(✉)]

Nanjing University of Science and Technology, Nanjing, China
huisama@njust.edu.cn

Abstract. Audio source separation is an important but challenging problem for many applications due to the only available single channel mixed signal. This work proposes a novel Non-Local Multi-scale Multi-band DenseNet model termed as NLMMDenseNet for audio source separation by jointly exploring the long-term dependencies and recovering the missing information around bands' borders. Specifically, to well leverage the long-term dependencies among the audio spectrogram, we propose a new non-local model by incorporating the non-local layer into MMDenseNet. It enables the proposed model to capture different audio sources features. Besides, the proposed model can also capture cross-band features, which are used to recover the missing information around bands' borders. The proposed model outperforms state-of-the-art results on the widely-used MIR-1K and DSD100 datasets by taking advantages of global information and bands' border information.

Keywords: Source separation · Deep learning · Signal processing

1 Introduction

Audio source separation is to identify one or more audio sources from one observed mixed signal, which is important for many applications, such as music transcription, audio noise reduction and music remixing. However, it is very challenging due to that many audio sources need to be recovered but only one observed mixed signal is available. Therefore, it is necessary and urgent to develop intelligent methods to automatically separate audio sources.

To solve this problem, many methods have been proposed including Nonnegative Matrix Factorization (NMF) [17], low-rank models [4], Bayesian models [25] and sparse models [4]. These methods are mainly based on some simple hypothesis to simplify the problem. However, these methods have poor generalization ability, which should be addressed for performance improvement.

Recently, Deep Neural Networks (DNNs) have shown good generalization power in visual applications because of their ability to learn the complex structures of the observed data [11]. Due to the superior performance of DNN, some

© Springer Nature Switzerland AG 2019
Z. Cui et al. (Eds.): IScIDE 2019, LNCS 11936, pp. 53–64, 2019.
https://doi.org/10.1007/978-3-030-36204-1_4

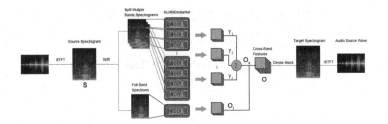

Fig. 1. The architecture of NLMMDenseNet with cross-band features.

deep methods have been developed based on Convolutional Neural Network (CNN) [12], Recurrent Neural Network (RNN) [5] and ensemble learning [21] for audio source separation. MMDenseNet [20] is a DenseNet-based model proposed for solving audio source separation problem and gets fine performance recently. The success of MMDenseNet results from dense block connection and parallelly built sub-Nets for multiple bands of the spectrogram, but there are problems of missing global information caused by convolutional layers in the denseblock and information around the bands' borders.

Towards this end, this paper proposes a new non-local deep neural network based on MMDenseNet, termed as NLMMDenseNet, by simultaneously exploring the global information and recovering the missing information around bands' borders for audio source separation, as illustrated in Fig. 1. The target audio source's magnitudes are estimated by using multiple-denseblock-based networks from the mixture magnitudes. To explore the long-term dependencies among the feature maps, the non-local scheme is used within the proposed MMDenseNet-based deep network by incorporating non-local layers. It can well estimate the global information from the feature maps extracted from the previous layers, which enables the network to better capture different audio sources features. Besides, some information around the bands' borders disappears due to the magnitude splitting. To deal with it, we propose to introduce new and generic features named cross-band features into the architecture to recover the missing information. To show the effectiveness of the proposed method, extensive experiments are conducted on two widely-used MIR-1K and DSD100 datasets. The proposed method achieves the best results compared with the state-of-the-arts.

2   Related Work

2.1   RNN/CNN-Based Approaches

RNN/CNN-based approaches are popular models to address audio source separation. By exploring the related prior information, and the sequential relativity between the mixture and each audio source, they usually estimate the sources directly from the mixture and approximate the original audio sources.

CNN-based models suffer from local receptive fields of each convolution operator. If the model heavily relies on convolutional feature maps, the long-term dependencies can be modeled only after passing through several convolutional layers [26]. That is, the model cannot capture the long-term dependencies if the scale of the model is not large enough. The RNN-based architecture has some innate weaknesses for long sequence [18]. It is difficult to learn good parameters when the length of the input sequence is long.

Different from above methods, we propose a deep model with the non-local mechanism, which can be deemed as a complementary to RNN/CNN by effectively modeling long range and multi-level dependencies.

2.2 Generative Approaches

Generative approaches are also widely used for audio source separation. The Bayesian-based methods perform well when we have prior information about the audio source. However, it is difficult to manually model prior information for thousands of audio sources. Thus, more constraints on the sources are usually needed to solve this problem. Traditional methods often introduce many simple and strong assumptions about the data generation process [10], which can simplify some difficult problems of posterior inference. But it always leads to unsatisfying separation results. Recently NMF-based generative methods are popular for separation [17]. They can integrate prior information about audio sources by adapting spectral templates to improve the performance. NMF is also limited on the assumption that the spectral context is factorized independently of time, and many spectral bases are required to represent complex instruments.

Currently generative approaches either are largely limited on many constraints for simplifying problems of posterior inference, or need expert knowledge to model the prior information. In contrast, the proposed method can learn source information from data without any assumption about the audio source.

3 Method

3.1 Preliminary

A mixed audio signal can be described as follows:

$$m[t] = \sum_{i=1}^{n} s_i[t] + o[t], t = 1, 2, 3, ..., k \tag{1}$$

where $m[t]$ denotes the observed mixed signal, $s_i[t]$ is the signal from the i-th audio source, $o[t]$ denotes the noisy signal at sampling tick t. n is the number of sources, and k is the total sampling number. The target of the audio source separation is to recover each audio source signal from the mixture by finding a set of functions $F = \{f_1(m), f_2(m), f_3(m), ..., f_n(m)\}$.

$$f_i(m[t]) \approx s_i[t], i = 1, 2, 3, ..., n \tag{2}$$

This problem is so-called "Cocktail Party Problem".

It is hard to manually design the function set F, and recent researches have shown that DNN-based methods enable to model complex structure effectively and also have fine generalization performance. Consequently, we propose an audio source separation method based on DNN to learn $f_i(m)$ by simultaneously considering MMDenseNet, cross-band features and non-local schemes into one unified framework, as shown in Fig. 1. The input mixed audio signal is converted to the frequency domain using the discrete STFT. The phase part of the STFT result is not processed and directly restored into the output.

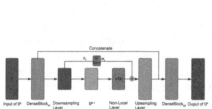

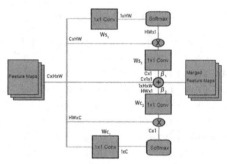

Fig. 2. Non-local MMDenseNet unit structure of k-th scale U^k.

Fig. 3. The structure of non-local layer. × denotes matrix multiplication and + denotes broadcast element-wise addition, which will first broadcasts features in each dimension to match the dimensions for two matrices with different dimensions.

3.2 Non-local MMDenseNet with Cross-Band Features

Recently, DenseNet [3] has shown fine performance on visual tasks among various types of CNNs. DenseNet is a network with dense concatenation where there is a direct connection between any two layers. This scheme optimizes the information flow between layers where the learned feature maps of one layer can be directly passed to all the following layers as the input to enable all the layers to receive the gradient directly and reuse features computed in the preceding layers.

MMDenseNet is a DenseNet-based CNN model which performs fine performance on audio source separation task. MMDenseNet has c scales which apply $c-1$ times downsampling and upsampling operations and has $2^{2(c-1)}$ times lower resolution compared to the original feature maps. The valuable idea of this architecture is that first it uses dense blocks at multiple scales which progressly downsample and upsample the feature maps to recover the target spectrogram with the input mixture spectrogram; and then it splits the full-band of the spectrogram into several sub-spectrograms and parallelly builds multiple sub-networks to respectively extract features from these sub-band-spectrograms and finally merge these features as the full feature. However, the basic MMDenseNet has also some problems as follows:

- The skeleton of basic MMDenseNet highly relies on denseblock which causes a number of feature maps in the finally full feature. Feature maps extracted by CNN can easily suffer from local receptive field of the convolution operator to miss the global information.
- The full-band is manually and roughly split into multiple bands and the CNN can hardly consider the missing information about the spectrograms around the border between 2 bands, even though a convolutional network is parallelly built for full-band spectrogram, which can be seemed as a work-around for the missing features around the band border.

To address these problems, we propose a new deep model NLMMDenseNet by jointly considering the non-local scheme and recovering the missing information around the band borders. To capture the global information from the feature maps extracted from the convolutional network, the non-local scheme is incorporated into the basic MMDenseNet as shown in Fig. 2. Furthermore, we propose a new and generic feature named cross-band features to recover the missing information around the band borders. It is worth noting that the features extracted by basic MMDenseNet can be deemed as a specific case of the cross-band features.

In addition, some new schemes are introduced to improve the performance. First, for the input of the downsampling layer, we append a residual structure between the output of the upsampling layer and the non-local layer to further optimize the flow of the gradient. Thus, the Non-local MMDenseNet unit structure U can be defined recursively as follows:

$$U^k(\mathbf{I}) = \begin{cases} DenseBlock_{k2}(Concat(Upsample(\alpha_k \mathbf{W}_k Q_k(\mathbf{I}) + \\ nl_k(U^{k-1}(Q_k(\mathbf{I})))), DenseBlock_{k1}(\mathbf{I}))) & k > 0 \\ DenseBlock_0(\mathbf{I}) & k = 0 \end{cases} \quad (3)$$

where $Q_k(\mathbf{I}) = Downsample(DenseBlock_{k1}(\mathbf{I}))$, $k = 1, 2...c$ denotes the k-th scale. $\mathbf{I}$ denotes the input of the unit, $\mathbf{W}_k$ denotes the 1×1 convolution layer, $nl_k(\mathbf{X})$ denotes the output of the non-local layer with input $\mathbf{X}$ (detailed in Sect. 3.3), and α_k is a trainable scalar variable with initial value 0, of the k-th scale. And we also change the upsampling layer from the deconvolution layer into the combination of interpolating upsampling layer and 1×1 convolutional layer to accelerate the training speed and reduce the number of parameters.

3.3 Non-local Layer

The non-local scheme has been verified that it enables to improve the performance of DNN-based model effectively [22]. For the audio source separation task, all we can analyze is just the spectrogram of the mixture channel. Thus, the relative information extracted from the mixture's spectrogram is essentially important. On the one hand, it is very difficult to have any available prior information during the processing. On the other hand, long-term dependencies are also important for CNN-based model, because the CNN-based model relies on a number of feature maps to recognize a specific audio source. Therefore, it is essential and applicable to figure out audio sources' features from the mixture by exploring the non-local scheme. It explores the internal relativity from the spectrogram, which can perform as the complementary to the local receptive field of the convolution operator [23]. Recent researches have shown that the traditional non-local mechanism can be simplified to reduce the calculation [1], so we have designed a light non-local layer in our network, besides, we have also introduced the dual-attention mechanism to further improve the performance [2].

The structure of the proposed non-local layer is shown as Fig. 3. For the feature maps $\mathbf{x}_p \in R^{C \times W \times H}$ extracted by the convolutional layer, we first transform $\mathbf{x}_p$ into the flatten feature maps $\mathbf{x} \in R^{C \times N} = \{\mathbf{x}_1, \mathbf{x}_2, ..., \mathbf{x}_N\}$ where $N = W \times H$. There are two calculation paths respectively for spatial attention $\mathbf{s} \in R^{C \times N} = \{\mathbf{s}_1, \mathbf{s}_2, ..., \mathbf{s}_N\}$ and channel attention $\mathbf{c} \in R^{N \times C} = \{\mathbf{c}_1, \mathbf{c}_2, ..., \mathbf{c}_C\}$.

$$\mathbf{s}_i = \mathbf{W}_{s_2} \sum_{j=1}^{N} \frac{e^{\mathbf{W}_{s_1} \mathbf{x}_j}}{\sum_{m=1}^{N} e^{\mathbf{W}_{s_1} \mathbf{x}_m}} \mathbf{x}_j, i = 1, 2, ..., N \tag{4}$$

where $\mathbf{W}_{s_1}$ and $\mathbf{W}_{s_2}$ denote 1×1 convolution layer. For the channel attention, we define $\tilde{\mathbf{x}} = \mathbf{x}^T \in R^{N \times C} = \{\tilde{\mathbf{x}}_1, \tilde{\mathbf{x}}_2, ..., \tilde{\mathbf{x}}_C\}$.

$$\mathbf{c}_{i'} = \mathbf{W}_{c_2} \sum_{j=1}^{C} \frac{e^{\mathbf{W}_{c_1} \tilde{\mathbf{x}}_j}}{\sum_{m=1}^{C} e^{\mathbf{W}_{c_1} \tilde{\mathbf{x}}_m}} \tilde{\mathbf{x}}_j, i' = 1, 2, ...C \tag{5}$$

where $\mathbf{W}_{c_1}$ and $\mathbf{W}_{c_2}$ denote 1×1 convolution layer. And the output of the non-local layer is a mergence between original feature maps $\mathbf{x}$ and spatial/channel attention feature maps $\mathbf{s}/\mathbf{c}$:

$$nl(\mathbf{x}) = Reshape(\beta_1 \mathbf{s} + \beta_2 \mathbf{c}^T + \mathbf{x}) \in R^{C \times W \times H} \tag{6}$$

where β_1 and β_2 are trainable scalar variables with initial value 0 to indicate the reliability of the attention features.

3.4   Cross-Band Features

For the basic multi-band method, supposing the full-band spectrogram is $\mathbf{S} \in R^{n \times m}$ where n is the number of frequency bins and m is the number of time frames, the interval $[0, n-1]$ of the spectrogram's band is split by a manually designed approach into multiple sub-bands (hereafter $\mathbf{F}$ means the band interval corresponding to $\mathbf{S}$):

$$C = \{\mathbf{F}_1, \mathbf{F}_2, ..., \mathbf{F}_m | \bigcap_{i \neq j} \mathbf{F}_i, \mathbf{F}_j = \phi, \bigcup_i \mathbf{F}_i = \mathbf{F}, i, j = 1, ..., m\} \tag{7}$$

Then we build $m+1$ parallel convolutional networks NET to extract the features of the input spectrograms.

$$NET = \{NET_1, NET_2, ..., NET_m, NET_f\} \tag{8}$$

$$\mathbf{O}_i = NET_i(\mathbf{S}_i), \mathbf{O}_f = NET_f(\mathbf{S}), i = 1, 2, ..., m, \tag{9}$$

These features on frequency dimension are concatenated:

$$\hat{\mathbf{O}} = Concat_{frequency}(\mathbf{O}_1, ..., \mathbf{O}_m) \tag{10}$$

Finally we get the full features extracted from the model by concatenating the full-band features on channel dimension:

$$\mathbf{O}_{basic} = Concat_{channel}(\mathbf{O}_f, \hat{\mathbf{O}}) \tag{11}$$

In the above basic multi-band method, there are no convolution networks paying attention to features around the border between neighbor bands. It can be seemed as a work-around for the missing features around the band borders by building the convolutional network for the full-band spectrogram. However, it will enlarge the input size and relatively reduce the local receptive fields of the convolutional network if the convolutional network is applied on the full-band spectrograms, and it will reduce the effectiveness of recovering the missing features around the band borders.

To well address the above problem, we propose a new cross-band approach to leverage the information around the borders. It applies multiple splitting strategies on the spectrogram, and parallelly builds multiple sub-CNNs corresponding to these strategies. All the features extracted by sub-CNNs will be merged as the full features.

For spectrogram $\mathbf{S} \in R^{n \times m}$, we split $\mathbf{S}$ into multiple bands spectrogram with $\lambda(\lambda > 0)$ strategies:

$$C_l = \{\mathbf{F}_{l1}, \mathbf{F}_{l2}, ..., \mathbf{F}_{lm_l} | \bigcap_{i \neq j} \mathbf{F}_{li}, \mathbf{F}_{lj} = \phi,$$
$$\bigcup_i \mathbf{F}_{li} = \mathbf{F}, i, j = 1, ..., m_l\}, l = 1, ..., \lambda \tag{12}$$

where C_l is an ordered set ordered by $LEFT(\mathbf{F}_{li})$ in ascending order. We defined the set of borders as

$$D_l = \{b_{li} | b_{li} = RIGHT(\mathbf{F}_{li}), i = 1, 2..., m_l - 1\} \tag{13}$$

and the minimum border overlapping size as $z \in N$. Here $LEFT(\mathbf{x})/RIGHT(\mathbf{x})$ means the left/right endpoint value of interval $\mathbf{x}$ respectively.

The splitting strategies subject to constraint as follows:

$$\forall b_{pi} \in D_p(1 \leq p \leq \lambda, 1 \leq i \leq m_p - 1)$$
$$\exists \mathbf{F}_{qj} \in C_q(q \neq p, 1 \leq q \leq \lambda, 1 \leq j \leq m_q)$$
$$b_{pi} \in \mathbf{F}_{qj} \tag{14}$$
$$while \ \forall b_{pi'} \in D_p(i' \neq i, 1 \leq i' \leq m_p - 1), b_{pi'} \notin \mathbf{F}_{qj}$$
$$and \ b_{pi} - LEFT(\mathbf{F}_{qj}) \geq z, RIGHT(\mathbf{F}_{qj}) - b_{pi} \geq z$$

For each strategy, we build convolutional networks:

$$NET_l = \{U_{l1}^c, U_{l2}^c, ..., U_{lm_l}^c\}, l = 1, .., \lambda \tag{15}$$

The output features of each sub-CNN are

$$O_l = Concat_{frequency}(U_{lr}^c(S_{lr})), l = 1, .., \lambda, r = 1, ..., m_l \qquad (16)$$

Finally, we merge all the sub-CNNs' features:

$$O_s = Merge(O_1, O_2, .., O_\lambda) = \sum_{i=1}^{\lambda} \gamma_i O_i \qquad (17)$$

where $\gamma_t(t = 2, ..., \lambda)$ are trainable scalar variables with initial value 0 while γ_1 is also a trainable scalar variable with initial value 1.

For improving the performance of our network, we still reserve the full-band convolutional network as U_f^c whose output features are still $O_f = U_f^c(S)$, which is deemed as an assistant network. And the final output features:

$$O = Concat_{channel}(O_f, O_s) \qquad (18)$$

which is named as cross-band features. The cross-band features ensure each band border and the spectrograms around it to be considered by at least one convolutional network without missing important information. Also, the features extracted by MMDenseNet can be deemed as a specific case of the cross-band features under the circumstance $\lambda = 1$.

4   Experiment

4.1   Dataset

Experiments are conducted on two widely-used datasets: MIR-1K and DSD100.

MIR-1K [15]. It is designed for the research of the singing voice separation. It has 1,000 songs clips with both music accompaniment and singing voice audio source tracks. Each clip is encoded at a sample rate of 16 kHz with durations from 4 to 13 s and extracted from 110 Chinese karaoke songs sung by both females and males. In experiments, each clip is preprocessed by mixing voice and accompaniment track into one track with equal energy.

DSD100 [21]. It contains 100 high quality songs. Each song has four different audio sources of vocals, bass, drums and other, and the mixture is provided which corresponds to the sum of all the signals. All those audio sources are stereophonic sound at the sampling rate of 44.1 kHz. In experiments, all the mixtures are preprocessed into one track. DSD100 provides songs with different styles, and each song has commercial grade quality.

4.2   Metrics and Setting

To evaluate the performance of the source separation methods, three widely-used metrics are utilized, i.e., Signal to Interface Ratio (SIR), Source to Artifacts Ratio (SAR), and Source to Distortion Ratio (SDR). The Normalized SDR (NSDR) is defined as follows:

$$NSDR(\hat{s}, s, x) = SDR(\hat{s}, s) - SDR(x, s) \qquad (19)$$

Table 1. Quantitative results of singing voice and accompaniments separation on the MIR-1K dataset.

Type	Vocals			Accompaniments		
Model	GNSDR	GSIR	GSAR	GNSDR	GSIR	GSAR
DRNN [6]	7.45	13.08	9.68	–	–	–
ModGD [16]	7.50	13.73	9.45	–	–	–
U-Net [7]	7.43	11.79	10.42	7.45	11.43	10.41
SH-4Stack [14]	10.51	16.01	12.53	9.88	14.24	12.36
MMDenseNet [20]	10.70	17.01	12.66	10.69	15.05	12.48
NLMMDenseNet	**10.93**	**17.21**	**12.84**	**10.82**	**15.25**	**12.61**

where $\hat{s}$ is the estimated audio source data, s is the target audio source data and x is the mixture data. In experiments, the Global Normalized SDR (GNSDR), Global SAR (GSAR) and Global SIR (GSIR) respectively defined as means of NSDRs, SARs and SIRs weighted by themselves' lengths are used. The bigger the value is, the better the result is.

In experiments, there are some settings to be set in advance. For the MIR-1K dataset, one male and one female's singing recording are used as the training set which contains 175 clips, and the remain 825 clips are used as the testing set, and for the DSD100 dataset, 100 songs are divided into 50 training sets and 50 testings sets [14]. For the STFT operation, we use 2048-point STFT with 50% overlap on the two datasets. For splitting strategies, considering the balance between effectiveness and scale of the model, we set $\lambda = 2$, $z = \lceil \frac{1}{6}n \rceil$ where $\mathbf{C}_1 = \{[0, \lceil \frac{1}{2}n \rceil - 1], [\lceil \frac{1}{2}n \rceil, n-1]\}$, $\mathbf{C}_2 = \{[0, \lceil \frac{1}{3}n \rceil - 1], [\lceil \frac{1}{3}n \rceil], [\lceil \frac{2}{3}n \rceil - 1], [\lceil \frac{2}{3}n \rceil, n-1]\}$. And the scale $c = 3$. The full-band network is the same as MMDenseNet's, and the growrate of all the denseblocks is set as 16, and each denseblock has 4 layers.

The Adam and SGD optimizers are used to train our network where we use Adam optimizer for main training (training duration until the loss gets stable) and SGD for manually fine-tuning after that, and we respectively train models for each audio source where the model with the best result is selected. The MWF (multi-channel Wiener Filter) is also used as post processing to further denoise the output signals to improve the quality of the output audio signals. The MSE (Mean Square Error) has been used as the loss function.

We also found that data augmentation is very important in our experiments and can effectively improve the performance of our network. In experiments, we adopt the following strategies [21].

– randomly shifting one or more audio sources' spectrograms and mix them up as the mixture.
– randomly scaling audio sources' amplitudes with ratio within $[0.5, 1.5]$ and mix them up as the mixture.
– randomly mixing up instruments from different songs as the mixture.

Table 2. Median SDR values for audio source separation on DSD100 dataset.

Model	Bass	Drums	Other	Vocals	Accompaniments
wRPCA [8]	–	–	–	3.92	9.45
NUG [13]	–	–	–	4.55	10.29
dNMF [24]	0.91	1.87	2.43	2.56	–
DeepNMF [9]	1.88	2.11	2.64	2.75	8.90
BLEND [21]	2.76	3.93	3.37	5.13	11.70
SH-4Stack [14]	1.77	4.11	2.36	5.16	12.14
MMDenseNet [20]	3.91	5.37	3.81	6.00	12.10
MMDenseLSTM [19]	3.73	5.46	**4.33**	6.31	**12.73**
NLMMDenseNet	**3.98**	**5.49**	4.17	**6.38**	12.39

Table 3. Median SDR values for audio source separation on DSD100 dataset for ablation experiments. NL denotes the usage of non-local scheme, CB denotes the usage of cross-band features and CB-FB denotes the usage of cross-band features without the full-band convolutional network.

Model	Bass	Drums	Other	Vocals	Accompaniments
MMDenseNet (baseline)	3.91	5.37	3.81	6.00	12.10
NLMMDenseNet (NL)	3.90	5.44	**4.19**	6.29	12.30
NLMMDenseNet (CB)	3.96	5.40	4.12	6.20	12.24
NLMMDenseNet (CB-FB)	3.93	5.38	4.07	6.16	12.19
NLMMDenseNet (NL+CB)	**3.98**	**5.49**	4.17	**6.38**	**12.39**

4.3 Experimental Results

The results of the proposed method and several state-of-the-art methods on the MIR-1K dataset are shown in Table 1 including, DRNN, ModGD, U-Net, SH-4Stack and MMDenseNet. From the results, we can easily observe that compared to the best results of the previous methods, the proposed method gets all the best results on vocal and accompaniment separation tasks. The compared MMDenseNet for MIR-1K dataset is implemented by ourselves.

The results of the proposed method and several state-of-the-art methods including DeepNMF, wRPCA, NUG, BLEND, dNMF, MMDenseNet, SH-4Stack and MMDenseLSTM on the DSD100 dataset are shown in Table 2. The music audio source separation on DSD100 aims at separating audio sources of vocals, bass, drums, other and accompaniments from the mixture. For each experiment, we measured all the SDR values of the separated audio sources and calculated median values. From the results, we get the best results on bass, drums and vocals source separation tasks and second best results on other and accompaniments source separation tasks.

Besides, we conducted the ablation experiments on the DSD100 dataset. The ablation experiments validate whether both of the two main components of

NLMMDenseNet contribute to the separation result. The results are shown in Table 3. Compared to original MMDenseNet's results, it is indicated that 1. both of the non-local scheme and cross-band features are valid and effective, besides, the combination of these two components can further improve the performance of MMDenseNet; 2. the cross-band features without the full-band path perform better than the full-band convolutional network, especially on medium-high frequency domain; 3. the full-band convolutional network can further improve the performance as the assistant network. The results can well indicate the effectiveness of our motivation by considering the long-term dependencies and the information around band borders.

5   Conclusion

In this paper, we propose an improved MMDenseNet architecture for audio source separation by using cross-band features and non-local mechanism. This network can recognize audio sources features from the spectrogram and successfully separate each audio source from the mixture. Experimental results show the effectiveness of the proposed deep architecture for audio source separation. In future, we will take into account ensemble learning of multiple models and combination with other advanced network layers. Besides, how to process the phase part of STFT result is another important direction.

Acknowledgement. This work was partially supported by the National Key Research and Development Program of China under Grant 2017YFC0820601 and the National Natural Science Foundation of China (Grant No. 61772275, 61720106004 and 61672304).

References

1. Cao, Y., Xu, J., Lin, S., Wei, F., Hu, H.: Gcnet: Non-local networks meet squeeze-excitation networks and beyond. arXiv preprint arXiv:1904.11492 (2019)
2. Fu, J., et al.: Dual attention network for scene segmentation. In: Proceedings of the IEEE Conference on Computer Vision and Pattern Recognition, pp. 3146–3154 (2019)
3. Huang, G., Liu, Z., Van Der Maaten, L., Weinberger, K.Q.: Densely connected convolutional networks. In: Computer Vision and Pattern Recognition, pp. 4700–4708 (2017)
4. Huang, P.S., Chen, S.D., Smaragdis, P., Hasegawa-Johnson, M.: Singing-voice separation from monaural recordings using robust principal component analysis. In: Acoustics, Speech and Signal Processing, pp. 57–60 (2012)
5. Huang, P.S., Kim, M., Hasegawa-Johnson, M., Smaragdis, P.: Singing-voice separation from monaural recordings using deep recurrent neural networks. In: ISMIR, pp. 477–482 (2014)
6. Huang, P.S., Kim, M., Hasegawa-Johnson, M., Smaragdis, P.: Joint optimization of masks and deep recurrent neural networks for monaural source separation. IEEE/ACM Trans. Audio Speech Lang. Process. **23**(12), 2136–2147 (2015)

7. Jansson, A., Humphrey, E., Montecchio, N., Bittner, R., Kumar, A., Weyde, T.: Singing voice separation with deep u-net convolutional networks (2017)
8. Jeong, I.Y., Lee, K.: Singing voice separation using rpca with weighted l_1-norm. In: International Conference on Latent Variable Analysis and Signal Separation, pp. 553–562 (2017)
9. Le Roux, J., Hershey, J.R., Weninger, F.: Deep NMF for speech separation. In: Acoustics, Speech and Signal Processing, pp. 66–70 (2015)
10. Li, Z., Liu, J., Tang, J., Lu, H.: Robust structured subspace learning for data representation. IEEE Trans. Pattern Anal. Mach. Intell. **37**(10), 2085–2098 (2015)
11. Li, Z., Tang, J., Mei, T.: Deep collaborative embedding for social image understanding. IEEE Trans. Pattern Anal. Mach. Intell. **41**(9), 2070–2083 (2019)
12. Miron, M., Janer Mestres, J., Gómez Gutiérrez, E.: Generating data to train convolutional neural networks for classical music source separation. In: The 14th Sound and Music Computing Conference (2017)
13. Nugraha, A.A., Liutkus, A., Vincent, E.: Multichannel audio source separation with deep neural networks. IEEE/ACM Trans. Audio Speech Lang. Process. **24**(9), 1652–1664 (2016)
14. Park, S., Kim, T., Lee, K., Kwak, N.: Music source separation using stacked hourglass networks. arXiv preprint arXiv:1805.08559 (2018)
15. Rafii, Z., Pardo, B.: A simple music/voice separation method based on the extraction of the repeating musical structure. In: Acoustics, Speech and Signal Processing, pp. 221–224 (2011)
16. Sebastian, J., Murthy, H.A.: Group delay based music source separation using deep recurrent neural networks. In: Signal Processing and Communications, pp. 1–5 (2016)
17. Sprechmann, P., Bronstein, A.M., Sapiro, G.: Real-time online singing voice separation from monaural recordings using robust low-rank modeling. In: ISMIR, pp. 67–72 (2012)
18. Sutskever, I., Vinyals, O., Le, Q.V.: Sequence to sequence learning with neural networks. In: Advances in Neural Information Processing Systems, pp. 3104–3112 (2014)
19. Takahashi, N., Goswami, N., Mitsufuji, Y.: MMDenseLSTM: an efficient combination of convolutional and recurrent neural networks for audio source separation. In: International Workshop on Acoustic Signal Enhancement, pp. 106–110 (2018)
20. Takahashi, N., Mitsufuji, Y.: Multi-scale multi-band densenets for audio source separation. In: Applications of Signal Processing to Audio and Acoustics, pp. 21–25 (2017)
21. Uhlich, S., et al.: Improving music source separation based on deep neural networks through data augmentation and network blending. In: Acoustics, Speech and Signal Processing, pp. 261–265 (2017)
22. Vaswani, A., et al.: Attention is all you need. In: Advances in Neural Information Processing Systems, pp. 6000–6010 (2017)
23. Wang, X., Girshick, R., Gupta, A., He, K.: Non-local neural networks. In: Proceedings of IEEE Computer Vision and Pattern Recognition, pp. 7794–7803 (2018)
24. Weninger, F., Roux, J.L., Hershey, J.R., Watanabe, S.: Discriminative NMF and its application to single-channel source separation. In: Annual Conference of the International Speech Communication Association (2014)
25. Yang, P.K., Hsu, C.C., Chien, J.T.: Bayesian singing-voice separation. In: ISMIR, pp. 507–512 (2014)
26. Zhang, H., Goodfellow, I., Metaxas, D., Odena, A.: Self-attention generative adversarial networks. arXiv preprint arXiv:1805.08318 (2018)

A New Method of Metaphor Recognition for A-is-B Model in Chinese Sentences

Wei-min Wang$^{(\boxtimes)}$, Rong-rong Gu, Shou-fu Fu,
and Dong-sheng Wang

School of Computer Science, Jiangsu University of Science and Technology,
Zhenjiang 212003, China
wangweimin@knowology.cn

Abstract. Metaphor recognition is the bottleneck of natural language processing, and the metaphor recognition for A-is-B mode is the difficulty of metaphor recognition. Compared with phrase recognition, the metaphor recognition for A-is-B mode is more flexible and difficult. To solve this difficult problem, the paper proposes a feature-based recognition method. First, the metaphor recognition problem for A-is-B model is transformed into a classification problem, then four sets of features of upper and lower position, sentence model, class, and Word2Vec are calculated respectively, and feature sets are constructed by using these four sets of features. The experiment uses the SVM model classifier and the neural network classifier to realize the metaphor recognition for the A-is-B mode. The experimental results show that the method using neural network classifier method has better accuracy and recall rate, 96.7% and 93.1%, respectively, but it takes more time to predict a sentence. According to the analysis of the experimental results of the two classifiers, the improved method achieved good results.

Keywords: A-is-B · Metaphor recognition · Feature fusion · Hyponymy · Sentence patterns · Class vocabulary · Word2Vec · SVM · Neural network

1 Introduction

Metaphor is an indispensable part of human language. The British rhetorician Richards found that people use metaphor every three sentences in their daily conversation [1]. In the actual text, the A-is-B metaphor is the type that all metaphor theories pay attention to, and one of the most commonly used sentences in the text.

Compared with the phrase metaphor and verb metaphor (V+N), the A-is-B metaphor is more flexible and more difficult to identify. The main difficulties are as follows: First, the scope of the source domain is more extensive. Because of the finiteness of verbs, we can use the semantic definition of verbs to construct recognition features when identifying V+N metaphors. But in the metaphor of the A-is-B model, the scope of A and B is very broad. It has no fixed source domain scope, even any noun or noun phrase may appear as a metaphor in a metaphor. For example, "他是我们班的孙悟空" ("He is the stone monkey of our class") is a metaphor sentence, and "孙悟空" ("stone monkey") rarely appears in metaphorical phrases. Second, the context is more sensitive. The same A-is-B sentence will have different meanings in different contexts. For example, the phrase "男人是动物" ("man is an animal") is just a classified statement in a biological article. But in other contexts, the sentence generally represents a metaphor. For another example, "她是一位辛勤的园丁" ("She is a hardworking gardener.") Non-metaphoric sentence: "她是一位辛勤的园丁，每日打理花园。" ("She is a hardworking gardener, taking care of the garden daily.") Metaphor sentence: "她是一位辛勤的园丁，对学生很负责。" ("She is a hardworking gardener and she is very responsible to the students.") Therefore, when identifying metaphors, we need to consider the context to be better identified. Third, the influence of pronouns. In the A-is-B metaphor, A is often a pronoun (I | you | he | she | it). For example, "他是一只老貔貅" ("He is an old brave troops.") When identifying metaphors, it is necessary to convert pronouns into specific nouns or noun phrases, that is, refers to disambiguation. But referring to disambiguation itself is a difficult job. Fourth, the existence of death metaphor. Because the metaphorical meaning of some source domain words is too prominent, people have defaulted the source domain words to the metaphorical meaning during use, which causes the source domain words to become a polysemous word. For example, "包袱" ("package") has a "some burden" in the basic interpretation of Chinese. Such dead source domain words also include "算账" ("accounting"), "水分" ("moisture") and so on. Therefore, the metaphor recognition work of A-is-B mode has become a difficult point in the development of natural language processing and machine translation.

In the 1980s, Lakoff and Johnson proposed conceptual metaphor theory from a cognitive perspective [2]. Since then, many metaphor recognition models have emerged, such as concept mapping models and selection priority models. After ten years of research, metaphor recognition research methods are broadly divided into the following categories: traditional methods based on semantics, statistics and cognition, research methods based on machine learning, and research methods based on deep learning and neural networks.

Traditional metaphor recognition tasks include semantic-based metaphor recognition methods and statistical-based metaphor recognition methods. In 2007, according to the characteristics of metaphor, Xu [3] established a metaphor recognition model based on the principle of maximum entropy, and demonstrated the rationality of using statistical means to establish the model. In 2017, Liu [4] used the web platform-based semantic annotation tool Wmatrix to retrieve and identify metaphors in self-built corpora. The experimental results show that Wmatrix can provide new ideas for the recognition of metaphors at the conceptual level. In the same year, Liu Chaojian also proposed the idea of combining corpus methods with metaphor theory. In 2019, Su [5] tried to analyze and explore English metaphor recognition from the perspective of language, psychology and cognition based on the difference calculation. Imitate humans to obtain the characteristics of the concept, choose the classification angle, calculate the difference under the specific classification angle, and carry out the experiment of English noun metaphor recognition. The experimental accuracy rate reached 85.4%.

In recent years, more research has analyzed text based on machine learning methods. In 2011, Liu [6] selected 20 commonly used metaphors, using the 2001–2004 People's Daily corpus for metaphor recognition research, based on machine learning algorithms and knowledge acquisition methods, explored various machine learning algorithms for metaphor recognition. The experimental idea avoids the shortage of manual knowledge base and rule method. In 2016, Bai [7] proposed a recognition method based on the topic model, and applied the topic model LDA (Latent Dirichlet Allocation) to the Chinese verb metaphor recognition process. By using the topic distribution of sentences as the feature and the method of machine learning to identify the verb metaphor, the average correct rate is 76.46%. After adding the topic annotation features, the average correct rate reaches 80.42%. In 2016, Zeng [8] based on the machine learning of large-scale corpus, using the maximum entropy classification model, proposed a metaphor recognition algorithm for automatic extraction of optimal feature templates, and discussed three different levels of feature templates, it contains both classic simple features, long-distance context information across multiple words, and word similarity feature templates that introduce semantic information. Experimental results show that the algorithm improves the accuracy of metaphor recognition. In 2018, Fu [9] proposed a metaphor phrase acquisition method based on the combination of clustering and classification, and clustered the phrases containing the source domain words. The results of the clustering are taken as a class of features of the classification. Experiments show that the classifier trained by using the features generated by clustering not only can well identify the existence of source domain word data in the training corpus, but also can well identify the lack of source domain word data in the training corpus. The experiment has a high recall rate. In 2015, Huang [10] used the logistic regression in statistical learning theory as the calculation model, and used the word vector obtained by the neural network language model as the feature. By constructing the abstract lexicon to obtain the feature weight vector and calculating the abstraction degree of Chinese words, the result can better fit the human cognition, and the accuracy reaches 67.1%. However, the word vector that relies solely on words indicates that the abstract features of words have certain limitations. After that, Huang [11] vectorized the experimental data, combining part-of-speech features and keyword features as input to the convolutional neural network, the features are extracted by the convolutional layer and the pooled layer, and SVM is used for classification. Aiming at the incompleteness of feature sampling in the pooling layer of convolutional neural networks, an improved method combining MaxPooling and Mean Pooling is proposed. The accuracy of the results increased by 4.12%, 0.84% and 4.50% in English verbs, English adjectives - noun phrase corpus and Chinese metaphor corpus, respectively. However, a convolutional neural network with too few layers affects the integrity of the feature.

At present, most of the research on Chinese metaphor recognition focuses on the recognition of noun metaphors at the phrase level. However, the frequency of A-is-B metaphors in actual texts is also high, and most Chinese metaphor researchers ignore them. Therefore, in order to improve the recognition rate of Chinese metaphors, this paper proposes a new method for the metaphor of A-is-B model.

2   Feature Extraction

We translate the metaphorical recognition of A-is-B into a classification problem; that is, given an A-is-B pattern of sentences, to determine whether the sentence is a metaphor. The feature set of the classifier is constructed by using the features of upper and lower position, sentence model, class, and Word2Vec. Finally, the SVM model classifier is used to realize the metaphor recognition classification of A-is-B mode. When constructing hyponymy relation libraries, we use the synonym word forest and recursive relationship to expand and disambiguate on the basis of predecessors. When calculating the similarity feature based on Word2Vec, we set a threshold of 0.35. Finally, using the trained training set, combined with the SVM algorithm for training and prediction.

2.1   Recognition Feature Based on Hyponymy Relation Library

The most common semantics of a sentence in the form of A-is-B is a subordinate relationship, that is, A belongs to one or the other of class B. If it can be judged that A and B do have a subordinate relationship, the statement can determine that it is not a metaphor; otherwise, this sentence may be a metaphor, or it may also be a sentence pattern of a partial overall relationship. The following example illustrates this feature.

(1) "Xiao Ming is active, he is the sun of the big guy"; (2) "Notre Dame Cathedral is a magnificent church".

In (1), using the hyponymy relation library to determine the superior of "Xiao Ming" is (name | person), but the superior of "sun" is "celestial body", and there is no hyponymy relation, so it is very possible to be a metaphor.

In (2), using the hyponymy relation library to judge the superior of "Notre Dame Cathedral" is (building | landmark), and the superior of the "church" is also a building, so you can judge this sentence is a non-metaphor sentence.

Through the work of predecessors [12, 13], we already have 1552400 pairs of hyponymy relations. The upper and lower relation libraries are constructed with the following points:

First, using *TongYiCiCiLin* to expand the existing hyponymy relations, since the synonym has been considered within the scope of this hyponymy relations, we don't consider the synonyms any more in the experiment.

Second, considering the recursiveness of the hyponymy relation, the inferior can directly find all the superior in the relation library, i.e., if A is the inferior of B and B is the inferior of C, then A may be the inferior of C. E.g.:

$$（红富士，苹果），（苹果，水果）\rightarrow（红富士，水果）$$

(Red Fuji, apple), (apple, fruit) → (Red Fuji, fruit)

Of course, this reasoning has the possibility of error, such as:

$$（红富士，苹果），（苹果，手机）\rightarrow（红富士，手机）$$

(Red Fuji, apple), (apple, cellphone) → (Red Fuji, cellphone)

This erroneous reasoning often occurs in the case of polysemy. In the construction of the hyponymy relation library, the recursive relation of such errors is also disambiguate. So our classification rules are:

$$F1 = \begin{cases} 1 & \text{if hyponymy relation exists} \\ 0 & \text{else} \end{cases} \tag{1}$$

In the experiment, it was found that many hyponymy relations are missing in the relation library. The main reason is that A is often the name of a new entity, which is not exist when building the hyponymy relation library. This situation will be taken out and mark separately, and F1 will be extended to the following classification feature formula.

$$F1 = \begin{cases} 1 & \text{if hyponymy relation exists} \\ 0 & \text{if hyponymy relation not exists and A in entity library} \\ 2 & \text{else} \end{cases} \tag{2}$$

2.2 Recognition Features Based on Sentence Patterns

When A and B present some specific patterns (referred to as sentence patterns), often A and B are not metaphorical relation. The hyponymy sentence pattern is used as an example to illustrate the recognition feature based on sentence patterns.

"糖类是人体必须的有机物之一" (Sugar is one of the essential organic substances in the human body.)

" 糖类,脂类,蛋白质类,核酸等有机物，是构成生命的基础" (Organic substances such as sugars, lipids, proteins, and nucleic acids are the foundation of life.)

The hyponymy relationship <sugar, organics> appears in different sentences, but metaphorical sentences (such as "上海是一颗明珠" ("Shanghai is a pearl")) are difficult to appear in the sentence model "A*等 B" ("A*, etc. B").

Table 1. Example of non-metaphoric sentence pattern

Sentence pattern	Example	Semantic
A .〈等等|等〉. B	糖类, 脂类, 蛋白质类, 核酸等 有机物，是构成生命的基础	Hyponymy relation
A. 〈组成|构成|做成|提供|做〉. B	在部队，三个班组成一个排	Partial overall relation
A. 〈包括|包含|含|含有|〉. B	文教学应该包括听、说、读、 写四项,不可偏轻偏重	Partial overall or hyponymy relation
A. 〈属于〉. B	最后的胜利一定属于我们	Partial overall or hyponymy relation
B .〈例如|例如| 如〉[、]. A	锻炼身体的方法有很多种，例 如长跑	Partial overall or hyponymy relation
A. 〈成为|作为〉. B 〈之一〉	"十二五" 时期, 我国已成为 全球产业链最完整的钢铁工业体系 国家之一	Hyponymy relation
A. 〈又称|定义|别名〉. B	再保险（reinsurance）又称分 保或"保险的保险"	same reference relation(notice：A< 称为|被称≫ B may be a metaphor)
B	江苏科技大学在镇江市	Hyponymy relation(A contains B

To determine whether A and B appear in a certain sentence pattern, this paper uses the Web search method. We will splicing A and B in a feature pattern. Once we can find the sentence of the pattern in the search engine, we think that A and B can be presented in this type of sentence pattern.

However, if A and B appear in another hyponymy sentence mode, we can also determine that A and B can form a hyponymy relation, thus negating the metaphor relation of A and B. Such sentence patterns are summarized in Table 1 above.

Where A and B respectively represent nouns or noun phrases appearing in the sentence, "|" means the side-by-side relation, "." means any string, "<>" means that an element must appear in the sentence; "[]" Indicates that it may or may not appear in the sentence.

Similarly, when A and B can be a metaphor relation, they often conform to the sentence pattern in Table 2.

Table 2. Example of analogy sentence pattern

Sentence pattern	Example
A. <如|就如|恰如|正像|像|就像|正| 正像>. B. [评价|所述|所说|一样|所言| 所示|所讲|所见]	弯弯的月亮像一条小船挂 在夜空中.

Analogy is different from metaphor in sentence structure, there are fewer words like "像、如" (like, as). So once A and B are presented in the sentence like this pattern, generally A and B are not metaphor relation. Based on the sentence patterns described above, we construct the following features:

$$F2 = \begin{cases} 1 & \text{if matches non-metaphoric sentence pattern} \\ 0 & \text{else} \end{cases} \quad (3)$$

$$F3 = \begin{cases} 1 & \text{if matches analogy sentence pattern} \\ 0 & \text{else} \end{cases} \quad (4)$$

2.3   Recognition Features Based on Class-Word

class-word: word that can be used as superior in the hyponymy relation library. In the sentence of the A-is-B mode, if A is a class-word, it is often less likely to be a hyponymy relation, that is, more likely to be a metaphor sentence. If B is a single case word, after excluding the "same reference relation", it is more likely to be the expression of a metaphor. For example, "他总是乐于助人，是我们中的宋江" ("He is always helpful, it is Songjiang among us.") In the metaphor, "Song Jiang" is a personal name and belongs to a single case word.

First we need to get the class, then determine if A belongs to the class, and whether B belongs to the class. Through this feature, we construct the following features.

$$F4 = \begin{cases} 1 & \text{if A is class-word} \\ 0 & \text{else} \end{cases} \tag{5}$$

$$F5 = \begin{cases} 1 & \text{if A is class-word} \\ 0 & \text{else} \end{cases} \tag{6}$$

It can be seen from the above definition that the establishment of this features require the acquisition of class-word, and the acquisition of the class itself is a complicated task. As space is limited, the detailed of the way to obtain class-word won't be described here.

2.4 Similarity Feature Based on Word2Vec Calculation

In 2013, Mikolov et al. [14] established the Word2Vec model, which is a model based on deep learning and trained on large corpora [15]. First, Word2Vec divides the sentences into words and maps each word into an N-dimensional vector, this converts the similarity comparison of two words into a similarity comparison between the two vectors, and then calculated using a model with a three-layer neural network. Finally, the K-Means clustering method [16] is used to cluster similar word vectors to form the word clustering model of Word2Vec. Compared with the traditional word bag model, Word2Vec can better express semantic information.

This paper uses Word2Vec to calculate the similar distance between A and B to determine whether the sentence is a metaphor. If the sentence of A-is-B mode is a metaphor, the distance between A and B will be farther, because they are two different things, otherwise A and B will be closer. That is: sim (A, B) = Word2Vec (A, B). For example, two examples are given in Table 3. The first is a metaphor and the second is a non-metaphor. From these two examples, the distance between A and B in a metaphor is much closer than the distance between A and B in a non-metaphor.

Table 3. Example of Word2Vec

Category	Example	A	B	Word2Vec
Forestry	在沙漠，树是黄金	树	黄金	0.11
Feeding	土鸡是什么鸡	土鸡	鸡	0.69

It is found that when the calculation results of A and B are greater than 0.35, it is often not a metaphorical relationship. Therefore, the threshold T is set to 0.35.

In the field trip, many Chinese nouns did not appear in the Word2Vec mode, which made it impossible to obtain similar values of A and B. Specifically because:

(1) The word segmentation granularity of Word2Vec is inconsistent with this experiment;
(2) Some new entity words are also not in the Word2Vec library;
(3) A and B may be verb phrases, often not in Word2Vec. For example, "吸烟就是在吸生命" ("smoking is sucking life") and "吸生命" ("sucking life") does not exist in Word2Vec.

When the result is empty, if A and B can also subdivide the words, then calculate the similarity between A and B's ending words. Because in Chinese, the ending word is often the superior of the word, which can largely reflect the original meaning of the word. For example, "乾隆是清朝皇帝吗" ("Is Qianlong the emperor of the Qing Dynasty?") A = "乾隆" ("Qianlong"), B = "清朝皇帝" ("Qing Dynasty Emperor"), but B is not in the Word2Vec mode, so the calculation result is empty. But at this time B can be subdivided into "清朝/n 皇帝/n" ("Qing Dynasty/n Emperor/n"), extract the ending words "皇帝" ("Emperor") of B, calculate Word2Vec (乾隆, 皇帝) = 0.72.

There is also a case that the calculation result is empty even if the ending word is extracted, at this point, we calculate the classification feature, we set the value to 0. The specific classification features are defined as follows:

$$F6 = \begin{cases} 1, & \text{IF} \quad Word2Vec(A, B) \geq 0.35 \\ 2, & \text{IF} \quad Word2Vec(t(A), t(B)) \geq 0.35 \\ 0, & \text{ELSE} \end{cases} \tag{7}$$

Where t(A) represents the ending word of A, and t(B) represents the ending word of B.

3   Feature-Based Classifier

3.1   SVM Classifier

Support Vector Machine (SVM) is an advanced machine learning method. Its theory is mainly based on the VC theory of statistical learning theory and the principle of structural risk minimization. The main advantage of SVM in machine learning is solving small sample, nonlinear and high dimensional pattern recognition problems. Another advantage is that the support vector machine is a relatively mature machine learning method. The theory is mature and the mathematical model is easy to understand.

We get six sentence features through the four algorithms in Sect. 2.1 above for each sentence, which we record as $[F_1, F_2, F_3, F_4, F_5, F_6]$ and then use the SVM classifier to get the final result. The experimental framework is shown in Fig. 1.

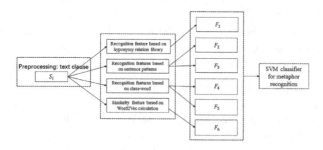

Fig. 1. SVM metaphor identification framework

3.2   Neural Network Classifier

In recent years, more studies have analyzed text based on deep learning methods. Kalchbrenner et al. [17] applied convolutional neural networks to natural language processing and designed a dynamic convolutional neural network (DCNN) model to process text of different lengths. At present, the related research of metaphor recognition is mainly based on machine learning and simple neural network, but there is still little research on Chinese metaphor recognition based on deep learning.

This paper uses expert knowledge to propose a new method for A-is-B metaphor recognition through the neural network (NN) of expert scoring. First, the text is divided into clauses, which we record as $\{S_1, S_2, S_3 \cdots S_n\}$, and the word segmentation and word vector representation are performed on S_i. The word segmentation is calculated by the lexical analyzer ICTCLAS, and the word vector is derived from the Wiki Chinese word vector. Then use the above four algorithms to get the characteristics of the clause S_i, which is recorded as $[F_1, F_2, F_3, F_4, F_5, F_6]$. Finally, it is judged whether the clause is an A-is-B metaphor by integrating the scoring system of expert experience. If the sentence is an A-is-B metaphor, it is recorded as 1, if it is not an A-is-B metaphor, then recorded as 0.

The specific algorithm steps for each of the clauses S_i are as follows:

S1: Any sentence S_i obtains the characteristics of the clause by the four algorithms introduced in Sect. 2.1, records the features obtained as $[F_1, F_2, F_3, F_4, F_5, F_6]$.
S2: Performing word segmentation and word vector representation on S_i.
S3: Using LSTM to extract sentence features from each sentence S_i, obtaining the sentence features of the sentence, and output is a 64-dimensional vector a_i.
S4: Concatenating the vector a_i is with the sentence features $[F_1, F_2, F_3, F_4, F_5, F_6]$, obtaining a 70-dimensional vector b_i.
S5: Using the vector b_i as an input, entering the Dense layer to reduce the dimension, and the output result is a probability y_i'.
S6: Calculate the loss value by using y_i' and y_i with two-class cross entropy to train the parameters of the whole neural network. (y_i is the result of the expert scoring system).

The experimental framework is shown in the Fig. 2.

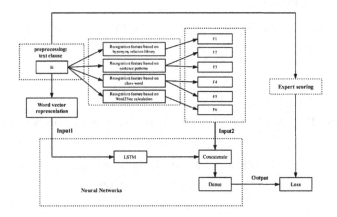

Fig. 2. Metaphor identification framework of fusion expert experience

The detailed structure of Neural Networks in the above figure is shown in the Fig. 3:

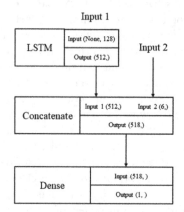

Fig. 3. Detailed structure of neural networks

4   Experiment and Evaluation Criteria

4.1   Construction of Metaphor Training Set

When constructing a training set, try to choose a case where A is not a pronoun. Because if A exists as a pronoun in the text, it is necessary to first work on the reference disambiguation. This is also a very complicated process. In order to more accurately evaluate the effect of metaphor recognition, the influence of ambiguity is temporarily circumvented.

The second level of the National Economic Industry Classification has 100 sub-sectors. But not all sub-sectors can find metaphors. At present, we have 650 metaphors and 1358 non-metaphors.

4.2   Experiment Platform

The experimental environment used in this paper is Win10 operating system, using I5-7500 processor, clocked at 3.4 GHz, memory is 16 GB, and the software programming environment is Python3.6.

4.3   Evaluation Criteria

The correct rate P, the recall rate R and the value F were used as the evaluation criteria of this experiment:

$$P = \frac{\text{Number of correctly identified metaphors}}{\text{Number of all metaphors}} \tag{7}$$

$$R = \frac{\text{Number of correctly identified metaphors}}{\text{Number of real metaphors}} \tag{8}$$

$$F = \frac{2 \times R \times P}{P + R} \tag{9}$$

4.4 Neural Network Training Process

In December 2014, two scholars, Diederik Kingma and Jimmy Ba, proposed the Adam optimizer, which is outstanding in the training optimization of neural networks. This article uses the parameter settings recommended in the paper: $\alpha = 0.001$, $\beta_1 = 0.9$, $\beta_2 = 0.9999$ and $\epsilon = 10^{-8}$.

The loss function chosen in this experiment is the cross entropy loss function, which is expressed as follows:

$$L = \sum_{i=1}^{n} \left[y_i \ln y_i' + (1 - y_i) \ln (1 - y_i') \right] \tag{10}$$

Where: y_i is the result of the expert scoring; y_i' is the probability value predicted by the neural network.

The training uses a ten-fold cross-validation and each time one tenth of the training data is reserved to evaluate the model. The model with the smallest L value is retained as the final model.

5 Experimental Results and Analysis

Using the types of feature calculations described above, and separately count the recognition effect of individual features, as shown in Table 4 for details.

Table 4. Results of feature recognition

Fusion name	Fusion ID	P	R	F
Hyponymy relation library	F_1	100%	20.5%	34.0%
Sentence pattern	F_2	100%	70.6%	82.8%
Sentence pattern	F_3	98.5%	38.6%	55.5%
Class-word	F_4	75.4%	76.5%	75.9%
Class-word	F_5	50.1%	48.2%	49.1%
Word2Vec	F_6	86.3%	83.5%	84.9%

From the above results, the judgment accuracy based on hyponymy relation library and sentence patterns is relatively high, but the recall rate is relatively low. The accuracy and recall rate based on Word2Vec is balanced. For example, the sentence "熬夜是健康的杀手" ("stay up late is a healthy killer"), Word2Vec (熬夜，杀手) = 0.35. The existence of A and B is a non-metaphorical relationship, but their Word2Vec values are relatively low. Another example, the sentence "清华大学是中华人民共和国教育部直属机构" ("Tsinghua University is directly affiliated to the Ministry of Education of the People's Republic of China"), Word2Vec (Tsinghua University, Institution) = 0.28. According to the analysis, although "Tsinghua University" and "institution" can constitute the hyponymy relation, However, because the upper level of the "institution" is relatively high, the scope of its use is relatively wide, so the value of Word2Vec will be lower when calculating.

The above calculation results are composed into the feature matrix, which is trained and classified by SVM classifier and neural network classifier respectively. In the prediction, we adopt the ten-fold cross-validation method. The specific results of the two experiments are shown in Table 5.

Table 5. Comparison table of experimental results

Method	P	R	F
SVM	95.5%	92.5%	94.0%
NN	96.7%	93.1%	94.6%

The experimental results show that the neural network classifier with expert experience has higher accuracy, recall rate and F value than the SVM classifier. Because the neural network of the expert scoring system combines the experience of experts, the experimental results are optimized without modifying the model structure. However, the training time and prediction time of this method have increased. The neural network classifier and the SVM classifier predict a sentence for 20 s and 6 s, respectively.

6 Conclusion

The metaphor of A-is-B mode is ubiquitous in people's lives. This paper analyzes the difficulties of A-is-B identification, uses industry classification to construct a descriptive training set, and combines the characteristics of Word2Vec, sentence patterns, hyponymy relation, and class-word, respectively using SVM model classifier and neural network classifier. The experimental results show that both methods have higher recognition accuracy and recall rate, and the neural network with expert experience can obtain better accuracy and recall rate (96.7% and 93.1%, respectively), but the prediction time of a single sentence also increases accordingly. This method provides a new solution for the problem of metaphor sentence of A-is-B mode recognition. In the future work, we should try to solve the problem of "reference ambiguity" and "death metaphor" as much as possible, and more consideration of metaphor recognition of other sentence types.

References

1. Koetter, R., Médard, M.: An algebraic approach to network coding. IEEE/ACM Trans. Network. **11**(5), 782–795 (2003)
2. Lakoff, G., Johnson, M.: Metaphors We Live By. University of Chicago Press, Chicago (1980)
3. Xu, Y.: Recognition of Chinese metaphor phenomenon based on maximum entropy model. Comput. Eng. Sci. (04), 95–97+103 (2007). (in Chinese)
4. Liu, C., Wang, J.: Research on metaphor recognition based on semantic annotation tool Wmatrix. Teach. Foreign Lang. **02**, 15–21 (2017). (in Chinese)
5. Su, C., Fu, Z., Zheng, F., Chen, Y.: Metaphor recognition method based on dynamic classification. J. Softw. (07), 1–15 (2019). (in Chinese)
6. Liu, J.: Research on metaphor recognition based on machine learning algorithm. Nanjing Normal University (2011). Instructor: Qu, W. (in Chinese)
7. Bai, Z.: Research on Chinese verb metaphor recognition method based on subject model. Hangzhou University of Electronic Science and Technology (2016). Instructor: Wang, X. (in Chinese)
8. Zeng, H., Zhou, C., Chen, Y., Shi, X.: Chinese metaphor computation based on feature automatic selection method. J. Xiamen Univ. (Nat. Sci.) **55**(03), 406–412 (2016). (in Chinese)
9. Fu, J., Wang, S., Cao, C.: A Chinese metaphor phrase recognition method based on combination of clustering and classification. J. Chin. Inf. Process. **32**(02), 22–28+49 (2018). (in Chinese)
10. Huang, X., Zhang, H., Lu, W., Wang, R., Wu, W.: A Chinese metaphor recognition method based on word abstraction. Mod. Libr. Inf. Technol. **04**, 34–40 (2015). (in Chinese)
11. Huang, X., Li, Y., Wang, R., Wang, X., Zhai, Z.: Metaphor recognition based on convolutional neural network and SVM classifier. Data Anal. Knowl. Discov. **2**(10), 77–83 (2018). (in Chinese)
12. Liu, L., Cao, C., Wang, H., Chen, W.: A subordinate concept acquisition method based on the "Yes" model. Comput. Sci. **09**, 146–151 (2006). (in Chinese)
13. Liu, L., Cao, C.: Verification method of upper and lower position relationship based on mixed features. Comput. Eng. (14), 12–13+16 (2008). (in Chinese)
14. Mikolov, T., et al.: Efficient estimation of word representations in vector space. Computer Science. arXiv (2013). http://arxiv.org/abs/1301.3781v3
15. Schmidhuber, J.: Deep learning in neural networks: an overview. Neural Netw. **61**, 85–117 (2015)
16. Han, J., et al.: Spatial clustering methods in data mining: a survey. Geogr. Data Min. Knowl. Discov. (2001)
17. Mikolov, T., Chen, K., Corrado, G., et al.: Efficient estimation of word representations in vector space [OL]. arXiv Preprint arXiv:1301.3781 (2013)

Layerwise Recurrent Autoencoder for Real-World Traffic Flow Forecasting

Junhui Zhao, Tianqi Zhu, Ruidong Zhao, and Peize Zhao[✉]

Beijing University of Posts and Telecommunications, Beijing, China
{zhaojunhui2016,zhutianqi,rdzhao,zhaopeize}@bupt.edu.cn

Abstract. Accurate spatio-temporal traffic forecasting is a fundamental task for wide applications in city management, transportation area and financial domain. There are many factors that make this significant task also challenging, like: (1) maze-like road network makes the spatial dependency complex; (2) the relationship between traffic flow and time brings non-linear temporal problem; (3) with the larger road network, the difficulty of flow forecasting grows. The prevalent and state-of-the-art methods have mainly been discussed on datasets covering relatively small districts and short time span. To forecast the traffic flow across a wider area and overcome the mentioned challenges, *Layerwise Recurrent Autoencoder* (LRA) is designed and proposed, in which a three-layer stacked autoencoder (SAE) architecture is used to obtain temporal traffic correlations in three different time scales and for each output of different time scales, a dedicate neural network is used for prediction. The convolutional neural networks (CNN) model is also employed to extract spatial traffic information within the road map for more accurate prediction. To the best of our knowledge, there is no effective method for traffic flow prediction which concerns traffic of city group and LRA is the first one. The experiment is completed on a large real-world traffic dataset to show the performance of the proposed. In the end, evaluations show that our model outperforms the state-of-the-art baselines by 6%–15%.

Keywords: Traffic forecasting · Neural networks · Stacked autoencoder

1 Introduction

Spatiotemporal traffic flow forecasting task is currently under a heated discussion and has attracted so many researchers. The application of this task is broad, including transportation anomaly detection, optimal resource allocation, logistic supply chain and city management. However, since the dynamic environment of traffic condition and the inherent complexity of large road networks, it is a challenge to make it accurate and efficient [9]. In this paper, we investigate the advantages from prevalent methods and propose a model that can solve the task with spatiotemporal modeling. Even in the dataset with large road network, the

© Springer Nature Switzerland AG 2019
Z. Cui et al. (Eds.): IScIDE 2019, LNCS 11936, pp. 78–88, 2019.
https://doi.org/10.1007/978-3-030-36204-1_6

model works well. The goal of the proposed framework is to predict the traffic flow in the next time slot at each sensor spot in the road network.

The main challenge of traffic flow prediction task is to find the appropriate spatiotemporal dependencies [2]. For two reasons, first, the traffic flow is dynamic, where the rush hours in the morning and evening generate a non-linear variate on the flow, and the information in different days of the week incurs more complex relationships; second, the space correlations between sensors in the road network are difficult to be determined. Figure 1 demonstrates a real-world example of the complexity in spatial dependency modeling. Point A and Point B are two sensors in a freeway network, and their geographic distance is close, but the driving distance is much farther than it seems to be. This example illustrates that the spatial distance in road network is not supposed to be Euclidean, but to be dominated by the road topology.

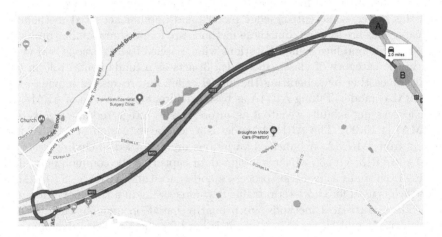

Fig. 1. Distance dependency is not simply directional and geographic. The Point A and B are geographically close, but they stand on opposite sides of the road, the driving distance is far. As a result, the spatial correlation between A and B should be low.

To overcome the challenges, we propose a deep-learning based *layerwise recurrent autoencoder* (LRA) for sequence-to-sequence traffic flow forecasting. The contributions of this framework are summarized as the following:

- Originally uses a very large traffic dataset[1], which covers three cities to evaluate and compare the performance of our model and other common used methods.
- Creatively employs flow information in three different time scales (flow sequence of today, flow sequence of this day in last week and in last four weeks) as three-layer inputs of LRA, from which LRA obtains the knowledge of time relationships and the periodical change of traffic flows.

---

[1] The dataset is from http://tris.highwaysengland.co.uk/detail/trafficflowdata.

– Innovatively exploits the driving distances between sensors in the road network to model spatial dependency, which is presented as a directed graph whose nodes are sensor spots and edge weights are spatial correlations.

2 Related Work

The history of traffic forecasting has been decades long, and many methods have been proposed. Especially in recent years, the instruments and infrastructures of sensors are developed, these sensors provide the possibility of accurate record of traffic volume within transportation network. The methods on this task can be mainly divided into two categories: classic statistical approaches and data-driven approaches. The classical time-series approaches are mainly based on queuing theory and statistic theory [5]. While the data-driven methods focus on curriculum learning and have recently attracted plenty of attentions.

In this paper, some cutting-edge models and popular accepted methods in both categories are studied, but these methods are found share similar problems in experiments and face some limitations when applied to large scope real-world data. In the category of classic methods, [3] acts as a fundamental role in this area of forecasting by generating the model called autoregressive moving average (ARMA) model. Taking ARMA as basis, an integrated version of ARMA for traffic forecasting is built, we cite it as autoregressive integrated moving average (ARIMA) [4,15,20]. The ARIMA model is a general extension of ARMA, and starting from ARIMA, a bunch of variations are proposed, including seasonal ARIMA (SARIMA) [23], which is designed for capturing the common periodical features from many time-series processes and space-time ARIMA (STARIMA) [24], which is used for short-term traffic flow forecasting in a transportation network. These statistical methods are primarily based on queuing theories and mathematical simulations, as a result, it is hard for them to learn dynamic and complex patterns in large scope use. As a result, though they perform satisfactorily on short-term and small-scope datasets in some research areas [17], when apply to real-time large-scope scenarios, their performance is barely satisfactory [6]. In this paper, we select the two most representative methods, ARMA and ARIMA as the baselines.

In data-driven learning category, most of models are built by neural networks, and have gotten high performances that surpass the methods in statistic category. The works in [14,26], apply recurrent neural networks (RNN) to study time series prediction. And convolutional neural networks (CNN) are also chosen for traffic flow forecasting in [19,27]. Besides, the echo state networks (ESN) [12] are deployed on some light applications for forecasting tasks. In [18], the authors use stacked autoencoder (SAE) model, which is built by autoencoder blocks, to learn time series features for prediction. See in [25], a temporal regularized matrix factorization method is proposed and uses graph regularization connections to learn the spatial dependency between different sensors, but this work pays insufficient attention on nonlinear temporal relationships. In many methods, researchers exploit space models for traffic volume prediction, however, the distance dependency is extracted by controversial geographical distances [8, 22]. To solve this problem, in [16], the authors firstly obtain the space-time dependency with diffusion convolutional recurrent neural networks (DCRNN), the distance correlations are represented as a directed graph and the model relates traffic flow to a diffusion process. In this work, we select ESN, SAE, RNN and DCRNN as the representatives of data-driven models, the performance comparison between these models and the proposed one is shown in the latter part of this paper, the generalization and effectiveness of these methods are also demonstrated.

3 Layerwise Recurrent Autoencoder

In this section, the structure of LRA is introduced by order, and spatial-temporal modeling is formulated.

3.1 Temporal Dependency Modeling

To extract temporal relationships from the history traffic flow data and reduce dimension of data, we originally model this process as a three-layer structure with autoencoder as cells. An autoencoder is used to reproduce its inputs, in other words, the target output of autoencoder is its input. With sequences of traffic flows $\{x^{(1)}, x^{(2)}, x^{(3)}, ...\}$ as input, an autoencoder first encodes the input $x^{(i)}$ to a hidden representation, and then decodes the representation back to a reconstruction. To minimizing reconstruction error $L(X, Z)$, where X is the input matrix of the autoencoder and Z is the output matrix, we denote it as θ,

$$\theta = arg\min_{\theta} L(X, Z) = arg\min_{\theta} \frac{1}{2} \sum_{i=1}^{N} \left\| x^{(i)} - z(x^{(i)}) \right\|^2, \qquad (1)$$

where N is the length of the input sequence, and $z(\cdot)$ is the reconstruction.

When take the sparsity constrains into consideration, to achieve the sparse representation in hidden layer [18], we minimize the reconstruction error as

$$SAO = L(X, Z) + \gamma \sum_{j=1}^{H_D} KL(\rho \| \hat{\rho}_j), \qquad (2)$$

where γ is the weight of the sparsity term, H_D is the number of hidden units, ρ is a sparsity parameter and is typically a small positive value, $\hat{\rho}_j$ is the average activation of hidden units, and $KL(\rho \| \hat{\rho}_j)$ is the Kullback-Leibler (KL) divergence, defined as

$$KL(\rho \| \hat{\rho}_j) = \rho \log \frac{\rho}{\hat{\rho}_j} + (1 - \rho) \log \frac{1 - \rho}{1 - \hat{\rho}_j}. \qquad (3)$$

The SAE is created by hierarchically stacked autoencoders, in which the input of the kth layer is the output of the $(k - 1)$th layer and a logistic regression is applied on the top. In this paper, to extract more detailed temporal relationships in traffic history, we employ a three-layer SAE, with three-layer input as the flow sequences of current day, flow sequences of this day in last week and in last four weeks, relatively. For example, if today is Friday, time slot is set as 15 min and we want to predict the traffic flow from 9:15 to 9:30 (as long as the predefined time slot), the input is supposed to be 6 traffic flows from 7:45 to 9:15 (as long as 6 times of predefined time slot) in this morning, the traffic flows at the same time in last Friday and the Friday in last four weeks.

Following the layering SAE model, we employ a sequence-to-sequence RNN structure for predicting the traffic flow. The backpropagation algorithm is used to optimize this process. To avoid the vanishing gradient problem in long lasting dataset with the traditional RNN models, we use long short-term memory (LSTM) [10] in our model. The key of LSTM model is memory cell, which allows LSTM to remove or maintain the information. The memory cell can help for remembering the temporal relationships extracted from SAE model and outperform other RNN models when competing on large-scope long-span dataset [21].

3.2 Spatial Dependency Modeling

The correlations of spatial dependency are complex, and even more abstract than the temporal relationships especially in large-scope road networks, in this paper, we model the spatial dependency between sensors as a directed graph and extract the relationships with CNN.

As for the directed graph, it takes sensor spots as nodes and driving distance as edges. We denote the graph as $G = \langle V, E \rangle$. V is the set of nodes and $E \subseteq \{(u,v)|u \in V, v \in V\}$ is the set of edges. In the graph G, we calculate the closest driving distance between each node, and build a triangle matrix to present these distances, denoted as

$$M = \begin{bmatrix} m_{11} & m_{12} & \cdots & m_{1N} \\ m_{21} & m_{22} & \cdots & m_{2N} \\ \vdots & \vdots & \ddots & \vdots \\ m_{N1} & m_{N2} & \cdots & m_{NN} \end{bmatrix} \tag{4}$$

where $m_{ij} = (-\frac{dist(v_i,v_j)^2}{\sigma^2})$, $dist(v_i, v_j)$ is the closet driving distance from v_i to v_j, σ denotes the standard deviation of these distances and N is the number of sensor spots. Besides, if $dist(v_i, v_j)$ is less than threshold $\kappa = 0.2$, we regard $m_{ij} = 0$.

To extract spatial dependency for traffic prediction, we encode the matrix M with a graph convolution networks (GCN) model [13]. The core work of the GCN is to map the input M to the influential index of each sensors.

The convolutional propagation in our paper is defined as

$$H^{(l+1)} = \sigma\left(D^{-\frac{1}{2}} M D^{-\frac{1}{2}} H^{(l)} W\right), \tag{5}$$

where $D = \sum_i \sum_j M_{ij}$ and W is the parameter matrix. While $\sigma(\cdot)$ is activation function, we use $ReLU(\cdot)$ in this work. $H^{(l)}$ denotes the matrix of last layer and $H^{(0)}$ is input matrix M.

We also consider convolutions with a filter $g_\theta = diag(\theta)$ ($\theta \in \mathbb{R}^N$ in the Fourier Domain) in GCN as the multiplication of element $x \in \mathbb{R}^N$ [7], defined as

$$g_\theta(L)x = U g_\theta(\Lambda)U^T x, \tag{6}$$

where U is the matrix of eigenvectors of the normalized graph Laplacian $L = I_N - D^{-\frac{1}{2}}MD^{-\frac{1}{2}} = U\Lambda U^T$ with a diagonal matrix of eigenvalues Λ, $U^T x$ is the Fourier transform of the element x. Analysis on Eq. 6 indicates that computational cost on computing the eigenvalue decomposition of L is expensive. To promote efficiency, we exploit a truncated presentation as

$$g_{\bar{\theta}}(\Lambda) \approx \sum_{k=0}^{K} \bar{\theta}_k T_k(\hat{\Lambda}), \tag{7}$$

where $\hat{\Lambda} = \frac{2}{\lambda_{max}}\Lambda - I_N$, λ_{max} means the largest eigenvalue of L. While $T_k(\cdot)$ is recursively defined as $T_k(x) = 2xT_{k-1}(x) - T_{k-2}(x)$, with $T_0 = 1$ and $T_1 = x$. See [11] for more discussion of the truncation.

Combine the knowledge of Eqs. 6 and 7, the filter is substituted by $g_{\bar{\theta}}$, and we have

$$g_{\bar{\theta}}(L)x \approx \sum_{k=0}^{K} \bar{\theta}_k T_k(\hat{L})x, \tag{8}$$

where $\hat{L} = \frac{2}{\lambda_{max}}L - I_N$. And the complexity of the computation decreases from $O(N^2)$ in Eq. 6 to $O(N)$ in Eq. 8.

With spatiotemporal modeling and LSTM network, LRA is built with three parts for extracting spatial-temporal dependencies, whose schematic architecture is shown in Fig. 2. Note that the architecture of LRA is much more complex than this schematic figure, for detail description of the LRA, please refer to here[2].

4   Experiment

In this paper, we complete the experiments with a large real-world dataset, **ENG-HW**: This dataset contains traffic flow information from inter-city highways between three cities, recorded by British Government. We conduct the experiments with 249 sensors and collect a whole year of data ranging from January 1st 2014 to December 31st 2014. In the experiments, we slice traffic flow information into 15 min as time slot, where 70% of data is for training, 10% for validation and remaining 20% for testing.

---

[2] The code and implemented details of LRA have been uploaded to https://github.com/bbklk/LRA-for-traffic-flow-forecasting.

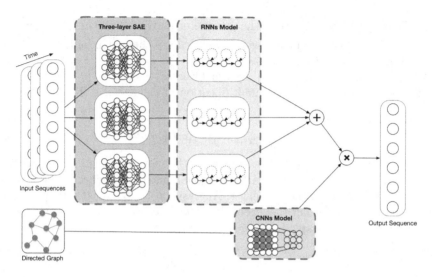

Fig. 2. The schematic architecture of LRA, designed for traffic flow forecasting. The temporal relationships are extracted by three SAEs, whose outputs are fed to LSTM model for prediction. Then the final outputs of LSTM model and GCN model are multiplied, and system output sequence is generated. The LRA is trained for minimizing the cross-entropy loss by backpropagation. (Color figure online)

4.1   Experimental Settings

In the experiments, we compare the performance of LRA with popularly used methods and state-of-the-art model, including: (1) **ARMA**: which provides a parsimonious description of a weakly stationary stochastic process, consists of two polynomials, one for autoregression and the second for moving average. (2) **ARIMA**: which is widely used in statistics and econometrics, especially in time series analysis. (3) **ESN**: a kind of RNN model with a sparsely connected hidden layer, which is fixed and randomly assigned. (4) **SAE**: a deep neural network model that uses autoencoder as cell, good for time series forecasting tasks since the capability of extracting temporal dependency. (5) **LSTM**: a variant of RNN model which is popular for classifying, processing and predicting tasks based on time series data. (6) **DCRNN**: one of the cutting edge deep learning models for forecasting, which uses a diffusion process during training stage to learn the representations of spatial dependency.

   We build all the above neural network based models by Tensorflow [1] and use the default parameters in the corresponding papers.

4.2   Experiment Results of Traffic Flow Forecasting

The algorithms are evaluated by three popularly accepted metrics in transportation area, including (1) Mean absolute error (MAE); (2) Root mean squared

Table 1. Performance comparison between LRA and other approaches for traffic flow forecasting. From the results of our experiments, LRA has the best performance.

	T	Metric	ARMA	ARIMA	ESN	SAE	LSTM	DCRNN	**LRA**
ENG-HW	15 min	MAE	40.88	40.17	33.43	30.22	25.67	24.82	**22.21**
		RMSE	69.81	68.23	48.25	52.52	43.93	42.26	**40.11**
		MAPE	15.9%	15.1%	15.8%	15.7%	15.2%	14.3%	**12.5%**
	30 min	MAE	41.44	42.98	35.08	35.12	28.34	29.27	**25.19**
		RMSE	74.49	73.24	56.06	59.66	50.91	50.19	**43.97**
		MAPE	18.4%	18.6%	17.2%	16.3%	16.7%	16.6%	**14.4%**
	60 min	MAE	54.89	58.45	50.19	42.34	37.78	32.47	**27.58**
		RMSE	78.65	78.62	59.76	61.47	53.21	55.76	**49.09**
		MAPE	22.3%	20.1%	20.2%	17.9%	17.4%	17.7%	**15.9%**

error (RMSE); (3) Mean absolute percentage error (MAPE). Note that comparisons across different datasets are invalid, since all the three metrics are scale-dependent.

Table 1 records the performances of different methods for three forecasting horizons (time scale) in the dataset. From the table, we notice the following facts that: (1) The deep learning based methods, including ESN, SAE, LSTM, DCRNN and LRA, outperform the statistical methods. (2) The deeper and more complex models are supposed to perform better than lighter ones, however there is an exception that the performances of DCRNN and LSTM are compatible, we guess it may because the diffusion convolutional layers in DCRNN extracts insufficient spatial correlations. (3) LRA achieves the best performance in any metrics and horizons, which reflects the generalization and effectiveness of the proposed model.

4.3 Experiment Results of Spatial and Temporal Dependency

In this part, we design experiments to proof the effect of spatial and temporal dependencies modeling by comparing the performance between LRA with two variants: (1) LRA-NoSAE, which feeds input time series sequences directly to LSTM model (the green part in Fig. 2) in the LRA and cancels encode-decode process of SAE model, this variant system is supposed to get fewer time relationships from inputs. (2) LRA-NoConv, which abandons CNN model (the pink part in Fig. 2) in the LRA, the outputs of LSTM model become the final output sequences, this mutation of LRA is supposed to get less sensitive to space correlations. Figure 3 shows the learning curve of the above two variants and LRA with regard of MAE, we keep the parameters of all three models as similar as possible. From the learning curve, LRA reaches the lowest MAE value, meanwhile, LRA-NoConv has a much higher MAE value, which illustrates the effect of our spatial dependency modeling. Besides, the learning curve of LRA-NoSAE almost gets the same level of the LRA, but the speed of convergence is much slower, this fact proves the effect of time relationship modeling. The Table 2 shows the comparison results of these three models and their convergent speed. Combine the observations of Fig. 3 and Table 2, the effect of spatial dependency modeling is proved for helping promote accuracy of prediction, while the aim of temporal dependency modeling is to boost the training progress.

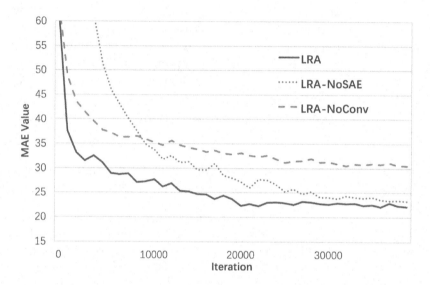

Fig. 3. Learning curve of LRA, LRA-NoConv and LRA-NoSAE on the ENG-HW dataset.

Table 2. Comparison for LRA, LRA-NoSAE and LRA-NoConv on the dataset ENG-HW in prediction horizon of 15 min. Note that the *Convergence Speed* means the number of iteration before models get convergent state (numerical fluctuation < 10%).

	MAE	Convergence speed
LRA	22.21	19000
LRA-NoSAE	23.27	26000
LRA-NoConv	30.49	19000

5   Conclusion

Traffic flow forecasting is an essential problem in many areas. There have been some methods performs well in specific conditions, however, a universal method for such problem is absent, especially in large-scope road network. In this paper, we modeled the spatial-temporal dependencies and formulated such task by proposing the *layerwise recurrent autoencoder* (LRA) model. This model originally uses driving distance for modeling space dependencies and works well for large-scope flow prediction. Meanwhile, the superiority and universality of our model are evaluated on a large real-world dataset with comparison to other common and state-of-the-art baselines. For the future work, we will investigate the following topics: (1) adding weather factors into LRA model for more accurate prediction; (2) implementing the proposed model on other real-world datasets and apply the LRA in more areas, e.g., pedestrian volume forecasting and audience distribution prediction.

References

1. Abadi, M., et al.: Tensorflow: a system for large-scale machine learning. In: OSDI, vol. 16, pp. 265–283 (2016)
2. Atwood, J., Towsley, D.: Diffusion-convolutional neural networks. In: Advances in Neural Information Processing Systems, pp. 1993–2001 (2016)
3. Box, G.E., Jenkins, G.M.: Time Series Analysis: Forecasting and Control. Wiley, Hoboken (1970)
4. Box, G.E., Jenkins, G.M., Reinsel, G.C., Ljung, G.M.: Time Series Analysis: Forecasting and Control. Wiley, Hoboken (2015)
5. Cascetta, E.: Transportation Systems Engineering: Theory and Methods, vol. 49. Springer, Heidelberg (2013)
6. Cheng, X., Zhang, R., Zhou, J., Xu, W.: Deeptransport: Learning spatial-temporal dependency for traffic condition forecasting. arXiv preprint arXiv:1709.09585 (2017)
7. Defferrard, M., Bresson, X., Vandergheynst, P.: Convolutional neural networks on graphs with fast localized spectral filtering. In: Advances in Neural Information Processing Systems, pp. 3844–3852 (2016)
8. Deng, D., Shahabi, C., Demiryurek, U., Zhu, L., Yu, R., Liu, Y.: Latent space model for road networks to predict time-varying traffic. In: Proceedings of the 22nd ACM SIGKDD, pp. 1525–1534. ACM (2016)
9. Drew, D.R.: Traffic flow theory and control. Technical report (1968)
10. Gers, F.A., Schraudolph, N.N., Schmidhuber, J.: Learning precise timing with LSTM recurrent networks. J. Mach. Learn. Res. 3(Aug), 115–143 (2002)
11. Hammond, D.K., Vandergheynst, P., Gribonva, R.: Wavelets on graphs via spectral graph theory. Appl. Comput. Harmon. Anal. 30(2), 129–150 (2011)
12. Ilies, I., Jaeger, H., Kosuchinas, O., Rincon, M., Sakenas, V., Vaskevicius, N.: Stepping forward through echoes of the past: forecasting with echo state networks. Technical report (2007)
13. Kipf, T.N., Welling, M.: Semi-supervised classification with graph convolutional networks. In: International Conference on Learning Representations (ICLR) (2017)
14. Laptev, N., Yosinski, J., Li, L.E., Smyl, S.: Time-series extreme event forecasting with neural networks at uber. In: International Conference on Machine Learning, vol. 34, pp. 1–5 (2017)
15. Lee, S., Fambro, D.: Application of subset autoregressive integrated moving average model for short-term freeway traffic volume forecasting. Transp. Res. Rec.: J. Transp. Res. Board **1678**, 179–188 (1999)
16. Li, Y., Yu, R., Shahabi, C., Liu, Y.: Diffusion convolutional recurrent neural network: Data-driven traffic forecasting. In: International Conference on Learning Representations (ICLR) (2018)

17. Lippi, M., Bertini, M., Frasconi, P.: Short-term traffic flow forecasting: an exper-
    imental comparison of time-series analysis and supervised learning. IEEE Trans.
    Intell. Transp. Syst. **14**(2), 871–882 (2013)
18. Lv, Y., Duan, Y., Kang, W., Li, Z., Wang, F.Y., et al.: Traffic flow prediction
    with big data: A deep learning approach. IEEE Trans. Intell. Transp. Syst. **16**(2),
    865–873 (2015)
19. Ma, X., Dai, Z., He, Z., Ma, J., Wang, Y., Wang, Y.: Learning traffic as images:
    a deep convolutional neural network for large-scale transportation network speed
    prediction. Sensors **17**(3), 818 (2017)
20. Moorthy, C., Ratcliffe, B.: Short term traffic forecasting using time series methods.
    Transp. Plann. Technol. **12**(1), 45–56 (1988)
21. Seo, Y., Defferrard, M., Vandergheynst, P., Bresson, X.: Structured sequence mod-
    eling with graph convolutional recurrent networks. arXiv preprint arXiv:1612.07659
    (2016)
22. Sun, S., Zhang, C., Yu, G.: A bayesian network approach to traffic flow forecasting.
    IEEE Trans. Intell. Transp. Syst. **7**(1), 124–132 (2006)
23. Williams, B.M., Hoel, L.A.: Modeling and forecasting vehicular traffic flow as a
    seasonal arima process: Theoretical basis and empirical results. J. Transp. Eng.
    129(6), 664–672 (2003)
24. Williams, B.M., Hoel, L.A.: Space-time modeling of traffic flow. Comput. Geosci.
    31(2), 119–133 (2005)
25. Yu, H.F., Rao, N., Dhillon, I.S.: Temporal regularized matrix factorization for high-
    dimensional time series prediction. In: Advances in Neural Information Processing
    Systems, pp. 847–855 (2016)
26. Yu, R., Li, Y., Shahabi, C., Demiryurek, U., Liu, Y.: Deep learning: a generic
    approach for extreme condition traffic forecasting. In: Proceedings of the 2017
    SIAM International Conference on Data Mining (SDM), pp. 777–785 (2017)
27. Zhang, J., Zheng, Y., Qi, D.: Deep spatio-temporal residual networks for citywide
    crowd flows prediction. In: AAAI, pp. 1655–1661 (2017)

Mining Meta-association Rules for Different Types of Traffic Accidents

Ziyu Zhao, Weili Zeng[✉], Zhengfeng Xu, and Zhao Yang

College of Civil Aviation, Nanjing University of Aeronautics and Astronautics,
Jiangjun Rd. no. 29, Nanjing 211106, China
zwlnuaa@nuaa.edu.cn

Abstract. Association rule method, as one of mainstream techniques of data mining, can help traffic management departments to identify the key contributing factors and hidden patterns in traffic accidents. However, there are still potential links between different accident attributes that have not been revealed, with poor universality of association rules obtained by current methods. In order to overcome the limitations of current methods, this paper proposes a new framework for mining universal rules over different types of traffic accidents, by accounting for the potential dependencies among varied rules suffered from the original methods, and improving the rule selection algorithm. First, different types of traffic accidents are classified and stored separately. Further, the strong association rules for each database are extracted, and then the frequent index approach is applied to organize a meta-rule set with universal applicability. Eventually, all traffic databases are excavated again with different thresholds to get association rules, and meta-rules are integrated into association rules to obtain the universal association rules in the form of a cell group. The proposed method is tested on real traffic databases of nine districts in Shenzhen, China. The results demonstrate that the improved association rules are more universal and representative than existing methods.

Keywords: Meta-association rules · Universal applicability · Traffic accidents · Data mining

1 Introduction

In recent years, with the growth of car ownership, the number of traffic accidents has increased rapidly. According to WHO's latest global road safety observation data [1], about 1.25 million people every year die from road traffic accidents in the world. In such a severe situation of traffic safety, how to prevent the occurrence of traffic accidents has become a difficult problem. However, the traffic system is a huge and complex system, with a number of uncertain factors in traffic accidents. It is not easy to find the internal mechanism of all kinds of traffic accidents. Fortunately, with the rapid development of cloud computing and data mining technology, the contributing factors to traffic accidents and potential patterns for the occurrence of traffic accidents can be find out, providing the basis for decision-making.

© Springer Nature Switzerland AG 2019
Z. Cui et al. (Eds.): IScIDE 2019, LNCS 11936, pp. 89–100, 2019.
https://doi.org/10.1007/978-3-030-36204-1_7

So far, a number of methods have been proposed for traffic accident analysis [2–4]. These methods can be divided into two categories: parametric models and non-parametric models. And the majority of studies are based on parametric models [5–8]. Considering the different crash levels and driver injury severities, a hierarchical Bayesian logistic model was applied to identify the contributing factors [9]. Tsubota et al. [10] aimed to reveal the relationship between different pavement types and traffic accident risks through Poisson regression analysis.

While parametric models are able to identify the most influential indicators between various factors and traffic accidents, most of them are failed to maintain stable performance than non-parametric models in actual test [11]. The reason is that parametric models have their own predetermined assumptions, which makes it possible to draw wrong conclusions if the hypothesis is broken.

Therefore, many studies of non-parametric models for traffic accident analysis have been done [12–17]. Chen et al. [18] employed support vector machine (SVM) models to identify injury severity patterns in crashes. Figueira et al. [19] used decision tree algorithms to detect contributing factors of different accidents. The coupled clustering method has been enhanced to deal with the occupational accidents data and present results with a readable and exportable form [20]. Getting rid of the constraints of predetermined assumptions, non-parametric models have the capability to discover significant patterns which have not been revealed in existing approaches.

The association rule mining method proposed by Agrawal et al. [21], as one of representative non-parametric models, has been widely used in traffic accident analysis at present. Marukatat [22] proposed a rule selection framework based on rule structure analysis to improve the readability and relevance of association rules. Weng et al. [23] proposed a method based on association rules to analyze the characteristics of work zone crash casualties. In order to improve the performance of association rules, some improved approaches have been proposed.

There are some shortages of existing methods using association rules for traffic accident analysis. It is worth noting that the methods require high threshold to ensure readability, which might lead to the fact that some hidden association rules cannot be embodied. For a large database, the contributing factors of the antecedent and consequent in the obtained association rules are overwhelming, lacking of precise description of correlation between each other. In addition, universality of association rules derived from the existing method is poor, that is to say, the association rules based on the database of a certain region can not apply to other regions.

The purpose of this paper is to propose a novel approach based on meta-rules, which can extract association rules with universality, readability and operability. The advantages of this method are as follows: (a) Hidden relevant information in classic association rules can be revealed through meta-rules. (b) It can be used to filter valuable rules and eliminate association rules that are not universally applicable. (c) It breaks down the limitations of overwhelming contributing factors in decision making, expressing resulting rules in a more readable and manageable way. This method provides decision support for traffic safety managers and helps decision-makers to discover the similarity among traffic accidents in different environmental conditions.

2 Methodology

2.1 Basic Idea

The main objective of this study is to mine association rules of different types of traffic accidents, with universality, readability and operability. The implementation process is shown in Fig. 1. The specific process is as follows: (a) Meta-rules extracting. Different types of traffic accidents in databases $D_1, D_2, \ldots, D_K$ should be divided in advance and stored separately. To extract universal strong rules from databases $D_1, D_2, \ldots, D_K$ (including K regions), firstly, frequent item sets of each database are excavated. Then, strong association rules of each database are generated by the frequent item sets. Finally, frequent index is used to filter strong association rules to get meta-rule set which is universally applicable; (b) Meta-rules mining. Considering that the traffic accident data stored in different databases, association rule mining of each database should be executed again. It is worth noting that meta-rules are considered together for mining, which is different from original mining process. The meta-rules are integrated into the association rules to get the resulting rules.

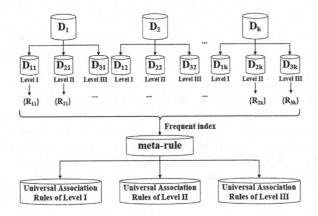

Fig. 1. Universal association rules mining process based on meta-rules.

2.2 Extracting Meta-rules

This section is to extract universally applicable meta-rules of the form $X \Rightarrow Y$ that relates the factor sets X and Y in K data sets $\{D_1, D_2, \ldots, D_K\}$. The left part X is known as the antecedent of the rule, and the right part as the consequent. The intensity of the rule is measured by support, confidence and lift, which is widely use in association rules mining [24, 25].

The support of a rule $X \Rightarrow Y$ is defined as

$$\text{Support}(X \Rightarrow Y) = \frac{|\{d \in D | X \cup Y \subseteq d\}|}{|D|} \tag{1}$$

where D denotes a set of traffic accident data, $|\cdot|$ represents cardinal number.
The confidence of a rule $X \Rightarrow Y$ is defined as

$$\text{Confidence}(X \Rightarrow Y) = \frac{|\{d \in D | X \cup Y \subseteq d\}|}{|\{d \in D | X \subseteq d\}|} \tag{2}$$

The confidence of association rules $X \Rightarrow Y$ is defined as

$$\text{Lift}(X \Rightarrow Y) = \frac{|\{d \in D | X \cup Y \subseteq d\}|}{|\{d \in D | X \subseteq d\}| \times |\{d \in D | Y \subseteq d\}|} \tag{3}$$

Assuming that the support threshold is MinSup, the frequent factor set F_i in data set D_i can be obtained by Apriori algorithm. For any data set D_i, the strong and interesting association rule set is generated as following steps:

(a) For each frequent factor set F_i, all non-empty subsets of F_i are generated;
(b) For each non-empty subset A, if $A \Rightarrow (F_i - A)$ satisfies:

$$\text{Confidence}(A \Rightarrow (F_i - A)) \geq MinConf \text{ and } \text{Lift}(A \Rightarrow (F_i - A)) \geq MinLift \tag{4}$$

where MinConf and MinLift are the minimum confidence threshold and the minimum lift threshold, respectively, and then the rule $A \Rightarrow (F_i - A)$ is the strong and interesting association rule in database D_i.

After strong and interesting association rules in all databases are obtained, meta-rules from these strong association rules can be extracted. Denoting $S = \{R_1, R_2, \ldots, R_N\}$, the set of strong association rules in all databases are established, in which N represents the total number of rules. In general, if the number of databases, which a rule is subordinate to, is more than that of another rule, then we can conclude that this rule is more general or universal than the other one. We will extract Meta-rules in terms of this idea. First, a binary data set is created by using Boolean value. If the rule R_i exists in the database D_i, the value is 1, otherwise 0. For any strong and interesting rule R_i, a frequent index of this rule is defined as

$$frequence(R_i) = \left(\frac{1}{K}\sum_{1}^{K} p_{ij}\right) \times 100\% \tag{5}$$

where p_{ij} is the Boolean value of the rule R_i in the database D_j. If $frequence(R_i)$ is greater than a given threshold, the rule R_i is meta-rule. Frequent index threshold should be based on the number of strong association rules in each database. If the threshold is too low, the universality of the meta-rules is poor. On the other hand, if the threshold is too high, the number of meta-rules is small and some frequent patterns are hidden. The above cases are not conducive to the analysis of the results. The determination of the final threshold needs to be tested and verified through practical experiments.

2.3   Mining Universal Association Rules

This section will Considering the meta-rules, we explore the contributing factors of traffic accidents and give the correlation among various factors. Although a number of studies have been conducted on the association rule analysis, the following form of association rules made up of different factors can only be excavated.

$$\{P_1, \ldots, P_i\} \Rightarrow \{L_1, \ldots, L_k\} \tag{6}$$

where P_i and L_k respectively denotes the contributing factors in the antecedent and the consequence. The information provided by Eq. 6 only reveals what factors lead to the specific traffic accident. It is difficult for the traffic management department to take some actions to prevent the potential traffic accident, especially when there exist a large number of contributing factors.

According to the structure of association rule as shown in Eq. 6, it is difficult to find the relationships between numerous factors mentioned in the antecedent. The actual implementation is rather difficult when we take precautions against traffic accidents on the basis of contributing factors. Therefore, we propose the following form of association rules:

$$\{P_1, (P_2, \ldots P_i) \Rightarrow P_j, \ldots, P_k \Rightarrow P_l\} \Rightarrow \{\{Q_1, \ldots, Q_t\} \Rightarrow Q_u, \ldots, Q_m\} \tag{7}$$

Nested form is applied in Eq. 7, which can include both meta-rules and single attributes, so that the resulting rules could contain more information. In practical application, the purpose of traffic accident prevention can be achieved by taking into account the association rules between the factors and focusing on the most critical factors.

For example, the resulting association rules with different approaches are shown in Table 1. There are three antecedents and three consequents, which have equal status in original approach. However, it is obvious that No.1225 illegal behavior is caused by time factors and the degree of perfection of marking, which lead to traffic accidents of level III with rainy days and male drivers. Male drivers are the main group of traffic accidents when it rains. According to the weather and the concerned radio programs of drivers with different genders, the specific broadcast reminding is carried out to reduce the rate of traffic accidents.

Table 1. Comparison of original approach and improved approach

	Antecedent	Consequent
Original approach	time = afternoon **AND** marking line = no **AND** illegal behavior = NO. 1225	Accident grade = level III **AND** Weather = rain **AND** gender = male
Improved approach	{(marking line = no, time = afternoon)⇒ illegal behavior = NO. 1225}	{Weather = rain ⇒ gender = male} **AND** Accident grade = level III

On the basis of combining the meta-rules, the database $D_1, D_2, \ldots, D_K$ is excavated with different thresholds once more, and association rules of different traffic accidents are obtained. In addition, cell pattern output is composed of multi component rules including the meta-rules and the single attribute. The resulting association rules are universally applicable and interpretative.

In this paper, the detailed steps of traffic accident factors mining based on meta-rules are as follows:

Step1: data preprocessing of databases $\{D_1, D_2, \ldots, D_K\}$.

Step2: Divide databases $\{D_1, D_2, \ldots, D_K\}$ by accident type, and frequent item sets of each database are excavated.

Step3: Based on the identical parameters MinSup, MinConf and MinFreq, data mining is executed through Apriori algorithm, and the strong association rules of each database are obtained.

Step4: A binary data set between the strong association rules and databases is constructed. Then, according to the frequent index of strong association rules, the meta-rules are extracted.

Step5: By integrating meta-rules and strong association rules, the databases are scanned again to extract universal association rules from multiple sets of data.

Step6: Based on the classification of traffic accidents, the cell pattern output rules of multiple traffic accident types are exhibited respectively.

3 Experimental Results and Analysis

3.1 Data Preparation

The experimental data were collected from the traffic accident database system (http:// rsrp.tmri.cn), covering traffic accidents of 9 districts in Shenzhen, China, from 2014 to 2016, with 237255 records. Each recorded data includes five types of information: (i) Essential information of accident (type, time, place, casualty, etc.); (ii) Responsible person information (gender, age, occupation, driving behavior, etc.) (iii) Vehicle information (vehicle type, performance, etc.) (iv) Road condition information (road level, road condition, alignment, marking line, etc.) (v) Environmental information (weather, lighting, etc.). In this paper, the attribute with a vacancy of less than 30% can be retained.

3.2 Parameter Selection

The method proposed in this paper involves three parameters: the support threshold MinSup, the confidence threshold MinConf and the frequent index threshold MinFreq. These parameters are mainly used to control the number of the resulting rules which determines the comprehensibility. In other words, a large number of rules are not convenient for traffic management and planning departments, and some rules are difficult to interpret. Conversely, some interesting association rules may be hidden with higher thresholds. Considering the coupling of these parameters, it is difficult to select the adaptive algorithm in practical application. In this section, we discuss the parameter selection through experimental analysis.

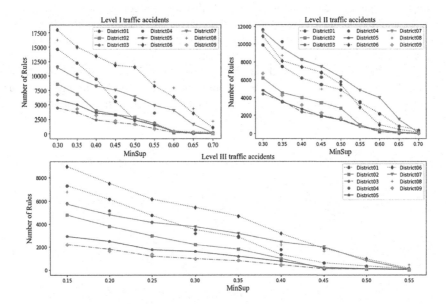

Fig. 2. Trend graphs of rule number with support threshold.

The support threshold is set first. From Fig. 2 (Level I traffic accidents), it can be seen that when the threshold ranges from 0.5 to 0.6, there are more than 7500 rules in 2 districts and more than 1000 in the other districts, and the rate of change in the number of rules is more balanced among the districts, and the K curve of the maximum item set of level I accidents is more stable (Fig. 3). The support threshold of level I traffic accidents is set as 0.5. According to Figs. 2 and 3, we can also get the support thresholds of level II and level III traffic accidents, which are 0.45 and 0.3.

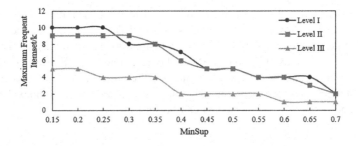

Fig. 3. Trend graph of maximum K for different types of traffic accidents with support threshold.

For different types of traffic accident data sets, different support threshold and confidence threshold are set up to compare and analyze the bubble relation diagram of the rule distribution and threshold setting (Fig. 4) which weigh the selection range of the support threshold and confidence threshold. The abscissa corresponds to the support threshold and the vertical coordinate corresponds to the confidence threshold. The bigger bubble, the more rules contained, in which the Benchmark bubble of Level I traffic accidents contains 15761 rules corresponding to (0.12, 0.15). For level I accidents, when the support threshold ranges from 0.5 to 0.7, according to the corresponding bubble graph, the confidence threshold is set as 0.65. According to similar analysis, the confidence thresholds of level II and level III traffic accidents can be determined as 0.55 and 0.5 respectively.

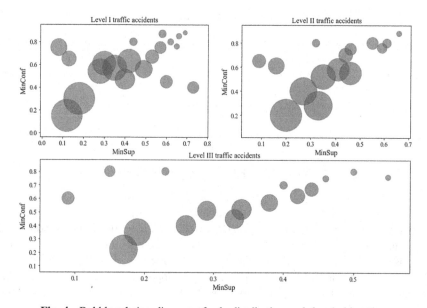

Fig. 4. Bubble relation diagram of rule distribution and threshold setting.

The frequent index threshold MinFreq is used to filter meta-rules with universal applicability. In Fig. 5, the number of rules obtained is close to the number of rules when the frequent index threshold is 0.6, with better universally applicable. However, when the frequent index threshold is 0.55, the redundancy of the rules is less, the universality and the analysis are better, so the MinFreq threshold of the frequent index of level I accidents is 0.55. According to similar analysis, the frequent index thresholds of level II and level III traffic accidents can be set to 0.55.

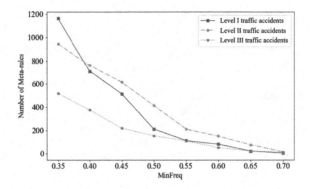

Fig. 5. The trend diagram of meta-rules under different frequent index thresholds.

3.3   Rule Generation and Analysis

After the selection of parameter threshold, the strong association rules that meet the support threshold interval and confidence threshold interval are generated by Apriori algorithm for different regional databases. Then the meta-rule sets are screened out according to the determined frequent index, and the above steps are repeated to get the cell pattern output rules which are composed of multi component rules.

Resulting rules of level I traffic accidents. Table 2 contains the resulting rules of Level I traffic accidents obtained when the support threshold of level I accidents is 0.5 and confidence threshold is 0.65.

Table 2. Resulting rules of level I traffic accidents.

Antecedent	Consequent	*Sup*	*Conf*	*Lift*
{Driving-Behavior = "1094 ⇒ Road-Grade = "General urban roads"} **AND** Weather = "Rain" **AND** Date = "Monday" **AND** Perfection-of-Mark-Line = "No"	{Age = "30–35"⇒ Driving-Vehicle = "Car"} **AND** Accident-Grade = "Level I"	0.51	0.74	2.38
{(Time = "Early evening (17:00–19:59)", Weather = "Rain")⇒ Road-Grade = "General urban roads"} **AND** Driving-Vehicle = "Car" **AND** Gender = "Male" **AND** Perfection-of-Mark-Line = "No"	{(Age = "19-23", Driving-Experience = "1-3")⇒ Driving-Behavior = "1225"} **AND** Accident-Grade = "Level I"	0.59	0.70	1.42

Resulting rules of level II traffic accidents. Table 3 contains the resulting rules of Level II traffic accidents obtained when the support threshold of level II accidents is 0.4 and the confidence threshold is 0.55.

Table 3. Resulting rules of level II traffic accidents.

Antecedent	Consequent	Sup	Conf	Lift
Time = "Late evening (20:00–22:59)" **AND** {Age = "19–23"⇒ Driving-Behavior = "1225"} **AND** Gender = "Male"	{(Driving-Vehicle = "Car", Perfection-of-Mark-Line = "No")⇒ Number-of-Injuries ≤ "4"} **AND** Accident-Grade = "Level II"	0.49	0.61	2.44
Perfection-of-Mark-Line = "No" **AND** {(Road-Grade = "Grade A highway", Month = "July, August")⇒ Driving-Behavior = "1225"}	{Weather = "Rain"⇒ Number-of-Injuries > 4} **AND** Accident-Grade = "Level II"	0.47	0.72	1.41

Resulting rules of level III traffic accidents. Table 4 contains the resulting rules of level III traffic accidents obtained when the threshold of support for level III accidents is 0.3 and the confidence threshold is 0.5.

Table 4. Resulting rules of level III traffic accidents.

Antecedent	Consequent	Sup	Conf	Lift
{Perfection-of-Mark-Line = "No"⇒ Road grade = "Grade A highway"} **AND** Lighting-Conditions = "Lighting at night" **AND** Time = "Late evening (20:00–22:59)"	{(Driving-Behavior = "1225", Age = "24–29")⇒ Gender = "Male"} **AND** Accident-Grade = "Level III"	0.31	0.60	1.94
Weather = "Rain" **AND** Perfection-of-Mark-Line = "No" **AND** {Driving-Vehicle = "Walking"⇒ Road-Grade = "General urban roads"} **AND** Gender = "Male"	{Gender = "Male"⇒ Driving-Behavior = "1225"} **AND** Accident-Grade = "Level III"}	0.42	0.56	1.37

The following conclusions can be drew from Tables 2, 3 and 4: (a) Imperfect marking is the most common factor and the main cause of accidents. (b) Both male drivers in early evening and female drivers in early morning have a high incidence of level I traffic accidents, but the former has a higher support than the latter, that is to say, male drivers in early evening have a higher chance of causing accidents. However, female drivers are the majority of level I traffic accidents on sunny days, and the female drivers driving on the express road with imperfect marks have the highest confidence in level I traffic accidents. (c) In July and August, the grade A highway with an imperfect marking line, under the rainy weather conditions, the illegal behavior of No. 1225 has a higher risk of serious injuries, accompanied by more than 4 people injured. (d) Male drivers are the main causes of level III traffic accidents. When the weather is rainy, we should pay more attention to the male drivers aged between 24 and 29 driving on highway with imperfect marking lines and straight road in late evening.

4 Conclusions

In this paper, a data mining approach based on meta-rules was proposed to analyze the causes of traffic accidents. The main advantage of this approach is that resulting rules are rules of relevance, readability and universal applicability through frequent index screening and meta-rules integration. It can guide the discovery of regular pattern and improve the readability and relevance of mining results, which is conducive to the analysis of decision-making level. By applying the idea and technology of data mining based on meta-rules to specific traffic accident data sets, the experimental research is carried out and resulting rules for different types of traffic accidents are obtained, and the corresponding analysis and improvement suggestions are put forward for such strong association rules. The results show that the mining mode is effective and feasible. And the application of this approach in other safety research topics will be very meaningful. Further, the application of data mining mode in the dynamic traffic system analysis model will be considered, hoping to provide effective support for the future intelligent transportation system mode.

Acknowledgements. This paper is supported by the Fundamental Research Funds for the Central Universities (NO. NS2018044), the Innovation Research Funds for Nanjing University of Aeronautics and Astronautics (NO. kfjj20180717), and the National Natural Science Foundation of China (NO. 51608268).

References

1. WHO Homepage. http://www.who.int/gho/road_safety/en/. Accessed 5 Jan 2018
2. Cerwick, D.M., Gkritza, K., Shaheed, M.S., Hans, Z.: A comparison of the mixed logit and latent class methods for crash severity analysis. Anal. Methods Accid. Res. **3–4**, 11–27 (2014)
3. Kwon, O.H., Rhee, W., Yoon, Y.: Application of classification algorithms for analysis of road safety risk factor dependencies. Accid. Anal. Prev. **75**, 1–15 (2015)
4. Zou, X., Yue, W.L.: A Bayesian Network Approach to Causation Analysis of Road Accidents Using Netica. J. Adv. Transp. **2017**, 1–18 (2017)
5. Ihueze, C.C., Onwurah, U.O.: Road traffic accidents prediction modelling: an analysis of Anambra state. Nigeria Accid. Anal. Prev. **112**, 21–29 (2018)
6. Lee, J., Chae, J., Yoon, T., Yang, H.: Traffic accident severity analysis with rain-related factors using structural equation modeling – a case study of Seoul city. Accid. Anal. Prev. **112**, 1–10 (2018)
7. Lord, D., Mannering, F.: The statistical analysis of crash-frequency data: a review and assessment of methodological alternatives. Transp. Res. Part Policy Pract. **44**(5), 291–305 (2010)
8. Xu, C., Li, H., Zhao, J., Chen, J., Wang, W.: Investigating the relationship between jobs-housing balance and traffic safety. Accid. Anal. Prev. **107**, 126–136 (2017)

9. Chen, C., et al.: Driver injury severity outcome analysis in rural interstate highway crashes: a two-level Bayesian logistic regression interpretation. Accid. Anal. Prev. **97**, 69–78 (2016)
10. Tsubota, T., Fernando, C., Yoshii, T., Shirayanagi, H.: Effect of road pavement types and ages on traffic accident risks. Transp. Res. Proc. **34**, 211–218 (2018)
11. Chang, L.-Y., Chien, J.-T.: Analysis of driver injury severity in truck-involved accidents using a non-parametric classification tree model. Saf. Sci. **51**(1), 17–22 (2013)
12. Hoglund, M.W.: Safety-oriented bicycling and traffic accident involvement. IATSS Res. **42**(3), 152–162 (2018)
13. Lavrenz, S.M., Vlahogianni, E.I., Gkritza, K., Ke, Y.: Time series modeling in traffic safety research. Accid. Anal. Prev. **117**, 368–380 (2018)
14. Shafabakhsh, G.A., Famili, A., Bahadori, M.S.: GIS-based spatial analysis of urban traffic accidents: case study in Mashhad. Iran. J. Traffic Transp. Eng. Engl. Ed. **4**(3), 290–299 (2017)
15. Weng, J., Meng, Q., Wang, D.Z.W.: Tree-based logistic regression approach for work zone casualty risk assessment: tree-based logistic regression approach. Risk Anal. **33**(3), 493–504 (2013)
16. Zhang, Y., Liu, T., Bai, Q., Shao, W., Wang, Q.: New systems-based method to conduct analysis of road traffic accidents. Transp. Res. Part F Traffic Psychol. Behav. **54**, 96–109 (2018)
17. Zhang, Z., He, Q., Gao, J., Ni, M.: A deep learning approach for detecting traffic accidents from social media data. Transp. Res. Part C Emerg. Technol. **86**, 580–596 (2018)
18. Chen, C., Zhang, G., Qian, Z., Tarefder, R.A., Tian, Z.: Investigating driver injury severity patterns in rollover crashes using support vector machine models. Accid. Anal. Prev. **90**, 128–139 (2016)
19. da Figueira, A.C., Pitombo, C.S., de Oliveira, P.T.M.E.S., Larocca, A.P.C.: Identification of rules induced through decision tree algorithm for detection of traffic accidents with victims: a study case from Brazil. Case Stud. Transp. Policy. **5**(2), 200–207 (2017)
20. Comberti, L., Demichela, M., Baldissone, G.: A combined approach for the analysis of large occupational accident databases to support accident-prevention decision making. Saftey Sci. **106**, 191–202 (2018)
21. Agrawal, R., Imieliński, T., Swami, A.: Mining association rules between sets of items in large databases. ACM SIGMOD Rec. **22**(2), 207–216 (1993)
22. Marukatat, R.: Structure-based rule selection framework for association rule mining of traffic accident data. In: International Conference on Computational Intelligence and Security, pp. 781–784. IEEE, Guangzhou (2006)
23. Weng, J., Zhu, J.-Z., Yan, X., Liu, Z.: Investigation of work zone crash casualty patterns using association rules. Accid. Anal. Prev. **92**, 43–52 (2016)
24. Wang, R., Ji, W., Liu, M.: Review on mining data from multiple data sources. Pattern Recogn. Lett. **109**, 120–128 (2018)
25. Ruiz, M.D., Romero, J.G., Solana, M.M.: Meta-association rules for mining interesting associations in multiple datasets. Appl. Soft Comput. **49**, 212–223 (2016)

Reliable Domain Adaptation
with Classifiers Competition

Jingru Fu and Lei Zhang$^{(\boxtimes)}$

School of Microelectronics and Communication Engineering, Chongqing University,
Chongqing 400044, China
{jrfu,leizhang}@cqu.edu.cn

Abstract. Unsupervised domain adaptation (UDA) aims to transfer labeled source domain knowledge to the unlabeled target domain. Previous methods usually solve it by minimizing joint distribution divergence and obtaining the pseudo target labels via source classifier. However, those methods ignore that the source classifier always misclassifies partial target data and the prediction bias seriously deteriorates adaptation performance. It remains an open issue but ubiquitous in UDA, and to alleviate this issue, a Reliable Domain Adaptation (RDA) method is proposed in this paper. Specifically, we propose double task-classifiers and dual domain-specific projections to align those easily misclassified and unreliable target samples into reliable ones in an adversarial manner. In addition, the domain shift of both manifold and category space is reduced in the projection learning step. Extensive experiments on various databases demonstrate the superiority of RDA over state-of-the-art unsupervised domain adaptation methods.

Keywords: Domain adaptation · Source domain · Target domain

1 Introduction

Many algorithms in computer vision areas are based on a fundamental assumption that the training and test data are drawn from the same distribution [13]. However, this assumption generally does not hold in many real-world scenarios such that the trained model often does not fit the test data, since training and test images are obtained under very different conditions (e.g., different camera device parameters, varying illuminations, and viewpoints, etc.) [25].

To address this issue, domain adaptation was proposed to exploit the rich labeled source domain data to facilitate the learning of a different but semantic related unlabeled target domain [7,11,19]. This is a *unsupervised domain adaptation* (UDA) problem. A common strategy to handle such unsupervised scenario is to align the distributions across the source and target domain. Maximum Mean Discrepancy (MMD) is a favorite principle to measure the discrepancy between two distributions [15]. Pan et al. proposed to learn a transferred subspace across domain by using MMD to measure the marginal distribution of domains [18].

© Springer Nature Switzerland AG 2019
Z. Cui et al. (Eds.): IScIDE 2019, LNCS 11936, pp. 101–113, 2019.
https://doi.org/10.1007/978-3-030-36204-1_8

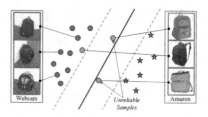

Fig. 1. The motivation of the proposed method. Source samples and target samples are denoted in blue and green, respectively. The classifier (solid line) is trained on the source samples (2 classes with different symbols for simplification). Target samples with large domain discrepancy have low classification confidence (within two dotted lines), which we define as unreliable samples. (Color figure online)

However, the source label information with rich semantics is ignored. To solve it, Long et al proposed to jointly minimize both the marginal and conditional distributions [15]. Since there is no target label, an iterative pseudo target label updating strategy was used to compute the conditional distribution. Many works [2,9,30,31] have experimentally demonstrated that the pseudo target labels can significantly boost the performance of UDAs. However, none of them take into account the misclassified target samples and the prediction bias, which we view as *unreliable* target samples. In fact, the unreliable target samples deteriorate the clustering performance of adaptation, due to that incorrect target labels cannot well account for the class distribution discrepancy. As shown in Fig. 1, our motivation is inspired by the fact that an easily misclassified sample generally closes to decision boundary and thus holds low confidence for a classifier. Apart from that, most of these methods assume there exists a common subspace between domains, which usually fails to extract domain-specific information from each domain.

To alleviate the pseudo target label prediction bias problem and preserve domain-specific information, we propose an RDA model composed of double task-classifiers and dual projections. The double task-classifiers are used to discover those unreliable target samples. Then two domain-specific projections are used to seek a reliable feature embedding that transforms those unreliable samples into reliable ones, in the meantime, they are forced to close to each other in order to reduce the distance across domains in the *Grassmann* manifold space [1]. Note that these two steps are trained in an adversarial manner.

Toward this end, we propose a **Reliable Domain Adaptation (RDA)** method for unsupervised domain adaptation, by discovering unreliable target samples with double classifiers and transforming the samples into new feature spaces, in an adversarial manner. We summarize the contributions of this paper as follows:

– We propose a RDA model to discover the unreliable target samples (i.e., easily-biased samples) via double task-classifiers and further transform the

unreliable samples into reliable feature embedding, which effectively alleviates the clustering bias resulted from the incorrect pseudo target labels.
- We propose the dual subspace projections to reduce the discrepancy between domains in manifold space and preserve domain-specific information across domains.
- Extensive experiments on challenging benchmark datasets demonstrate that our method achieves the best performance by comparing to state-of-the-arts including shallow and deep learning methods.

2 Related Work

In this section, some related works are divided into three aspects:

Subspace-Driven Methods. Subspace alignment (SA) [8] aims at learning a linear mapping for aligning subspaces spanned by eigenvectors using principal component analysis (PCA) across domains. Geodesic flow kernel (GFK) [1] characterized the changes of geometric and statistical properties across domains by integrating numerous subspaces. CORAL [24] alleviated the domain shift by aligning the second-order statistics (e.g., covariance) between two domains. Those methods aligned the statistical features over domains in manifold space, where the global property of domains is well represented. The tolerance of noise is then improved. However, they ignored the distribution alignment.

Data-Driven Methods. Transfer component analysis (TCA) [18] learned the transfer components between domains using Maximum Mean Discrepancy (MMD). Domain invariant projection (DIP) [2] proposed to construct the MMD in the manifold space. Statistically invariant embedding (SIE) [3] used Hellinger distance on statistical manifolds to approximate the geodesic distance. Transfer Joint Matching (TJM) [16] matched the feature representations by re-weighting the instances. However, none of them utilized the semantic information that is beneficial to the discrimination of the model. So, joint distribution alignment (JDA) [15] proposed to reduce both the marginal distribution and conditional distribution measured by using MMD and pseudo target labels. However, due to the clustering bias, the predicted pseudo target labels are not reliable.

Adversarial Learning Methods. Generative adversarial networks (GAN) [10] was the first proposal for adversarial learning. It can be seen as a distribution matching method, for matching the generated data (i.e. generator) with the target data, supervised by a domain classifier (i.e. discriminator). Tzeng et al. [26,27] proposed adversarial domain adaptation models by enhancing the domain feature confusion, supervised by a domain classifier. Motivated by the theory proposed in [5], Saito et al. [22] considered the decision boundaries between classes for the first time and aimed at aligning the distribution between classes. These methods are structured based on convolutional neural network (CNN), that well accelerates the discriminative feature representation. Our approach is based on a statistical learning framework that also uses the adversarial idea to achieve reliable unsupervised domain adaptation.

3   Reliable Domain Adaptation

In this section, we introduce the proposed method in detail. First, the problem and notations are defined, then the overall method is presented, and details of model and solution are finally introduced. Note that our approach is a statistical learning framework, *not* a CNN-based deep network.

3.1   Problem Definition

Given a labeled source domain $\mathcal{D}_s = \{(x_i, y_i)\}_{i=1}^{n_s}, x_i \in \mathbb{R}^D$ and an unlabeled target domain $\mathcal{D}_t = \{(x_j)\}_{j=1}^{n_t}, x_j \in \mathbb{R}^D$, where n_s and n_t indicate the number of samples in source and target domain, and D is the dimension of the original samples. We assume the label spaces between domains are the same, i.e. $\mathcal{Y}_s = \mathcal{Y}_t$, and the label space $\mathcal{Y}$ is a C-cardinality label set. $X_s \in \mathbb{R}^{D \times n_s}$ and $X_t \in \mathbb{R}^{D \times n_t}$ are domain-specific datasets drawn from distribution $\mathcal{P}_s(\cdot)$ and $\mathcal{P}_t(\cdot)$, respectively.

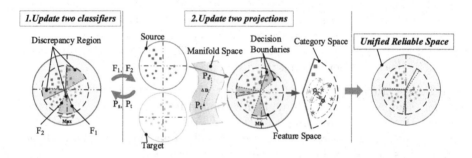

Fig. 2. Overview of the RDA approach. Two classifiers F_1 and F_2 (dotted line and solid line) are presented to discover the unreliable target samples by maximizing the discrepancy region (shadow region). Dual projections P_s and P_t aim to seek new reliable features by minimizing the shadow region and, simultaneously, align domains in both manifold and category spaces. The ultimate goal is to find a reliable space where clustering of the same class across domains is as good as possible. Note that solid circles define the ranges of subspaces, dotted circles define the ranges of features' distributions. (Color figure online)

3.2   Model Formulation

Clearly in Fig. 2, two fundamental steps are included in RDA in an adversarial manner:

Step 1. *Train two classifiers.* We introduce two *task-classifiers* aiming at discovering the unreliable target samples in this step, which are hard to be classified by the source-classifier due to distribution mismatch. Note that two *task-classifiers* are initially trained on source data, and are forced to classify source data as accurately as possible during the whole training processing. The double task-classifiers are trained by maximizing the discrepancy region, so that the target samples which are close to classifier boundaries (unreliable samples, green sample in Fig. 2) can be discovered as much as possible.

We suppose the input features of classifiers to be $z_s = g_s(x_s)$ for source samples and $z_t = g_t(x_t)$ for target samples, respectively. $g_s(\cdot)$ and $g_t(\cdot)$ indicate the function of dual *domain-specific projections*. We train both classifiers $f_1(\cdot)$ and $f_2(\cdot)$ to classify the source samples as correct as possible and, simultaneously, maximize the discrepancy over classification outputs. The objective function of the first step is as follows:

$$\min_{f_1,f_2} \mathcal{L}_f(z_s, y_s) - \lambda \mathcal{L}_{adv}(z_t) \tag{1}$$

where $\mathcal{L}_f$ represents the classifiers' loss function for source features, $\mathcal{L}_{adv}$ represents the adversarial loss formulated as the discrepancy between two classifiers' outputs on target samples, and λ is the trade-off parameter.

Step 2. *Train dual projections.* The ultima goal of learning is to obtain dual generators that can generate reliable transferred features. We firstly propose to align conditional feature distribution over domains to ensure that the same classes of both domains can be clustered as well as possible and then introduce the pseudo label prediction strategy. The dual *domain-specific projections* are used to map data of both domains, respectively. The involved clustering bias is solved by minimizing the discrepancy in this step to align such unreliable target features. We define the pseudo target labels as $\hat{y}_t$. The objective function of the second step can be formulated as:

$$\min_{g_s,g_t} \mathcal{L}_{feat}(x_s, x_t, y_s, \hat{y}_t) + \alpha \mathcal{L}_{sub} + \lambda \mathcal{L}_{adv}(x_t) \tag{2}$$

where $\mathcal{L}_{feat}$ and $\mathcal{L}_{sub}$ represent the feature-align loss and the subspace-align loss, respectively, α is trade-off parameter. The subspace-align and adversarial loss are treated as the regularization terms in our model.

These two steps are updated alternately, and ultimately, a unified reliable transferred space can be obtained, where samples of the same category in two domains are clustered. In the next subsection, the technical details of RDA are presented.

Details in Step 1. This subsection explains the specific implementation of Step 1, which aims to train double *task-classifiers*. We formulate double classifiers as the coefficient vector $F \in \mathbb{R}^{d \times C}$, and dual projections as $P_s \in \mathbb{R}^{D \times d}$ and $P_t \in \mathbb{R}^{D \times d}$ according to the representer theorem [4], d donates the dimension of features.

(1) Classification loss $\mathcal{L}_f$ on source domain. The loss function of the source classifiers is formulated as a regularized least-square loss:

$$\mathcal{L}_f(z_s, y_s) = \sum_{i=1}^{2} (\sum_{j=1}^{n_s} (f_i(z_s^j) - y_s^j)^2 + \eta \|f_i\|^2)$$
$$= \sum_{i=1}^{2} (\|Z_s^T F_i - Y_s\|_F^2 + \eta \|F_i\|_F^2), \tag{3}$$

where $Z_s = [z_s^1, z_s^2, ..., z_s^{n_s}] \in \mathbb{R}^{d \times n_s}$ (Note that $Z_s = P_s^T X_s$) is the source domain feature set, η is trade-off parameter. $\|M\|_F = \sqrt{tr(M^T M)}$ is the Frobenius norm of matrix M, $tr(\cdot)$ is trace operator. We define the constructed source label matrix as $Y_s = [y_s^1, y_s^2, ..., y_s^{n_s}]^T \in \{-1, 1\}^{n_s \times C}$, and $y_s^i(c) = 1$ means that the i-th source sample is associated with the c-th class.

(2) Adversarial loss $\mathcal{L}_{adv}$ on target domain. The unreliable target features are samples which close to classifier boundaries, inspired by a CNN-based deep method [22], which utilizes classifiers' difference to represent classifier boundaries, we also formal the outputs' discrepancy as adversarial loss term which can be formulated as:

$$\mathcal{L}_{adv}(z_t) = \sum_{i=1}^{n_t} (f_1(z_t^i) - f_2(z_t^i))^2$$
$$= \|Z_t^T F_1 - Z_t^T F_2\|_F^2, \tag{4}$$

where $Z_t = [z_t^1, z_t^2, ..., z_t^{n_t}] \in \mathbb{R}^{d \times n_t}$ ($Z_t = P_t^T X_t$) is target domain feature set. From Fig. 1 we find that when we force to increase the difference between the two task-classifiers' outputs, target samples that are closing to the decision boundary can fall into the region between the two classifiers' decision boundaries and can then be found.

Details in Step 2. This subsection explains the specific implementation of Step 2.

(1) Feature-align loss $\mathcal{L}_{feat}$. The proposed feature alignment loss aims at clustering the same class of the source and target domain in category space, such that the disparity between the conditional distributions across domains is reduced. The semantic guided MMD alike feature alignment loss is used to measure the dissimilarity of conditional distributions [15, 28, 30]. The pseudo target label is represented as $\hat{y}_t$. Then the feature-align loss is as:

$$\mathcal{L}_{feat} = \sum_{c=1}^{C} \|\frac{1}{n_s^{(c)}} \sum_{x_i \in \mathcal{D}_s^{(c)}} P_s^T x_i - \frac{1}{n_t^{(c)}} \sum_{x_j \in \mathcal{D}_t^{(c)}} P_t^T x_j\|^2, \tag{5}$$

where $\mathcal{D}_s^{(c)} = \{x_i | x_i \in \mathcal{D}_s | y_s(x_i) = c\}$ ($\mathcal{D}_t^{(c)} = \{x_j | x_j \in \mathcal{D}_s | \hat{y}_t(x_j) = c\}$) is the set of source (target) samples (a total of $n_s^{(c)}$ ($n_t^{(c)}$) samples) with respect to class c, $y_s(x_i)$ is the true source label of sample x_i.

The feature-align loss aims to reduce the class-wise distance between domains. As illustrated in Fig. 2, the data points with the blue circle and yellow circle represent the class-wise center of target and source domain in the *category space* of the F_1 classifier.

(2) Subspace-align loss $\mathcal{L}_{sub}$. Similar to [8], our goal is to decrease the distance (i.e. $\triangle D$ in Fig. 2) between two domain-specific projections. [30] confirmed that the shift of subspace geometries can be reduced in this way. For better non-parameter learning, instead of learning an additional mapping function, we propose to minimize the following smooth subspace-align loss directly:

$$\mathcal{L}_{sub} = \|P_s - P_t\|_F^2 , \tag{6}$$

(3) Adversarial loss. For correcting the unreliable target samples found in Step 1, we expect to reduce the discrepancy in an adversarial way. Note that the dual projections (P_s, P_t) instead of classifiers (F_1, F_2) are trained to minimize the classifiers' difference. The following adversarial loss function is minimized:

$$\mathcal{L}_{adv}(x_t) = \sum_{i=1}^{n_t} (f_1(g_t(x_t^i)) - f_2(g_t(x_t^i)))^2 \tag{7}$$
$$= \left\| (P_t^T X_t)^T F_1 - (P_t^T X_t)^T F_2 \right\|_F^2 ,$$

where the two task classifiers F_1 and F_2 have been solved in Step 1.

A deep adaptation method MCD_DA that is relevant to this paper was proposed by Saito et al [22], in which two classifiers are considered for solving UDA. Here, we briefly highlight the main differences between MCD_DA and RDA as following: (1) MCD_DA just tries to align source-unsupported target samples without considering to align conditional feature distribution, while RDA takes it into consideration. (2) MCD_DA only trains one shared generator between domains, but RDA considers the domain-specific generators (projections), and they are beneficial to reduce domain shifts in the manifold space. (3) MCD_DA is a deep adaptation method while RDA is a statistical learning framework. The necessity and effectiveness of the first two items are verified in the *ablation analysis* part.

Overall Model of RDA. The ultimate model of RDA consist of two objectives:

$$\min_{F_1, F_2} \mathcal{L}_f(z_s, y_s) - \lambda \mathcal{L}_{adv}(z_t) \tag{8}$$

$$\min_{P_s, P_t} \mathcal{L}_{feat}(x_s, x_t, y_s, \hat{y}_t) + \alpha \mathcal{L}_{sub} + \lambda \mathcal{L}_{adv}(x_t) \tag{9}$$

where all terms in the minimax optimization model have been presented above.

In the optimization of the RDA model, we adopt the alternating optimization strategy, i.e., fix the projections in training the two task-classifiers and fix the task-classifiers in training the two projections. The predicted pseudo-labels of target data are updated in each loop. For each step, ADMM algorithm is considered [6]. The optimization of RDA is summarised in Algorithm 1.

Algorithm 1. Reliable Domain Adaptation

Input: Data and source labels: X_s, X_t, y_s; Parameters: $d = 20$, $\eta = 1$, α, λ, T.
Output: Projection matrices: P_s and P_t; Predicted target labels: $\hat{y}_t$.

1: Initialize P_s and P_t using existing method. e.g. SA [8], PCA, etc.
   While iteration $t < T$ **do**
2: Update $\hat{y}_t$ using a base classifier, there is
   $\hat{y}_t = classifier(P_s^T X_s, y_s, P_t^T X_t)$.
3: Fix P_s and P_t, and update F_1 and F_2 by solving (8).
4: Fix F_1 and F_2, and update P_s and P_t by solving (9), calculate $Z_s = P_s^T X_s$, $Z_t = P_t^T X_t$.
5: $t = t + 1$.
   End while
6: **return** P_s, P_t, $\hat{y}_t$

---

4 Experiment

A number of experiments are conducted to evaluate the performance of RDA for unsupervised scenarios, which is closer to real-world applications. We compare our methods with state-of-the-art: (1) *Subspace-driven methods*: SA [8], GFK [1] and CORAL [24]; (2) *Data-driven methods*: JDA [15], DIP [2], JGSA [30] and TJM [16]; (3) *Adversarial learning methods*: Deep Domain Confusion (DDC) [27]; (4) *Deep transfer learning methods*: Domain Adaptation Networks (DAN) [14] and Residual Transfer Network (RTN) [17]. Notice that, it is unfair for RDA to compare directly against the deep DA methods, since RDA is a statistical shallow learning method. Therefore, deep features extracted using a pre-trained CNN are fed into RDA, and expect to further reduce the discrepancy of deep representation.

4.1 Data Preparation

In experiments, five different visual benchmark datasets are exploited and tested.

(1) **Office-10+Caltech-10 (4DA)** [1]: The Office data [21] contains three real-world object domains, including **A**mazon, **W**ebcam and **D**SLR. Caltech-256 [12] is a standard database for object recognition. 4DA is formulated with 10 shared categories between Office and Caltech datasets. Two kinds of features, i.e. hand-crafted SURF feature and deep CNN features, are used. *First*, the SURF features [1] that are encoded with 800-dimension BoW features are used as the shallow feature. *Second*, the features extracted from a deep model (the FC7 activations of VGG-VD-16 model) [23] are exploited as the deep feature. By randomly selecting two different domains as the source and target domain, a total of 12 cross-domain tasks are constructed.

(2) **MSRC+VOC2007** [16]: Six shared semantic classes from both datasets are formulated, and 1,269 images in MSRC and 1,530 images in VOC2007 are selected for domain adaptation. The 128-dimensional dense SIFT (DSIFT) features were extracted using the VLFeat open-source software package, and K-means clustering was used to obtain the 240-dimensional codebook. Following

the experimental setting as [29], two cross-domain tasks are constructed: M vs. V and V vs. M.

(3) **COIL20** [20]: Dataset contains 20 objects with 1440 gray scale images. Each image has 32×32 pixels and 256 gray levels per pixel. In experiments, the dataset is divided into two subsets COIL1 and COIL2 by following [29]. Specifically, the COIL1 (C1) and COIL2 (C2) contain the images taken in the directions of $[0°, 85°] \bigcup [180°, 265°]$ and $[90°, 175°] \bigcup [270°, 355°]$, respectively.

Table 1. Recognition accuracies (%) on Office+Caltech10 dataset with the deep feature from VGG-VD-16 model. * denotes deep transfer learning methods. Red: ranks the 1st; **Blue**: ranks the 2nd; Green: ranks the 3rd.

Task	Raw	SA	JDA	GFK	JGSA	CORAL	DIP	TJM	DDC*	DAN*	RTN*	RDA
C→A	91.5	93.2	93.7	93.6	94.2	91.6	93.3	93.9	91.9	92.0	**94.4**	96.0
C→W	83.7	86.4	94.6	86.8	93.3	78.9	86.2	92.0	85.4	90.3	**96.6**	99.0
C→D	89.9	95.0	93.2	91.0	**94.4**	87.6	91.4	90.8	88.1	90.5	92.9	94.3
A→C	81.7	77.1	**90.1**	85.3	87.2	80.1	86.0	86.4	85.0	85.1	88.5	93.2
A→W	74.8	80.4	91.5	85.8	95.7	75.7	74.1	87.3	86.1	93.8	**97.0**	98.6
A→D	77.2	89.6	91.3	85.5	94.1	76.2	83.4	89.9	89.0	92.4	**94.6**	96.8
W→C	77.3	77.9	86.7	81.3	82.3	77.6	81.2	81.4	78.0	84.3	**88.4**	92.6
W→A	85.5	87.3	93.8	90.2	**94.9**	90.7	88.4	91.1	84.9	92.1	93.1	96.0
W→D	99.0	98.0	96.1	98.0	96.1	98.0	98.0	97.6	100	100	100	**99.4**
D→C	75.0	78.6	84.8	82.3	**85.2**	73.1	81.0	81.8	81.1	82.4	84.3	91.3
D→A	83.6	83.8	91.7	90.8	93.8	84.5	90.0	91.4	89.5	92.0	95.5	**94.5**
D→W	95.8	97.0	89.2	97.3	96.4	94.9	95.2	96.8	98.2	**99.0**	98.8	99.7
Average	84.6	87.0	91.4	89.0	92.3	84.1	87.4	90.0	88.2	91.2	**93.7**	96.0

4.2 Experimental Setting

We strictly follow the experimental configuration for UDA as [1,16,29]. SVM is trained on the labeled source data for generating pseudo-target-labels. Three trade-off parameters: α, λ and η are involved in the proposed method. We set $\eta = 1$ for all experiments to simplify the tuning steps. For fairness, α and λ are only tuned from the parameter set [0.1, 1, 10]. We empirically set the subspace dimension $d = 20$ for all experiments.

4.3 Experimental Results

The recognition accuracies of RDA are shown in Tables 1, 2, and 3, respectively. From those results, we observe that RDA outperforms the state-of-the-art in a number of cross-domain tasks (21/28 tasks). Moreover, we achieve at least the second-best performance except for the three tasks: C→D, A→W and D→W. The average classification accuracy of RDA on the total 28 tasks is **74.3%**, which is **3.3%** higher than the state-of-the-art JDA (71.0%). Notice that, the results

Table 2. Recognition accuracies (%) on Office+Caltech10 dataset with SURF features.

Task	Raw	SA	JDA	GFK	JGSA	CORAL	DIP	TJM	RDA
C→A	50.1	54.4	59.8	56.6	55.1	45.9	56.4	54.4	**59.4**
C→W	43.1	45.8	50.1	48.1	49.7	37.8	**51.2**	44.0	57.6
C→D	**47.8**	40.9	44.1	42.9	46.0	31.8	46.9	38.4	51.6
A→C	42.8	44.8	**44.9**	44.3	40.8	37.1	41.4	42.4	49.2
A→W	37.0	44.1	**47.0**	42.7	59.0	37.9	44.8	39.5	45.1
A→D	37.2	37.7	44.2	39.9	**49.4**	38.5	47.8	45.6	50.3
W→C	29.5	32.3	29.8	32.0	29.7	32.5	30.0	**33.3**	40.9
W→A	34.2	**43.3**	42.0	38.3	34.6	39.4	33.8	39.5	45.6
W→D	80.6	70.3	86.3	78.7	78.5	80.9	79.6	**83.6**	78.3
D→C	30.1	31.1	**34.4**	30.8	30.2	27.8	29.3	32.3	36.6
D→A	32.1	40.8	**44.6**	40.4	39.0	31.9	31.6	37.1	46.5
D→W	72.2	74.4	**83.3**	80.3	75.1	69.4	67.5	83.7	80.7
Average	44.7	46.7	**50.9**	47.9	48.9	42.6	46.7	47.8	53.5

Table 3. Recognition accuracies (%) on MSVC-VOC2007 and COIL20 datasets. * denotes the results of GFK based on 1-nearest neighbor (1-NN) classifier.

Task	Raw	SA	JDA	GFK*	JGSA	CORAL	TJM	RDA
M→V	37.1	31.8	38.2	28.8	**38.7**	33.9	38.3	39.7
V→M	55.5	46.0	**59.3**	48.9	49.3	54.1	54.1	62.3
C1→C2	82.7	86.7	**88.7**	72.5	85.1	84.9	83.1	93.5
C2→C1	84.0	90.6	93.1	74.2	83.9	87.9	88.5	**91.8**
Average	64.8	63.8	**69.8**	56.1	64.3	65.2	66.0	71.8

are obtained from a number of benchmark visual datasets, which can effectively demonstrate that RDA is capable of reducing the domain shift for UDA.

Second, for local comparisons, RDA generally outperforms the subspace-driven methods (i.e., SA, GFK and CORAL) and data-driven methods (i.e., DIP, JDA and TJM). The reason is that those methods do not reduce the cross-domain discrepancy in both category space and domain-specific subspace. Our approach considers both aspects. Above all, compared to the methods using the pseudo label strategy (e.g., JDA and JGSA), our RDA alleviates the clustering bias resulted from unreliable pseudo target labels and guarantees the reliability.

Third, compared with shallow features (e.g., SURF features), deep features obtain significantly better results for all models. The proposed RDA shows significant improvement (**3.7%**) on average compared to the best shallow transfer method (i.e. JGSA), and **2.3%** comparing to the best deep transfer method (i.e. RTN) as shown in Table 1. The comparison shows that the proposed RDA, as a shallow learning method, is more effective but reliable.

4.4   Model Analysis and Discussion

Parameter Sensitivity. *First*, the recognition performance on several datasets with regard to the iterations T is shown in Fig. 3(a). We set the maximum number of iteration $T=10$ in all experiments. From the results, it can be observed that classification performance rises slowly and tends to be smooth on some tasks (e.g., $C \rightarrow A(VGG)$ and $M \rightarrow V$), but shows a clear upward trend on other tasks (e.g., $C \rightarrow A$ and $C1 \rightarrow C2$). Empirically, we are able to get relatively good results with $T=10$. *Second*, we investigate the sensitivity of subspace dimension d with a wide range of $d \in \{10, 20, ..., 100\}$ to illustrate the relationship between d and the classification accuracy in Fig. 3(b). From the results, it can be observed that RDA is robust and keeps stable with regard to the different numbers of d. *Third*, the two parameters are tuned from the given set $[10^{-1}, 10^{0}, 10^{1}]$. From the results on two tasks $C \rightarrow A$ and $M \rightarrow V$ shown in Fig. 3(c) and d), we can observe that the parameter α has a relatively larger impact on the performance, which represents the importance of the subspace alignment loss. Generally, a larger λ contributes much to the unreliable target sample discovery and rectification with the adversarial loss. In general, the parameters can be easily tuned in experiments.

Ablation Analysis. In RDA, three main components are involved: subspace alignment loss $\mathcal{L}_{sub}$ (SA), feature alignment loss $\mathcal{L}_{feat}$ (FA) and adversarial loss $\mathcal{L}_{adv}$ (Adv). For a better insight into the model, ablation analysis is presented.

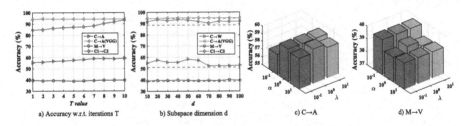

Fig. 3. Convergence and parameter sensitivity analysis of RDA model on several datasets. Note that the dashed lines in (b) show the best baseline results.

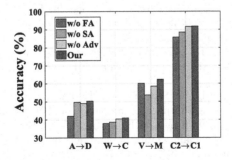

Fig. 4. Ablation analysis of RDA model.

We randomly select several tasks and report the results in Fig. 4 by using the model without (w/o) the associated loss terms. From the results, we observe that each loss is indispensable. In general, the FA term has the greatest impact on performance. This is because there exists a large distribution divergence between two domains and most of the samples are misclassified. The SA term is also important, since it verifies that using domain-specific projections is more effective than a shared projection. The results also demonstrate that the adversarial loss term can further boost performance by improving the reliability of the model. The effectiveness of the adversarial regularization term is verified.

5 Conclusion

In this paper, we proposed a new Reliable Domain Adaptation (RDA) approach for UDA. RDA tries to simultaneously align the manifold and category space across domains through two dual projections. In order to address the prediction bias problem involved by pseudo labels, an adversarial learning strategy is introduced. Firstly, RDA focuses on the discovery of unreliable samples by maximizing the discrepancy between the two task-classifiers. Secondly, RDA focuses on the correction of those unreliable target samples by minimizing the classifiers discrepancy. Comprehensive experiments validate the superiority of RDA over state-of-the-arts.

References

1. Gong, B., Shi, Y., Sha, F., Grauman, K.: Geodesic flow kernel for unsupervised domain adaptation. ICCV **157**(10), 2066–2073 (2012)
2. Baktashmotlagh, M., Harandi, M.T., Lovell, B.C., Salzmann, M.: Unsupervised domain adaptation by domain invariant projection. In: ICCV, pp. 769–776 (2013)
3. Baktashmotlagh, M., Harandi, M.T., Lovell, B.C., Salzmann, M.: Domain adaptation on the statistical manifold. In: CVPR, pp. 2481–2488 (2014)
4. Belkin, M., Niyogi, P., Sindhwani, V.: Manifold Regularization: A Geometric Framework for Learning from Labeled and Unlabeled Examples. JMLR.org (2006)
5. Ben-David, S., Blitzer, J., Crammer, K., Kulesza, A., Pereira, F., Vaughan, J.W.: A theory of learning from different domains. Mach. Learn. **79**(1–2), 151–175 (2010)
6. Boyd, S., Parikh, N., Chu, E., Peleato, B., Eckstein, J.: Distributed optimization and statistical learning via the alternating direction method of multipliers. Found. Trends Mach. Learn. **3**(1), 1–122 (2011)
7. Chu, W.S., Torre, F.D.L., Cohn, J.F.: Selective transfer machine for personalized facial expression analysis. IEEE Trans. Pattern Anal. Mach. Intell. **39**, 529–545 (2017)
8. Fernando, B., Habrard, A., Sebban, M., Tuytelaars, T.: Unsupervised visual domain adaptation using subspace alignment. In: ICCV, pp. 2960–2967 (2014)
9. Ghifary, M., Balduzzi, D., Kleijn, W.B., Zhang, M.: Scatter component analysis: a unified framework for domain adaptation and domain generalization. IEEE Trans. Pattern Anal. Mach. Intell. **39**(7), 1414–1430 (2017)
10. Goodfellow, I.J., et al.: Generative adversarial nets. In: NIPS, pp. 2672–2680 (2014)

11. Gopalan, R., Li, R., Chellappa, R.: Domain adaptation for object recognition: an unsupervised approach. In: ICCV (2011)
12. Griffin, G.S., Holub, A.D., Perona, P: Caltech-256 object category dataset. California Institute of Technology (2007)
13. Kan, M., Wu, J., Shan, S., Chen, X.: Domain adaptation for face recognition: targetize source domain bridged by common subspace. Int. J. Comput. Vis. **109**(1–2), 94–109 (2014)
14. Long, M., Cao, Y., Wang, J., Jordan, M.I.: Learning transferable features with deep adaptation networks. In: ICML, pp. 97–105 (2015)
15. Long, M., Wang, J., Ding, G., Sun, J., Yu, P.S.: Transfer feature learning with joint distribution adaptation. In: ICCV, pp. 2200–2207 (2014)
16. Long, M., Wang, J., Ding, G., Sun, J., Yu, P.S.: Transfer joint matching for unsupervised domain adaptation. In: CVPR, pp. 1410–1417 (2014)
17. Long, M., Wang, J., Jordan, M.I.: Unsupervised domain adaptation with residual transfer networks. In: NIPS (2016)
18. Pan, S.J., Tsang, I.W., Kwok, J.T., Yang, Q.: Domain adaptation via transfer component analysis. IEEE Trans. Neural Netw. **22**(2), 199–210 (2011)
19. Pan, S.J., Yang, Q.: A survey on transfer learning. IEEE Trans. Knowl. Data Eng. **22**(10), 1345–1359 (2010)
20. Rate, C., Retrieval, C.: Columbia object image library (coil-20). Computer(2011)
21. Saenko, K., Kulis, B., Fritz, M., Darrell, T.: Adapting visual category models to new domains. In: Daniilidis, K., Maragos, P., Paragios, N. (eds.) ECCV 2010. LNCS, vol. 6314, pp. 213–226. Springer, Heidelberg (2010). https://doi.org/10.1007/978-3-642-15561-1_16
22. Saito, K., Watanabe, K., Ushiku, Y., Harada, T.: Maximum classifier discrepancy for unsupervised domain adaptation. In: CVPR (2018)
23. Simonyan, K., Zisserman, A.: Very deep convolutional networks for large-scale image recognition. Comput. Sci. (2014)
24. Sun, B., Feng, J., Saenko, K.: Correlation alignment for unsupervised domain adaptation. In: Csurka, G. (ed.) Domain Adaptation in Computer Vision Applications. ACVPR, pp. 153–171. Springer, Cham (2017). https://doi.org/10.1007/978-3-319-58347-1_8
25. Torralba, A., Efros, A.A.: Unbiased look at dataset bias. In: CVPR, pp. 1521–1528 (2011)
26. Tzeng, E., Hoffman, J., Saenko, K., Darrell, T.: Adversarial discriminative domain adaptation. In: CVPR, pp. 7167–7176 (2017)
27. Tzeng, E., Hoffman, J., Darrell, T., Saenko, K.: Simultaneous deep transfer across domains and tasks. In: ICCV, pp. 4068–4076 (2017)
28. Wang, J., Feng, W., Chen, Y., Yu, H., Huang, M., Yu, P.S.: Visual domain adaptation with manifold embedded distribution alignment. In: ACM MM (2018)
29. Xu, Y., Fang, X., Wu, J., Li, X., Zhang, D.: Discriminative transfer subspace learning via low-rank and sparse representation. IEEE Trans. Image Process. **25**(2), 850–863 (2016)
30. Zhang, J., Li, W., Ogunbona, P.: Joint geometrical and statistical alignment for visual domain adaptation. In: CVPR, pp. 5150–5158 (2017)
31. Zhang, L., Zhang, D.: Robust visual knowledge transfer via extreme learning machine-based domain adaptation. IEEE Trans. Image Process. **25**(10), 4959–4973 (2016)

An End-to-End LSTM-MDN Network for Projectile Trajectory Prediction

Li-he Hou and Hua-jun Liu[(⊠)]

School of Computer Science and Engineering,
Nanjing University of Science and Technology, Nanjing, China
liuhj@njust.edu.cn

Abstract. Trajectory prediction from radar data is an example of a signal processing problem, which is challenging due to small sample sizes, high noise, non-stationarity, and non-linear. A data-driven LSTM-MDN end-to-end network from incomplete and noisy radar measurements to predict projectile trajectory is investigated in this paper. Traditional prediction algorithm usually uses Kalman Filter (KF) or the like to estimate target's position and speed, then uses the numerical integral, such as Runge-Kutta, Adams, etc. to extrapolate the launch point or impact point, which mainly relies on the accuracy of dynamic models. A Long Short-Term Memory (LSTM) network is designed to estimate the real position from sampled and noisy radar measurements series, and a Mixture Density Network (MDN) is developed for trajectory extrapolation and projectile launch point prediction. These two subnetworks are integrated into an end-to-end network, which is trained by the radar measurement samples of a projectile and the corresponding ground truth of its launch point. Compared with the traditional methods, amount of experiments show that our proposed method is superior to the traditional model-based methods, and its adaptability to the range of initial launch angle is significantly better than the traditional method.

Keywords: Trajectory prediction · Mixture Density Network · Long Short-Term Memory Network

1 Introduction

With the emergence of low-cost location sensing and related devices, coupled with the maturity of wireless transceiver technology, the trajectory information of mobile targets becomes one of the most important context information in many application scenarios, and it can provide important support for various services and applications, such as Intelligent Transportation System (ITS), intelligent navigation system, route planning, recommendation system [1], public facilities planning [2], and public safety [3], etc. At present, the analysis of time series data for trajectory prediction mainly focuses on pedestrians [4, 5], cars [6] on the road or drones in the air, aircrafts [7, 8], and various balls [9, 10] in some sports scenes and so on, and the analysis of the above objects mainly uses the video or image as input. For some other data source such as radar data [11] or sonar data, due to their small sample size, high noise, and non-stationary characteristics, trajectory prediction is more challenging than the visual video data.

© Springer Nature Switzerland AG 2019
Z. Cui et al. (Eds.): IScIDE 2019, LNCS 11936, pp. 114–125, 2019.
https://doi.org/10.1007/978-3-030-36204-1_9

Reconnaissance radars usually use the observed motion trajectory of a non-maneuverable projectile target to obtain the impact point or launch point of the target through filtering and extrapolation. Based on a few segments of sampled and incomplete measurement data with high noise, some prior aerodynamic models and optimal estimation algorithms are designed to filter the noise, to reconstruct the trajectory and to predict the launch point in the traditional pipeline.

According to the external ballistic theory [12], the flight path of the non-maneuverable projectile target is a non-linear dynamic process. Therefore, the filtering technique of non-linear system is needed to reduce random noise and to estimate the real state (position and velocity, etc.) of a target. Currently, the extended Kalman filter (EKF), unscented Kalman filter (UKF) algorithm, and the Markov chain (MC) model are widely used in this field.

Specifically, in the projectile trajectory prediction domain, as the initial launch angle of the projectile target decreases, the prediction accuracy based on the traditional EKF method would also decrease because of the smaller sample size and the stronger nonlinearity. Especially in the low angle field where the initial launch angle is less than 20 degrees, we need a more robust algorithm to improve the prediction accuracy. For non-maneuverable projectile target trajectories with different launch angles and sampling rate, it is difficult to establish a general purpose dynamic model, which are several equations with uncertain parameters, for trajectory extrapolation and prediction.

With the rise of deep learning, many researchers began to use Recurrent Neural Network (RNN) methods for trajectory prediction. Such as Al-Molegi et al. [5] proposed the Spatial Temporal Featured-RNN (STF-RNN) algorithm, which is trained by the target trajectory data to predict its specific position at a certain moment, and achieved good results. However, RNN cannot handle the trajectory prediction problem of long-term sequences, because an excessively long sequence will cause the RNN to produce gradient exploding or gradient vanishing during training. The improved LSTM model can avoid similar phenomena. Park et al. [6] used the LSTM Encoder-Decoder-based method to generate K most likely trajectory candidates for car trajectory prediction through beam search technique. Shi et al. [7] proposed a flight trajectory prediction model based on a LSTM network. By applying sliding windows in LSTM to maintain the continuity and avoid dynamic compromising dependencies of adjacent states in the long-term sequences. However, the above algorithms did not consider the case of data with small sample sizes and high noise.

In response to the above problems, a data-driven LSTM-MDN end-to-end network from radar measurements with high-noise and small size sample for predicting projectile trajectory is investigated in this paper. Compared with the traditional EKF based method, merits of this method lies in that it does not need to establish the precise projectile motion equation in advance, and it is a data-driven and model-free architecture, and the trained neural network parameters can approximate the model with extremely low error in essence if we can get amount of training samples. Moreover, it is a less parameter and end-to-end architecture, except several hyper-parameters, most parameters of the whole neural network can be learnable through training samples, so it's more practical for all kind of projectile trajectories prediction. LSTM is a dynamic model for predicting and generating time series data. It belongs to a branch of the RNN. Compared with ordinary RNN, LSTM has better performance in long-term sequences.

Our experimental results show that compared with the existing methods, the LSTM +MDN method has a significant improvement in the accuracy of low-angle trajectory prediction.

To summarize, the main contributions of this paper can be listed as the following three aspects:

(1) An end-to-end deep neural network architecture for projectile trajectory prediction with incomplete and noisy radar measurement data is proposed.
(2) A mixed density network used to predict target trajectories and launch point of the non-maneuverable projectile target is one of the earliest proposals in this domain.
(3) Time-stamp for enriching the data dimension and training samples interpolation for data densification is used for data augmentation on the training dataset, which can improve the prediction accuracy.

The organization of the remainder is shown as follows: Sect. 2 describes the related work, and Sect. 3 introduces details of our proposed end-to-end trajectory predict system. The experimental results and analysis will be described in Sect. 4. Finally, the conclusion is given in Sect. 5.

2   Related Work

The fire launch point prediction is one of the main functions of a reconnaissance radar. The schematic diagram of the traditional EKF prediction method is shown in Fig. 1. It usually uses a radar to capture the section of moving trajectories of projectile targets and can acquire targets discrete flight arc; then it performs target's tracking and filtering to reduce the noise; subsequently, it performs ballistic extrapolation to estimate the fire location, which is exactly the weapon's launch point. Usually, this general process can be divided into three steps:

(1) The radar detects flight path of moving target in its ascending arc and obtains 3-D position coordinates and time-stamps with a certain sampling rate.
(2) Using a simplified ballistic model combined with the tracking and filtering algorithms to obtain the targets ballistic parameters as accurately as possible.
(3) A precise ballistic model and numerical integral algorithms are used for the trajectory reconstruction and extrapolation to predict the target's launch point.

As shown in Fig. 1, the *ab* segment in the figure is in a radar tracking stage which is an observable segment by radar, the *bc* segment is the ballistic extrapolation segment which cannot be detected by radar, and the point *a* is the extrapolation starting point, and *c* is the fire launch point which we want to extrapolate through the observable segment.

For different projectile trajectories' prediction and extrapolation, traditional EKF methods often need to establish different mathematical models. As an improvement of EKF method, Mousavi et al. [13] introduced a fading factor based on the traditional extended Kalman filter to form a fading memory Kalman filter, which improved the filtering accuracy and convergence.

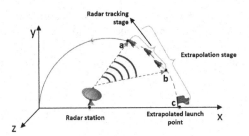

Fig. 1. Traditional extrapolation method

However, the same mathematical model tends to be inferior when moving to other different scenarios for extrapolation. Therefore, some researchers advocate the Markov model to solve this problem. The Markov model uses the transition probability between targets to establish the transfer matrix, and then uses the transfer matrix to predict the position of the next target [14, 15]. Later, Rendle et al. [16] proposed a factorized personalized MC (FPMC) model, which combines both a common Markov chain and the normal matrix factorization model, and obtain better experimental results on the actual data. There are other further studies on the position prediction of Markov models, and some scholars such as Mathew et al. [17] used the hidden Markov model (HMM) to solve such problems. However, the drawback of using these Markov models is that the computational complexity increases exponentially as the data grows, so it cannot meet the needs of real-time prediction.

In recent years, neural networks, especially RNNs, have been widely used in time series processing [18, 19]. However, standard RNNs are not very competent when dealing with long-term sequences. As one of the extensions of RNN, LSTM has better performance on long-term sequence prediction [20].

Figure 2 shows the ability of learning a sin curve based on an LSTM network. As the training round increases, the LSTM network learns the variation of the sin curve and can predict the next change of the sin curve well.

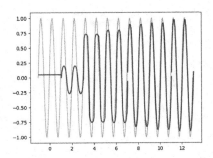

Fig. 2. LSTM learning a sine wave

However, for the incomplete and noisy long-term sequence of radar detection data, LSTM will accumulate the error of each prediction step when making predictions, as a result, the prediction accuracy of the target's trajectory will eventually become unacceptable. The MDN can replace the unimodal distribution of a typical neural network output with a linear combination of kernel functions to simulate an arbitrary output distribution, making it better for trajectory prediction. For most cases, the kernel function generally selects a Gaussian function whose probability density can be expressed as:

$$p(y|x; w) = \sum_{k=1}^{K} \pi_k(x; w) \varnothing \left(\mu_k(x; w), \sigma_k^2(x; w) \right) \tag{1}$$

where, $\sum_{K=1}^{K} \pi_k(x) = 1$, $0 \le \pi_k(x; w) \le 1$, this K here is the number of mixed distributions, and w is the weight matrix of the neural network. $\pi_k(x; w)$, $\mu_k(x; w)$ and $\sigma_k^2(x; w)$ are the output of the MDN, which represent the mixing coefficients, mean and variance of the Gaussian mixture distribution respectively. Finally, these mixed values are integrated to establish a mixed Gaussian distribution model, and the final prediction trajectory can be obtained by a probability weighting mechanism.

3   LSTM-MDN Prediction Network

The framework of this section can be roughly divided into three parts. In Sect. 3.1, the details of the LSTM network [21] whose input at this moment is its output previous moment are listed. In Sect. 3.2, the second part of the model, a MDN network for trajectory prediction will be introduced, its output will be used to synthesize a Gaussian mixture distribution representing the target trajectory. In Sect. 3.3, a fully connected (FC) network paralleling with the MDN network will be introduced, it takes the state of the last moment of the LSTM output as input, and it will be used to predict the launch point. Finally, the total loss function of the end-to-end network will be discussed, it is the intersection where two parallel networks will meet. By minimizing the total loss, we can get a convergent trajectory prediction and launch point prediction at the same time. The network structure of our proposal is shown below (Fig. 3):

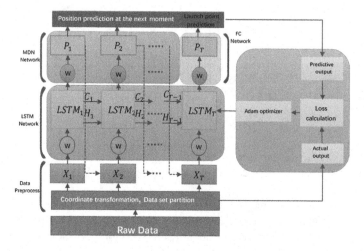

Fig. 3. The LSTM-MDN prediction network structure

3.1 LSTM

The RNN has a good performance in sequence processing [22]. But, in terms of long-term sequence prediction, LSTM has the advantage that ordinary RNN can't match, because of its more parameters and more complex network structure [23]. As a variant of RNN, LSTM has the same input and output as a normal RNN network. However, in LSTM, due to its input gates, forgetting gates and output gates, information can be selectively passed, thereby reducing the phenomenon of gradient exploding or gradient vanishing.

The mathematical equations of LSTM are:

$$\begin{cases} f_t = \sigma\left(W_f[h_{t-1}, x_t] + b_f\right) \\ i_t = \sigma\left(W_i[h_{t-1}, x_t] + b_i\right) \\ \tilde{C}_t = tanh(W_C[h_{t-1}, x_t] + b_C) \\ C_t = f_t * C_{t-1} + i_t * \tilde{C}_t \\ o_t = \sigma(W_o[h_{t-1}, x_t] + b_o) \\ h_i = o_t * \tanh(C_t) \end{cases} \tag{2}$$

In our case, x_t represents the input data, which is a 1-dimensional vector with 4 elements, which are the coordinates of the x, y, and z axes and the time-scale information t. W_f, W_i, W_C and W_o are the weight matrix, while b_f, b_i, b_C, and b_o are the bias matrix. And f_t represents the forgetting gate which determines whether information can be held in the cell state. It outputs a number between 0 and 1 to determine the persistence of information, where 1 means "completely reserved" and 0 means "completely discarded".

3.2 MDN

The MDN is different from ordinary neural networks. In our structure, its input is the LSTM's output after subtracting the final time step output element. MDN's output are the parameters that make up the Gaussian Mixture Model. The operation of the MDN can be described as follows, where h_t is the first Ts-1 term of the LSTM output (Ts represents the time steps).

$$y_t = \left\{ \left(\tilde{\mu}_t^c, \tilde{\pi}_t^c, \tilde{\sigma}_t^c, \tilde{\rho}_t^c\right) \right\}_{c=1}^{C} = W_y * h_t + b_y \tag{3}$$

where the c represents the c-th Gaussian component. Each of the parameters $\tilde{\mu}_t^c, \tilde{\pi}_t^c, \tilde{\sigma}_t^c, \tilde{\rho}_t^c$ can be seen as a function of input h_t.

There is a restriction that the sum of π_t^c must add up to one, to ensure the Gaussian Mixture function integrates to 100%. At the same time, σ_t^c must be strictly positive. Therefore, the mathematical equation of the MDN is defined as follows.

$$\begin{cases} \mu_t^c = \tilde{\mu}_t^c \\ \pi_t^c = \dfrac{\exp(\tilde{\pi}_t^c)}{\sum_{i=1}^{C} \exp(\tilde{\pi}_t^i)} \\ \sigma_t^c = \exp(\tilde{\sigma}_t^c) \\ \rho_t^c = \tanh(\tilde{\rho}_t^c) \end{cases} \tag{4}$$

Then, the Gaussian mixture distribution is generated by the processed parameters, and the predicted value of the XYZ coordinates at each moment is obtained.

3.3 Fully Connected Network

A FC sub-network is designed to predict the final launch point of the projectile. The FC sub-network has three layers with Rectified Linear Unit (ReLU) as its activation function. The number of FC network's nodes are 32, 16, and 3 respectively. The FC network Takes the output of the last moment of the LSTM network as its input, and its output is the extrapolated projectile launch point.

3.4 Loss Function

In our structure, the weighted sum of the MDN and FC sub-network's cost functions is defined as the total loss function. Because the whole network has two aims: predicting the projectile trajectory and its launch point simultaneously.

In MDN, the output is an entire description of the probability distribution, so the min square error L2 loss function cannot be simply used. A more suitable loss function would like to minimize the logarithm of the likelihood of the distribution with the training data, so in our proposal, the loss function for i-th trajectory p^i can be defined as:

$$L_{mdn}(p^i, y^i) = \sum_{t=0}^{T_i-1} -\log \left[\sum_{c=1}^{C} \pi_c^t \varnothing(y_t^i; \mu_t^c, \sigma_t^c, \rho_t^c) \right] \tag{5}$$

where T_i represent time steps we choose and $\varnothing$ is the Gaussian distribution.

In the FC network, the Root Mean Square (RMS) error between the predicted launch point and the launch point ground truth is compared statistically as its loss function. The FC network's loss function can be defined as:

$$L_{fc}(\hat{y}^i, y^i) = \left(\sum_{t=0}^{T_i-1} (\hat{y}_t^i - y_t^i)^2 \right)^{1/2} \tag{6}$$

Two branches cost functions are calculated in parallel, then the sum of them is calculated as the total cost functions. The final total loss function is defined as follows:

$$L = (1 - \lambda) \cdot L_{mdn}\left(p^i, y^i\right) + \lambda \cdot L_{fc}\left(\hat{y}^i, y^i\right) \tag{7}$$

Where λ represents the weight coefficient of the loss of the MDN network and the FC network, and $0 < \lambda < 1$, increase λ will improve the accuracy of the network prediction about launch point, but the performance of the trajectory prediction will be degraded. When λ decreases, vice versa. Usually, we set $\lambda = 0.5$ for a balanced accuracy.

The loss function is optimized using the Backpropagation Through Time (BPTT), and prior to this, a gradient clipping step is also needed to reduce the possibility of a gradient exploding. Eventually, the LSTM, MDN and FC network are jointly trained from randomly-initialized parameters.

4 Experiment Results and Analysis

In this section, the source of the experimental data, and the pre-processing process will be introduced. At the same time, the actual performance of the proposed method will be verified on the experimental data, compared with the traditional method. The experimental results show that the method is superior to the traditional nonlinear filtering and integral extrapolation method in accuracy.

4.1 Dataset

The dataset used in this paper consists of 16,000 sets of trajectory data from non-maneuverable projectile target trajectory simulation program, where the training and test division ratio is 3:1. The launch conditions (initial velocity, angle, ballistic coefficient, etc.) and noise factors (wind direction, wind speed, radar system error, etc.) of these trajectories vary. Because the data generated by the simulator is the polar coordinate system data centered on the radar station, it is not convenient for calculation and processing. Therefore, before training, the coordinate conversion is needed first, and the polar coordinate data will be converted to Cartesian coordinate system data:

$$\begin{cases} x = RsinAcosE \\ y = RcosAcosE \\ z = RsinE \end{cases} \tag{8}$$

where, R, A, and E are the range, azimuth angle, and elevation angle of the generated analog data respectively.

4.2 Network and Training

The network is implemented based on a very popular deep learning framework TensorFlow, version 1.13.0. The experimental models were all trained with the NVIDIA Titan Xp GPU. The model with the CPU is E5-2630 V2.

To ensure the optimization function closer to the optimal solution, the method of exponential decay learning rate is adopted. The specific parameters are as follows:

The initial learning rate is set to 1e-4, the decay period is 1000 iterations, and the stepwise decay method is used with a decay rate of 0.95.

Batch gradient descent is chosen to be our optimization algorithm using Adaptive Moment Estimation(Adam) [24] as its Optimizer. The batch size used in the model is 40. To avoid overfitting, early stop is used, which stops training when the model's performance on the validation set begins to drop.

As the first part of the model, the basic structure of LSTM contains three layers, each layer has 64 units, and its input is a vector consisting of X, Y, Z and time stamp t after coordinate transformation. The input sequence length is 19 of each time, then the coordinates of the corresponding point's launch point are trained as a label. The LSTM output with last item removed will be fed to the MDN for trajectory prediction, and the last item will be fed to the FC network to predict the location of the launch point.

Three Gaussian distributions were used to simulate the target object motion trajectory. Each Gaussian distribution contains eight parameters, which are the mean μ and variance σ of the three-dimensional Gaussian distribution, and two correlation coefficients, so the final output of the MDN has 24 elements.

The FC network responsible for predicting the launch point has three layers, and the number of nodes per layer are 32, 16, and 3 respectively. To avoid over-fitting, the drop rate for each layer has been set to 0.7.

4.3   Performance Metric

The percentage of the circular error probability (CEP) and the target's range has been used as a metric for launch point prediction precision. CEP is often used to measure the accuracy of the projectile hitting, and it is calculated by launching multiple shells at the same target under the same conditions. In general, the impact point will be evenly distributed around the target. Centered on the aiming point, the radius of the circle containing 50% of the impact point is called the CEP of the projectile.

The formula for calculating the probability of a circle is:

$$R = \sqrt{ln2}\left(\left(D_1^2 + D_2^2 + \cdots D_n^2\right)/n\right)^{1/2} \tag{9}$$

where, $D_i = \sqrt{(X_i - \bar{X})^2 + (Y_i - \bar{Y})^2}$, in this paper, X_i, Y_i represent the coordinates of the i-th launch point predicted by the model, while $\bar{X}, \bar{Y}$ represent the coordinates of the corresponding actual launch point.

4.4   Results and Analysis

In this section, the traditional EKF-based prediction method, vanilla LSTM method and the LSTM + MDN method (proposed in this paper) are compared on the same testing data. The testing samples can be grouped by the key parameters, such as initial launch angle A and the noise covariance E of simulated data. The first group is configured with a fixed initial launch angle of $A = 20°$ and a varying noise covariance. The second one is configured with a fixed noise covariance of $E = 5$ and a varying initial launch angle.

The performance curve of the Loss and the CEP with training iterations in our LSTM + MDN method when $A = 20°$ and $E = 5$ are shown in Figs. 4 and 5.

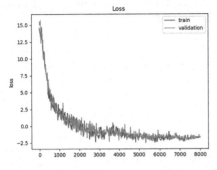

Fig. 4. Curves of Loss with iterations

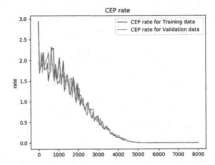

Fig. 5. Curves of CEP rate with iterations

Table 1. Performance comparison with different initial launch angles (E = 5)

Evaluation criteria (°)	Method	A = 10	A = 15	A = 20	A = 30	A = 40
CEP rate	LSTM + MDN	**0.332**	**0.297**	**0.285**	**0.277**	0.271
(%)	LSTM	0.398	0.347	0.342	0.330	0.319
	Traditional EKF	0.434	0.377	0.332	0.306	**0.262**

Table 2. Performance comparison with different noise covariance (A = 20°)

Evaluation criteria	Method	E = 5	E = 10	E = 15	E = 20	E = 25
CEP rate	LSTM + MDN	**0.285**	**0.384**	0.591	0.746	1.004
(%)	LSTM	0.342	0.441	0.650	0.836	1.145
	Traditional EKF	0.332	0.407	**0.531**	**0.684**	**0.855**

The qualitative performances of the traditional EKF based method, vanilla LSTM method and the LSTM + MDN method on different testing samples with $E = 5$ and $A = 20°$ are listed in Tables 1 and 2.

From Tables 1 and 2, it can be shown that the CEP rate of LSTM + MDN is superior to the traditional EKF and LSTM method overall, especially in the cases of low launch angles and low noise covariance, advantages of our proposed are apparent.

It can be analyzed qualitatively from Table 1 that when the noise covariance is constant the performance of LSTM + MDN will not change obviously with the initial launch angle decreasing, which indicates that our proposed is sensitive less to initial launch angle variation compared with the traditional EKF-based method. It can be seen when A = 40°, traditional EKF method's performance gets a significant improvement and performs better than LSTM+MDN method.

As can be seen from Table 2, when the initial launch angle is fixed, the LSTM+MDN method is superior to the traditional EKF method at the cases of low noise covariance. But at the cases of larger noise covariance, the difficulty of learning the projectiles' trajectory features by the vanilla LSTM and LSTM+MDN method will increase rapidly, and some noise factors will be learned as a feature by the network model. However, since the projectile motion equation is established in advance, the traditional EKF method can still predict the approximate trajectory of the projectile accurately.

In summary, the performance of the LSTM+MDN trajectory prediction algorithm proposed by this paper is superior to vanilla LSTM method in most cases, and it's more sensitive to the noise covariance, but it is more accurate in most cases with varying initial launch angle than traditional EKF based method.

5   Conclusion

An end-to-end method for trajectory prediction based on incomplete and noisy time series data is proposed in this paper. A LSTM, a MDN and a FC sub-network are integrated into our end-to-end network. The 4-D radar measurement data (R, A, E, t) is taken as input for our network, after training, it can directly extrapolate the launch point which cannot be directly detected by the radar.

The comparison experiments with traditional EKF-based methods are done on a large amount of testing data, and it can be concluded from analysis that the LSTM +MDN network can be learnable completely from noisy and incomplete radar data without any prior aerodynamic models and any tuned parameters; the LSTM+MDN network has achieved better performance than the traditional method on the same testing data; Moreover, the LSTM+MDN network can be more adaptable to the initial launch angle variation than the traditional method.

In the future, a method on the Meta-Learning [25, 26] based on less training samples will be our next work.

References

1. Bao, J., Zheng, Y., Mokbel, M.F.: Location-based and preference-aware recommendation using sparse geo-social networking data. In: Proceedings of the 20th International Conference on Advances in Geographic Information Systems, pp. 199–208. ACM (2012)
2. Bao, J., He, T., Ruan, S., et al.: Planning bike lanes based on sharing-bikes' trajectories. In: Proceedings of the 23rd ACM SIGKDD International Conference on Knowledge Discovery and data MINING, pp. 1377–1386. ACM (2017)
3. Feng, Z., Zhu, Y.: A survey on trajectory data mining: techniques and applications. IEEE Access **4**, 2056–2067 (2016)
4. Amirian, J., Hayet, J.B., Pettré, J.: Social ways: learning multi-modal distributions of pedestrian trajectories with GANs. In: Proceedings of the IEEE Conference on Computer Vision and Pattern Recognition Workshops (2019)
5. Al-Molegi, A., Jabreel, M., Ghaleb, B.: STF-RNN: space time features-based recurrent neural network for predicting people next location. In: 2016 IEEE Symposium Series on Computational Intelligence (SSCI), pp. 1–7. IEEE (2016)
6. Park, S.H., Kim, B.D., Kang, C.M., et al.: Sequence-to-sequence prediction of vehicle trajectory via LSTM encoder-decoder architecture. In: 2018 IEEE Intelligent Vehicles Symposium (IV), pp. 1672–1678. IEEE (2018)

7. Shi, Z., Xu, M., Pan, Q., et al.: LSTM-based flight trajectory prediction. In: 2018 International Joint Conference on Neural Networks (IJCNN), pp. 1–8. IEEE (2018)

8. Li, Z., Li, S.H.: General aircraft 4D flight trajectory prediction method based on data fusion. In: 2015 International Conference on Machine Learning and Cybernetics (ICMLC), vol. 1309–315. IEEE (2015)

9. Shah, R., Romijnders, R.: Applying deep learning to basketball trajectories. arXiv preprint arXiv:1608.03793. (2016)

10. Wang, K.C., Zemel, R.: Classifying NBA offensive plays using neural networks. In: Proceedings of MIT Sloan Sports Analytics Conference, vol. 4 (2016)

11. Liu, H., Xia, L., Wang, C.: Maneuvering target tracking using simultaneous optimization and feedback learning algorithm based on Elman neural network. Sensors **19**, 1–19 (2019)

12. Wang, N., Li, C., Lu, M.: Research on the simulation method of external Ballistic flight speed of fuze. In: 2017 36th Chinese Control Conference (CCC), pp. 10252–10258. IEEE (2017)

13. Mousavi, M.S.R., Boulet, B.: Estimation of the state variables and unknown input of a two-speed electric vehicle driveline using fading-memory Kalman filter. IEEE Trans. Transp. Electrif. **2**(2), 210–220 (2016)

14. Gambs, S., Killijian, M.O., del Prado Cortez, M.N. : Next place prediction using mobility markov chains. In: Proceedings of the First Workshop on Measurement, Privacy, and Mobility, vol. 3. ACM (2012)

15. Krumm, J.: A Markov model for driver turn prediction. In: Proceedings of the Society of Automotive Engineers World Congress, Detroit, pp. 1–7, 14–17 April 2008. SAE World Congress, Warrendale (2016)

16. Rendle, S., Freudenthaler, C., Schmidt-Thieme, L.: Factorizing personalized markov chains for next-basket recommendation. In: Proceedings of the 19th international conference on World wide web, pp. 811–820. ACM (2010)

17. Mathew, W., Raposo, R., Martins, B.: Predicting future locations with hidden Markov models. In: Proceedings of the 2012 ACM Conference on Ubiquitous Computing, pp. 911–918. ACM (2012)

18. Karpathy, A., Johnson, J., Fei-Fei, L.: Visualizing and understanding recurrent networks. arXiv preprint arXiv:1506.02078. (2015)

19. Graves, A.: Generating sequences with recurrent neural networks. arXiv preprint arXiv: 1308.0850. (2013)

20. Gao, Ya., Jiang, G., Qin, X., Wang, Z.: Location prediction algorithm of moving object based on LSTM. J. Front. Comput. Sci. Technol. (2019)

21. Schmidhuber, J.: Deep learning in neural networks: an overview. Neural Netw. **61**, 85–117 (2015)

22. Zhang, H., Liu, H., Wang, C.: Learning to multi-target tracking in dense clutter environment with JPDA-recurrent neural networks. J. Phy.: Conf. Series **1207**(1), 012011 (2019). IOP Publishing

23. Liu, H., Zhang, H., Mertz, C.: DeepDA: LSTM-based Deep Data Association Network for Multi-Targets Tracking in Clutter. arXiv preprint arXiv:1907.09915. (2019)

24. Kingma, D.P., Ba, J.: Adam: a method for stochastic optimization. arXiv preprint arXiv: 1412.6980. (2014)

25. Andrychowicz, M., Denil, M., Gomez, S., et al.: Learning to learn by gradient descent by gradient descent. In: Advances in Neural Information Processing Systems, pp. 3981–3989 (2016)

26. Finn, C., Abbeel, P., Levine, S.: Model-agnostic meta-learning for fast adaptation of deep networks. In: Proceedings of the 34th International Conference on Machine Learning, vol. 70, pp. 1126–1135. JMLR. org (2017)

DeepTF: Accurate Prediction of Transcription Factor Binding Sites by Combining Multi-scale Convolution and Long Short-Term Memory Neural Network

Xiao-Rong Bao, Yi-Heng Zhu, and Dong-Jun Yu[(⊠)]

School of Computer Science and Engineering,
Nanjing University of Science and Technology,
200 Xiaolingwei, Nanjing 210094, China
njyudj@njust.edu.cn

Abstract. Transcription factor binding site (TFBS), one of the DNA-protein binding sites, plays important roles in understanding regulation of gene expression and drug design. Recently, deep-learning based methods have been widely used in the prediction of TFBS. In this work, we propose a novel deep-learning model, called Combination of Multi-Scale Convolutional Network and Long Short-Term Memory Network (MCNN-LSTM), which utilizes multi-scale convolution for feature processing, and the long short-term memory network to recognize TFBS in DNA sequences. Moreover, we design a new encoding method, called multi-nucleotide one-hot (MNOH), which considers the correlation between nucleotides in adjacent positions, to further improve the prediction performance of TFBS. Stringent cross-validation and independent tests on benchmark datasets demonstrated the efficacy of MNOH and MCNN-LSTM. Based on the proposed methods, we further implement a new TFBS predictor, called DeepTF. The computational experimental results show that our predictor outperformed several existing TFBS predictors.

Keywords: Transcription factor binding site · Multi-scale convolution · Long Short-Term Memory Network · Multi-nucleotide one-hot

1 Introduction

Accurate prediction of transcription factor binding site (TFBS) from DNA sequences is critical for regulation of gene expression and drug design [1]. Traditionally, researchers identified TFBS through biochemical methods, such as ChIP-seq [2] and ChIP-chip [3]. However, these methods are time-consuming and laborious, that cannot keep up with relevant research advances in the postgenomic era. Hence, many intelligent computational methods, such as traditional-machine-learning-based ones and deep-learning-based ones, have been proposed and achieved outstanding performance in the task of predicting TFBS [4].

© Springer Nature Switzerland AG 2019
Z. Cui et al. (Eds.): IScIDE 2019, LNCS 11936, pp. 126–138, 2019.
https://doi.org/10.1007/978-3-030-36204-1_10

Traditional-machine-learning-based methods [5–7] led to trends in the field of early TFBS prediction. Wong *et al.* [6] proposed kmerHMM, which used Hidden Markov Models (HMMs) to derive precise TFBS. Fletez-Brant *et al.* [5] developed kmer-SVM, which used a support vector machine with k-mer sequence features to identify predictive combinations of short TFBS. Nevertheless, there are more and more datasets related to transcription factors in the postgenomic era, and traditional-machine-learning-based methods have been unable to quickly and effectively predict TFBS in DNA sequences due to that they are designed for small-scale dataset [8].

Recently, deep-learning-based models [9, 10] have achieved good performances in prediction of TFBS. For example, Alipanahi *et al.* [9] implemented DeepBind, ascertaining sequence specificities from experimental data with Convolutional Neural Network (CNN) [11], to identify TFBS in DNA sequences. Next, CNN-Zeng was presented by Zeng *et al.* [10], which adopted CNN to capture the sequence-based features which are critical to accurate characterization of TFBS.

Although these methods have achieved remarkable performance in prediction of TFBS [12], there is still room for further enhancing the performances as the following two points. Firstly, these methods only take consideration of the independent relationship among nucleotides in TFBS and ignore the interdependence of nucleotides in the sequence. Taking DeepBind [9] as an example, which used one-hot to represent DNA sequences, it ignores the relationship between adjacent nucleotides. However, studies [13] have shown that considering the correlation between nucleotides in adjacent positions can effectively promote the prediction performance. Secondly, many CNN-based models employ a fixed-size convolution kernel to identify TFBS in DNA sequences. For example, CNN-Zeng take the fixed kernel size 24 when do prediction. However, it has a limitation that the length of the TFBS in the model is fixed. It is well known that different kinds of TFBS have different binding lengths [14]. Therefore, it will be useful to employ multiple size of convolutional kernels to capture multi-scale features.

In this work, we first propose a new encoding method, multi-nucleotide one-hot (MNOH), which takes account of the correlation between nucleotides in adjacent positions. Moreover, we design a new deep learning architecture, called Combination of Multi-Scale Convolutional Network and Long Short-Term Memory Network (MCNN-LSTM), which uses multi-scale convolution layers for feature processing, and then uses Long Short-Term Memory Network (LSTM) [15] as a recurrent model that recognizes TFBS in DNA sequences [16]. Based on above, we further implement a TFBS predictor, called DeepTF. Stringent cross-validation and independent tests on benchmark datasets have demonstrated the efficacy of our proposed methods.

2 Materials and Methods

2.1 Benchmark Datasets

We used 690 ChIP-seq datasets, constructed in CNN-Zeng [10] as benchmark datasets to evaluate the proposed method. The positive set consisted of the centering 101 bps that overlaps one TFBS, which have more than 40 000 binding events, while the negative set was 101bp sequences, which do not bind to transcription factor. For each dataset, we use 70% for training, 10% for validating and the remaining 20% for testing.

2.2   Feature Representation

We use the MNOH encoding method to convert DNA sequences into feature matrixes. Specifically, for one DNA sequence $S = (s_1, s_2, \ldots s_n)$ with the length of n, we have $s_i \in A(adenine), C(cytosine), G(guanine), T(thymine)$, which means s_i can be any one of the 4 nucleotide bases in DNA. MNOH encodes it as a matrix, denoted as Z_{MNOH}, as follows:

$$Z_{MNOH} = [Z_1 \ Z_2 \ \cdots \ Z_{n-h+1}]$$

$$= \begin{bmatrix} z_{1,1} & z_{1,2} & \cdots & z_{1,4^h} \\ z_{2,1} & z_{2,2} & \cdots & z_{2,4^h} \\ \vdots & \vdots & & \vdots \\ z_{n-h+1,1} & z_{n-h+1,2} & \cdots & z_{n-h+1,4^h} \end{bmatrix} \quad (1)$$

Where h represents the degree of MNOH, Z_i represents an one-hot vector of the i-th $(i = 1, 2, \ldots, n - h + 1)$ nucleotide in DNA sequence, and $z_{i,j}$ is the value of j-th position in vector Z_i, which is defined as follows:

$$z_{i,j} = \begin{cases} 1 & when \ j = 4^{h-1}f(s_i) + 4^{h-2}f(s_{i+1}) + \ldots + 4^0 f(s_{i+h-1}) \\ 0 & others \end{cases} \quad (2)$$

Where $f(s_i)$ represents the function corresponding to the nucleotide at the i-th position in the DNA sequence. The specific definition of $f(s_i)$ is in Eq. (3):

$$f(s_i) = \begin{cases} 1 & when \ s_i = A \\ 2 & when \ s_i = C \\ 3 & when \ s_i = G \\ 4 & when \ s_i = T \end{cases} \quad (3)$$

Finally, each DNA sequence can be encoded into a $(n - h + 1) \times 4^h$ matrix. The one-hot encoding method is a special kind of the MNOH when $h = 1$. That means we only consider each individual nucleotide, and each nucleotide in the DNA sequence is denoted as one of the 4 one-hot vectors, i.e., [1, 0, 0, 0], [0, 1, 0, 0], [0, 0, 1, 0] and [0, 0, 0, 1]. Furthermore, when $h = 2$, MNOH takes the dependencies between two adjacent nucleotides into account, which has 16 dinucleotides totally, i.e., [AA AC AG AT CA CC CG CT GA GC GG GT TA TC TG TT]. As shown in Eq. (2), these dinucleotides can be denoted as one of the 16 one-hot vectors, i.e., [1, 0, 0, 0, 0, 0, 0, 0, 0, 0, 0, 0, 0, 0, 0, 0], [0, 1, 0, 0, 0, 0, 0, 0, 0, 0, 0, 0, 0, 0, 0, 0], ..., and [0, 0, 0, 0, 0, 0, 0, 0, 0, 0, 0, 0, 0, 0, 0, 1].

However, as the degree of MNOH increases, the dimension of the vector will rise at the same time. Hence, the GPU memories consumed by the model will increase and the calculations will drop. Due to the limitation of GPU memory, we only evaluate the performances of MNOH with degree less than five.

2.3   Model Architecture

In this section, we introduce a novel deep learning model, named MCNN-LSTM, by combining multi-scale CNN with LSTM. Figure 1 illustrates the workflow of MCNN-LSTM, which consists of three parts, *i.e.*, multi-scale convolutional layer, LSTM network layer and fully-connected (fc) layer. Firstly, the input DNA sequence is encoded into a $(n - h + 1) \times 4^h$ matrix, by MNOH. Then, this matrix is passed on to the multi-scale convolutional layer, and each sublayer scans the output of the previous sublayer through convolutional kernels of different sizes. In other words, this layer can be thought of as feature extractors, which can capture the features of TFBS of different lengths. Next, we use LSTM as a recurrent model that recognizes TFBS in DNA sequences. At last, the fc layer takes the outputs of the LSTM network and produces two classification results.

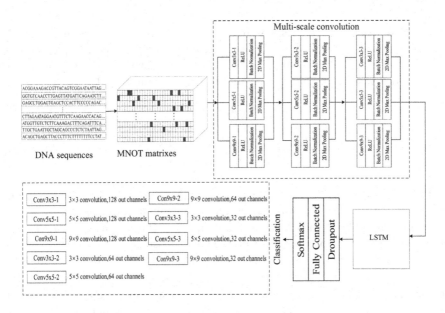

Fig. 1. The architecture of CNN-LSTM model.

Multi-scale CNN. The first part contains three multi-scale CNN layers, and each layer consists of three sublayers with different kernel sizes. Specifically, each sublayer contains convolution, rectified linear unit (ReLU) [17], batch normalization (BN) and max-pooling layer. Moreover, these sublayers separately capture different local features of the DNA sequences and produce multi-scale outputs. In light of this, we use the average of the three outputs as inputs for the subsequent layers. In addition, fused features from CNN layers with multiple kernels sizes will be more distinguishable than that from single convolutional kernel.

Here, the multi-scale CNN layers are introduced, which play an important role when processing DNA sequence features. After inputting the matrix Z_{MNOH}, the implementation of the convolutional part is as follows:

$$X_{i,k} = \sum_{j=1}^{m} \sum_{l=1}^{4^h} z_{i+j,l} M_{k,j,l} + b_k \tag{4}$$

Where $i \in [1, n+m-h]$ and $k \in [1, d]$, d is the number of convolution kernels, M is the kernel of convolution, b_k represents the bias term, m represents the size of the convolutional kernel, and h represents the degree of MNOH.

We then feed X into ReLU [17] activation function to alleviate the problem of overfitting. Next, we add BN to speed up the training of the model and improve the prediction accuracy of the model. The final layer is a max-pooling layer. This layer only picks out the maximum value of its respective previous layer outputs. The function of the max-pooling process is reducing the amount of computation.

LSTM Network Layer. As we know, the TFBS is a fragment in the DNA sequence. Hence, the TFBS prediction can be interpreted as a sequence prediction problem. In recent years, LSTM networks have been widely used for sequence tasks, such as speech recognition [18] and language translation tasks [19].

$$\begin{aligned}
f_t &= \sigma(W_f[h_{t-1}, x_t] + b_f) \\
i_t &= \sigma(W_i[h_{t-1}, x_t] + b_i) \\
C_t &= f_t C_{t-1} + i_t tanh(W_c[h_{t-1}, x_t] + b_C) \\
o_t &= \sigma(W_o[h_{t-1}, x_t] + b_o) \\
h_t &= o_t tanh(C_t)
\end{aligned} \tag{5}$$

The process of LSTM is above. Where x_t is the current input, i_t is the input gate at time t, h_t is the hidden state, C_t is the cell state at time t, f_t is the forget gate, σ is the sigmoid function and the o_t is the output gate, b_f, b_i, b_C and b_o is bias terms. Moreover, the weight matrix W subscripts have the obvious meaning. For instance, W_f is the forget gate matrix, W_i is the input gate matrix.

Fc Layer. The final part contains the fc layer with a dropout [20] regularization and a softmax activation function. The dropout is used to randomly mask portions of its output to avoid overfitting and softmax function is used to transform the output into a probability distribution over two classes.

2.4   Evaluation Metric

In this study, we choose Sensitivity (Sen), Specificity (Spe), Accuracy (Acc) and the Matthews correlation coefficient (MCC) to evaluate the performance of the method [21], which are defined as follows:

$$Sen = TP/(TP + FN)$$
$$Spe = TN/(TN + FP)$$
$$Acc = (TP + TN)/(TP + FP + TN + FN) \quad (6)$$
$$MCC = \frac{(TP \times TN - FP \times FN)}{\sqrt{(TP+FP) \times (TN+FN) \times (TP+FN) \times (TN+FP)}}$$

Where TP, TN, FN and FP means true positives, true negatives, false negatives and false positives.

We further employ another evaluation metric, the area under the receiver operating characteristic curve (ROC), called AUC [22]. The value of AUC is between 0 and 1. The closer AUC tends to 1, the better the predictor.

3 Results

3.1 Comparisons Among CNN-LSTM, CNN, and LSTM

Recently, CNN and LSTM have been adapted to the task of predicting TFBS. To demonstrate the efficacy of CNN-LSTM, we compare it with CNN and LSTM. Note that all of the three models use the simplest one-hot encoding method. Moreover, we randomly select five datasets from all 690 ChIP-seq datasets, namely A549, GM12878, Hela, Hep-G2 and H1-hESC, as benchmark datasets. For each dataset, we separately train three models on the corresponding training dataset, and evaluate the performances of these models on the test datasets. In the remaining experiments, we take the same setting which afore-mentioned.

Table 1. Ablation study of CNN-LSTM, CNN and LSTM.

Dataset	Model	Sen (%)	Spe (%)	Acc (%)	MCC	AUC
A549	CNN-LSTM	59.3	**82.0**	**76.1**	**0.518**	**0.831**
	CNN	64.7	79.8	72.7	0.449	0.804
	LSTM	**66.5**	44.1	55.6	0.109	0.619
GM12878	CNN-LSTM	67.2	**77.3**	**77.2**	**0.545**	**0.854**
	CNN	69.5	69.3	74.1	0.480	0.826
	LSTM	**76.6**	59.8	60.7	0.213	0.645
Hela	CNN-LSTM	62.5	**84.6**	**80.3**	**0.604**	**0.801**
	CNN	66.9	80.7	69.9	0.387	0.773
	LSTM	**75.7**	33.9	56.2	0.112	0.567
Hep-G2	CNN-LSTM	72.6	**80.9**	**76.8**	**0.617**	**0.844**
	CNN	75.1	76.9	71.5	0.536	0.781
	LSTM	**76.9**	68.5	67.3	0.487	0.672
H1-hESC	CNN-LSTM	**82.2**	**77.7**	**80.0**	**0.599**	**0.885**
	CNN	79.2	71.4	75.5	0.481	0.793
	LSTM	70.5	72.8	68.4	0.364	0.619

From Table 1, we find that CNN-LSTM performs better than both CNN and LSTM. As for CNN, the four evaluation metrics (*i.e.*, Spe, Acc, MCC and AUC) of CNN-LSTM are 4.9%, 5.4%, 11.0% and 4.8% higher on average than CNN on the five datasets, respectively. Although the Sen of CNN-LSTM are lower than CNN on four out of five datasets (i.e., A549, GM12878, Hela, Hep-G2), the corresponding Spe, Acc, MCC and AUC values of CNN-LSTM are relatively higher. The underlying reason is that too many false positive example in CNN. The similar observations can also be obtained on LSTM. For example, the CNN-LSTM on Hela dataset achieves a Spe of 84.6%, an Acc of 80.3%, a MCC of 60.4% and an AUC of 80.1%, which are 50.7%, 24.1%, 49.2% and 23.4%, respectively, higher than LSTM.

To further investigate the effectiveness of the CNN-LSTM, we implement three models on all 690 ChIP-seq datasets. Figure 2 summarizes the distribution of AUCs of these models. The red dotted line indicates the median value of AUCs.

It is easy to see that the distribution of the AUCs of CNN-LSTM on benchmark datasets is significantly higher than both CNN and LSTM. Moreover, it can be clearly observed from Fig. 2 that the median AUC of CNN-LSTM is 0.812, while the median AUC of CNN is 0.764 and LSTM is 0.527. Considering that the benchmark datasets contain 690 ChIP-seq datasets, these experimental results show the superiority of the CNN-LSTM we proposed.

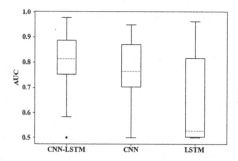

Fig. 2. The distribution of AUCs on 690 ChIP-seq datasets. (Color figure online)

These findings indicate that CNN-LSTM can effectively improve the prediction accuracy for predicting TFBS in this study. One side, CNN has made great achievements in task of predicting TFBS, by which the binding preferences of transcription factor can be well characterized. Moreover, CNN layers with kernel size of 24 scans the input sequence, which aims to extract features of the TFBS. On the another side, LSTM has proved to be effective in the processing of sequence features. The LSTM network layer can take its advantages to deal with sequence features to predict the TFBS in DNA sequences.

3.2   MNOH Helps to Improve the Prediction Performance

In this section, we will choose the optimal degree h of MNOH and prove the efficacy of MNOH. Specifically, we employ MNOH with four different degrees of MNOH, i.e. 1, 2, 3 and 4, denoted as 1NOH, 2NOH, 3NOH and 4NOH, respectively, to extract features from DNA sequences, which are used as the inputs of CNN-LSTM. Therefore, these CNN-LSTM models with inputs generated by different h are denoted as C-L-1NOH, C-L-2NOH, C-L-3NOH, and C-L-4NOH. Similarly, we select the five datasets above as benchmark datasets. Table 2 summarizes the contrast results of the models on five datasets.

Table 2. Ablation study of MNOH.

Dataset	Model	Sen (%)	Spe (%)	Acc (%)	MCC	AUC
A549	C-L-1NOH	59.3	**82.0**	76.2	0.518	0.831
	C-L-2NOH	**62.0**	79.6	**77.9**	**0.546**	**0.852**
	C-L-3NOH	60.3	81.1	77.2	0.526	0.849
	C-L-4NOH	61.6	80.1	76.9	0.520	0.841
GM12878	C-L-1NOH	67.2	77.3	77.2	0.545	0.854
	C-L-2NOH	69.9	**89.4**	**80.8**	**0.604**	**0.872**
	C-L-3NOH	**78.1**	82.6	80.4	0.577	0.863
	C-L-4NOH	75.9	80.3	78.2	0.582	0.867
Hela	C-L-1NOH	62.5	**84.6**	80.3	**0.604**	0.801
	C-L-2NOH	**74.4**	79.2	**83.4**	0.506	**0.822**
	C-L-3NOH	70.8	77.5	82.2	0.483	0.819
	C-L-4NOH	67.8	82.0	83.5	0.503	0.824
Hep-G2	C-L-1NOH	**72.6**	80.9	76.8	0.617	0.844
	C-L-2NOH	66.3	**83.6**	**78.6**	**0.635**	**0.857**
	C-L-3NOH	70.2	81.8	77.1	0.621	0.851
	C-L-4NOH	67.4	82.5	77.8	0.629	0.855
H1-hESC	C-L-1NOH	**82.2**	77.7	80.0	0.599	0.885
	C-L-2NOH	76.1	**83.5**	**81.4**	**0.613**	**0.898**
	C-L-3NOH	77.8	82.0	81.1	0.610	0.893
	C-L-4NOH	79.2	81.5	80.7	0.604	0.889

According to Table 2, we can draw two mainly conclusions as follows.

Firstly, 1NOH has the worst results. Using A549 dataset as an example, in terms of Acc, MCC and AUC three metrics, 2NOH achieves 1.7%, 2.8% and 2.1% better than 1NOH respectively. 3NOH is 1.0%, 0.8% and 1.8% better than 1NOH. 4NOH achieves 0.7%, 0.2% and 1.0% respectively, better than 1NOH. Similar observations can also be obtained on other datasets. The underlying cause of this phenomenon is that the higher h, the more comprehensive the sequence characteristics obtained. For example, 2NOH considers the correlation between nucleotides in two adjacent positions.

Secondly, 2NOH is a better choice in this study for encoding the DNA sequences. By revisiting Table 2, we observe that when h is higher than 2, the performance is not significantly improved, even deteriorate. For instance, 2NOH achieves the prediction results with an Acc of 83.4%, a MCC of 50.6% and an AUC of 82.2%, which are 1.2%, 2.3% and 0.3%, respectively, better than 3NOH on the Hela dataset. We speculate that as h increases, the prediction effect will deteriorate. The main reason for this phenomenon is that MNOH takes into account redundant information as h increases. Take the 4NOH as an example, a sequence, with the length of n, is encoded into a $(n-3) \times 256$ matrix, but 99.6% of the elements in the matrix are 0. This means the matrix will be too sparse and the prediction performance will be poor. Therefore, the more the relationship between nucleotides is considered, the prediction cannot be continuously improved. Based on conclusions above, we decide to use 2NOH in our study.

3.3   Multi-scale Convolution Outperforms in Prediction

In order to further explore the usefulness of multi-scale CNN layer, we compare the performances between MCNN-LSTM and CNN-LSTM on 5 datasets. Here, MCNN-LSTM denotes the CNN-LSTM, which contains multi-scale CNN layers. Note that both of the methods are based on 2NOH. Moreover, CNN-LSTM use only a fixed convolutional kernel size 24 to predict TFBS in DNA sequences, while MCNN-LSTM uses different convolutional layers with kernel size (3, 5 and 9) to capture multiple sequence features. Figure 3 demonstrates the performance comparison between the proposed MCNN-LSTM and CNN-LSTM. Figure 4 depicts the ROC curves along with the AUCs, listed in Fig. 3.

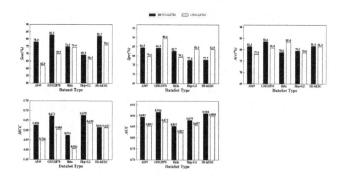

Fig. 3. Performance comparisons between MCNN-LSTM and CNN-LSTM.

As shown in these figures, it is straightforward to see that MCNN-LSTM outperform the CNN-LSTM. Compared with CNN-LSTM, MCNN-LSTM achieves the better values of Sen, Acc, MCC and AUC. For example, the four evaluation metrics of MCNN-LSTM achieve 78.0%, 81.0%, 63.2% and 89.0% on average,

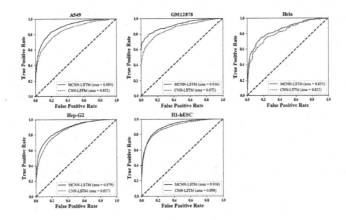

Fig. 4. The ROC curves of MCNN-LSTM and CNN-LSTM.

which are 7.7%, 0.6%, 5.1% and 3.0% higher than those of CNN-LSTM. In addition, it cannot escape our notice that the Spe value of MCNN-LSTM are lower than that for CNN-LSTM on GM12878, Hep-G2 and H1-hESC datasets. Maybe since that CNN-LSTM tends to predict positive samples as negatives.

According to the figures above, it is obvious that the multi-scale convolutional layer is effective to predict TFBS in DNA sequences. This phenomenon can be explained as follows. The binding lengths of the transcription factor are different, hence the performance of the model with a fixed-size convolutional kernel is lower than that of the model with a multiple size convolutional kernel. Specifically, when the model scans DNA sequences with multi-scale convolution, the features of TFBS tend to be more comprehensively captured.

3.4   Comparisons with Existing Predictors

In this section, we implement a new TFBS predictor, called DeepTF, by combining the new encoding method MNOH with MCNN-LSTM. To demonstrate the effectiveness of DeepTF, we compare it with the existing sequence-based predictors, including DeepBind [9] and CNN-Zeng [10]. Table 3 summarizes the prediction performance of DeepTF, DeepBind and CNN-Zeng on five datasets including A549, GM12878, Hela, Hep-G2 and H1-hESC dataset, selected in Sect. 3.1. Note that we separately train the models on the train dataset of each dataset, and evaluate the performances on the corresponding test datasets.

As shown in Table 3, it is obviously that DeepTF achieves better performance than DeepBind and CNN-Zeng with respect to all of the three evaluation metrics, i.e., Acc, MCC and AUC. Compared to DeepBind, DeepTF gets 1.0%, 2.0%, and 1.1% increases on A549 dataset in terms of Acc, MCC and AUC, respectively. Moreover, the Sen, Acc, MCC and AUC of DeepTF are, respectively, 2.1%, 0.1%, 0.4% and 0.3% higher than the values measured for CNN-Zeng on GM12878 datasets. As another example, the average AUCs are also 1.1% and 0.9%, respectively, higher than the corresponding methods on five datasets.

Table 3. Results comparisons of DeepTF, DeepBind and CNN-Zeng.

Dataset	Model	Sen (%)	Spe (%)	Acc (%)	MCC	AUC
A549	DeepTF	**78.0**	84.5	**81.4**	**0.626**	**0.893**
	DeepBind	76.2	84.2	80.4	0.606	0.882
	CNN-Zeng	74.9	**86.4**	81.0	0.615	0.874
GM12878	DeepTF	**82.9**	84.3	**83.6**	**0.672**	**0.916**
	DeepBind	73.4	**90.0**	82.5	0.642	0.904
	CNN-Zeng	80.8	86.0	83.5	0.668	0.913
Hela	DeepTF	**74.8**	82.7	**78.8**	**0.574**	**0.853**
	DeepBind	62.7	**90.1**	78.4	0.547	0.844
	CNN-Zeng	63.0	89.1	77.8	0.538	0.839
Hep-G2	DeepTF	69.3	**77.6**	**79.5**	**0.675**	**0.879**
	DeepBind	**71.6**	73.5	76.2	0.658	0.866
	CNN-Zeng	66.4	69.7	77.8	0.671	0.877
H1-hESC	DeepTF	**82.2**	77.7	**81.6**	**0.614**	**0.910**
	DeepBind	79.3	69.4	81.6	0.591	0.897
	CNN-Zeng	75.5	**78.6**	79.5	0.602	0.901

To further investigate the effectiveness of our predictor, we train the three predictors on all 690 ChIP-seq datasets. Figure 5 is the box plots of all AUCs of DeepTF and two other predictors. The red dotted line represents the median value of AUCs for all 690 datasets. Specifically, the median AUC of DeepTF is 0.880, which are 1.5% and 0.8% higher than DeepBind and CNN-Zeng, respectively, which indicates that DeepTF is superior to other predictors in terms of median AUC. Moreover, our predictor achieves the maximum value of the AUCs similar to the other two predictors, but the number of the AUCs close to 0.6 is less than both DeepBind and CNN-Zeng, which means that the DeepTF we proposed has good generalization ability and adaptability to different samples.

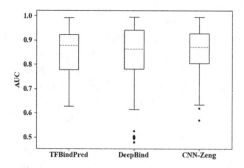

Fig. 5. The distribution of median AUCs on 690 ChIP-seq datasets. (Color figure online)

4 Conclusions

In this study, we propose a new sequence-based TFBS predictor, called DeepTF, by combining a new encoding method, called MNOH, with a new deep-learning model, called MCNN-LSTM. By comparison with several existing sequence-based TFBS predictors, on the 690 ChIP-seq datasets, the efficacy of the proposed predictor has been demonstrated. The superior performance of DeepTF is mainly attributed to MNOH and MCNN-LSTM. Specifically, MNOH takes account of the correlation between nucleotides in adjacent positions; MCNN-LSTM uses multi-scale convolution layers for feature processing, and then uses LSTM as a recurrent model that recognizes TFBS in DNA sequences.

Although DeepTF achieves some progress, there is still room to further improve its performance due to the following two points. First, we only use three multi-scale convolution layers in MCNN-LSTM model due to GPU memory limitations. In our future work, we try to use more multi-scale convolution layers for enhancing the performance of MCNN-LSTM. Second, DeepTF is specifically designed to predict TFBS from DNA sequences. In the future, we will further investigate the applicability of DeepTF to other types of binding site prediction problems, such as RNA-binding sites [23], ATP-binding sites [24].

Acknowledgments. This work was supported by the National Natural Science Foundation of China (No. 61772273, 61373062) and the Fundamental Research Funds for the Central Universities (No. 30918011104).

References

1. Lee, D., et al.: A method to predict the impact of regulatory variants from DNA sequence. Nat. Genet. **47**(8), 955 (2015)
2. Kharchenko, P.V., Tolstorukov, M.Y., Park, P.J.: Design and analysis of ChIP-seq experiments for DNA-binding proteins. Nat. Biotechnol. **26**(12), 1351 (2008)
3. Ji, H., Jiang, H., Ma, W., Johnson, D.S., Myers, R.M., Wong, W.H.: An integrated software system for analyzing ChIP-chip and ChIP-seq data. Nat. Biotechnol. **26**(11), 1293 (2008)
4. Siggers, T., Gordân, R.: Protein-DNA binding: complexities and multi-protein codes. Nucleic Acids Res. **42**(4), 2099–2111 (2013)
5. Fletez-Brant, C., Lee, D., McCallion, A.S., Beer, M.A.: kmer-SVM: a web server for identifying predictive regulatory sequence features in genomic data sets. Nucleic Acids Res. **41**(W1), W544–W556 (2013)
6. Wong, K.C., Chan, T.M., Peng, C., Li, Y., Zhang, Z.: DNA motif elucidation using belief propagation. Nucleic Acids Res. **41**(16), e153–e153 (2013)
7. Ghandi, M., Lee, D., Mohammad-Noori, M., Beer, M.A.: Enhanced regulatory sequence prediction using gapped k-mer features. PLoS Comput. Biol. **10**(7), e1003711 (2014)
8. Nutiu, R., et al.: Direct measurement of DNA affinity landscapes on a high-throughput sequencing instrument. Nat. Biotechnol. **29**(7), 659 (2011)
9. Alipanahi, B., Delong, A., Weirauch, M.T., Frey, B.J.: Predicting the sequence specificities of DNA-and RNA-binding proteins by deep learning. Nat. Biotechnol. **33**(8), 831 (2015)

10. Zeng, H., Edwards, M.D., Liu, G., Gifford, D.K.: Convolutional neural network architectures for predicting DNA–protein binding. Bioinformatics **32**(12), i121–i127 (2016)
11. LeCun, Y., Bengio, Y., Hinton, G.: Deep learning. Nature **521**(7553), 436 (2015)
12. Hassanzadeh, H.R., Wang, M.D.: DeeperBind: enhancing prediction of sequence specificities of DNA binding proteins. In: 2016 IEEE International Conference on Bioinformatics and Biomedicine (BIBM), pp. 178–183. IEEE (2016)
13. Zhou, J., Troyanskaya, O.G.: Predicting effects of noncoding variants with deep learning–based sequence model. Nat. Methods **12**(10), 931 (2015)
14. Siebert, M., Söding, J.: Bayesian Markov models consistently outperform PWMs at predicting motifs in nucleotide sequences. Nucleic Acids Res. **44**(13), 6055–6069 (2016)
15. Salekin, S., Zhang, J.M., Huang, Y.: Base-pair resolution detection of transcription factor binding site by deep deconvolutional network. Bioinformatics **34**(20), 3446–3453 (2018)
16. Hochreiter, S., Schmidhuber, J.: Long short-term memory. Neural Comput. **9**(8), 1735–1780 (1997)
17. Glorot, X., Bordes, A., Bengio, Y.: Deep sparse rectifier neural networks. In: Proceedings of the Fourteenth International Conference on Artificial Intelligence & Statistics, AISTATS, vol. 130, p. 297 (2011)
18. Graves, A., Jaitly, N., Mohamed, A.R.: Hybrid speech recognition with deep bidirectional LSTM. In: 2013 IEEE Workshop on Automatic Speech Recognition and Understanding, pp. 273–278. IEEE (2013)
19. Sutskever, I., Vinyals, O., Le, Q.V.: Sequence to sequence learning with neural networks. In: Advances in Neural Information Processing Systems, pp. 3104–3112 (2014)
20. Srivastava, N., Hinton, G., Krizhevsky, A., Sutskever, I., Salakhutdinov, R.: Dropout: a simple way to prevent neural networks from overfitting. J. Mach. Learn. Res. **15**(1), 1929–1958 (2014)
21. Hu, J., Zhou, X., Zhu, Y.H., Yu, D.J., Zhang, G.: TargetDBP: accurate DNA-binding protein prediction via sequence-based multi-view feature learning. IEEE/ACM Trans. Comput. Biol. Bioinform. (2019)
22. Zhu, Y.H., Hu, J., Song, X.N., Yu, D.J.: DNAPred: accurate identification of DNA-binding sites from protein sequence by ensembled hyperplane-distance-based support vector machines. J. Chem. Inf. Model. (2019)
23. Ren, H., Shen, Y.: RNA-binding residues prediction using structural features. BMC Bioinform. **16**(1), 249 (2015)
24. Chen, K., Mizianty, M.J., Kurgan, L.: ATPsite: sequence-based prediction of ATP-binding residues. In: Proteome Science, vol. 9, p. S4. BioMed Central (2011)

Epileptic Seizure Prediction Based on Convolutional Recurrent Neural Network with Multi-Timescale

Lijuan Duan[1,2], Jinze Hou[1,2], Yuanhua Qiao[3(✉)], and Jun Miao[4,5]

[1] Faculty of Information Technology, Beijing University of Technology, Beijing, China
[2] Beijing Key Laboratory of Trusted Computing, Beijing, China
[3] College of Applied Science, Beijing University of Technology, Beijing, China
qiaoyuanhua@bjut.edu.cn
[4] School of Computer Science, Beijing Information Science and Technology University, Beijing, China
[5] Beijing Key Laboratory of Internet Culture and Digital Dissemination Research, Beijing, China

Abstract. Epilepsy is a common disease that is caused by abnormal discharge of neurons in the brain. The most existing methods for seizure prediction rely on multi kinds of features. To discriminate pre-ictal from inter-ictal patterns of EEG signals, a convolutional recurrent neural network with multi-timescale (MT-CRNN) is proposed for seizure prediction. The network model is built to complement the patient-specific seizure prediction approaches. We firstly calculate the correlation coefficients in eight frequency bands from segmented EEG to highlight the key bands among different people. Then CNN is used to extract features and reduce the data dimension, and the output of CNN acts as input of RNN to learn the implicit relationship of the time series. Furthermore, considering that EEG in different time scales reflect neuron activity in distinct scope, we combine three timescale segments of 1 s, 2 s and 3 s. Experiments are done to validate the performance of the proposed model on the dataset of CHB-MIT, and a promising result of 94.8% accuracy, 91.7% sensitivity, and 97.7% specificity are achieved.

Keywords: EEG · Seizure prediction · Deep learning · Multi-timescale

1 Introduction

Epilepsy is caused by abnormal discharge of brain neurons, and EEG reflects the activity of neuron population [1, 2]. Therefore EEG analysis is promising for building seizure detection and prediction model. Seizure prediction is helpful in preventing patients from doing dangerous things before the seizure, like swimming, driving, therefore, building seizure prediction model is of great significance for epileptics [3].

Studies have shown that the EEG patterns change with time from a normal state to seizure. According to that, many methods have been proposed for seizure prediction [4,

© Springer Nature Switzerland AG 2019
Z. Cui et al. (Eds.): IScIDE 2019, LNCS 11936, pp. 139–150, 2019.
https://doi.org/10.1007/978-3-030-36204-1_11

5]. However, due to the complexity of epilepsy pathology and the differences among people, there are few robust methods for seizure prediction in the past few years.

Most of the existing seizure prediction model using EEG signals to distinguish preictal from interictal patterns is actually a binary classification. EEG signals include intracranial EEG (iEEG) and scalp EEG (sEEG). Compared with iEEG, sEEG is easy to collect but with lower signal to noise ratio, which makes the research based on sEEG more difficulty [5]. A lot of investigations have been made on seizure detection and prediction and Most of them extract features from EEG signals and then classify them with a classifier as KNN, SVM, Decision Tree, ADaboost, etc. [6–11]. Cho [7] uses filtering algorithms such as band pass filtering, empirical mode decomposition, noise-assisted multivariate empirical mode decomposition (NA-MEMD) etc. to decompose spectral components from the EEG data. Phase locking values are calculated and used to classify EEG signals with a SVM. Through NA-MEMD with 0-dB they achieve the best result of 82.44% sensitivity and 82.76% specificity. The spectral power of raw, bipolar and time-differential iEEG in nine bands are computed respectively in [10] and it reaches 97.5% sensitivity by Cost-sensitive SVM. Discrete wavelet transform [8] is applied for the analysis of EEG signals and the number of zero-crossings of detail coefficients in level 1 is calculated, which achieves 98% sensitivity, 88% specificity and 91% accuracy by self-organizing maps with polling mechanism. Cui propose a bag-of-wave method for EEG data investigation, and a dictionary of EEG segments is constructed by K-Means clustering algorithm, a classifier is used to learn the synchronization patterns of a series of continuous segments by mapping to the codebook elements, which achieves 88.24% sensitivity and 25% false prediction [9].

With rapidly development of deep learning in recent years, it has achieved excellent results in tasks such as image classification, object detection, and natural language processing [12–14]. Many methods based on deep learning for seizure detection and prediction have been proposed [15–24]. Truong uses the short-time Fourier transform to convert the data from time domain to frequency domain. Then a CNN is used to extract features and classify the data. It reaches 81.4%, 81.2%, 82.3% sensitivity and 0.06/h, 0.16/h, 0.22/h false prediction rate on three datasets respectively [16]. Mirowski calculates bivariate features of EEG synchronization, like cross-correlation, nonlinear interdependence, wavelet coherence etc., and different combinations of classifiers (including SVM, logistic regression and CNN) with different features are evaluated. CNN performs best with 71% sensitivity and without false alarms by using wavelet coherence [17]. Another model is proposed to capture subtle chaotic dynamics in fractional Fourier transform domain for epileptic signals and it applies an energy measure to determine appropriate fractional order used in time domain, and the modified largest Lyapunov exponent is calculated as artificial neural network's input. The results show good performance in identifying the preictal state with 89.5% sensitivity and 89.75% specificity. Since EEG signals are long-term-dependent, it is important to consider the temporal implicit relationship. Ma uses LSTM to process EEG signals, where the average spectral power in six frequency bands, the standard deviation of the six powers, the correlation in the time and frequency domain are computed as input for LSTM. The approach is evaluated in two patient cases and get an average accuracy of 89.36% [18]. In addition, for epileptic signal detection, [19] a sparse auto-encode is used to translate EEG fragments into words, and the neighboring words of the current one are referred to learn the contextual information. Then a binary classifier is applied to classify epileptic and non-epileptic signals.

Most of the above methods are patient-specific. Due to the complexity of epileptic EEG signals among different patients, patient-independent seizure prediction is more challenging [15]. The existing methods are generally based on multi kinds of features which increase the time complexity in feature extraction step. Moreover, using only single timescale features may result in information lost. In this paper, a multi-timescale seizure prediction method based on convolutional recurrent network in patient-independent is proposed. The correlation coefficients among electrodes are the only features we used, and multi-timescale feature extraction of EEG can mine more information.

The main contributions are as follows:

1. A specific neural network architecture is designed for extracting temporal EEG features.
2. Multi-timescale features are fused for seizure prediction and ablation experiments are done to validate the effectiveness of the proposed model.

2 Proposed Model

The flow chart of our model is shown in Fig. 1. The method contains three parts: data segmentation, feature extraction, and MT-CRNN construction.

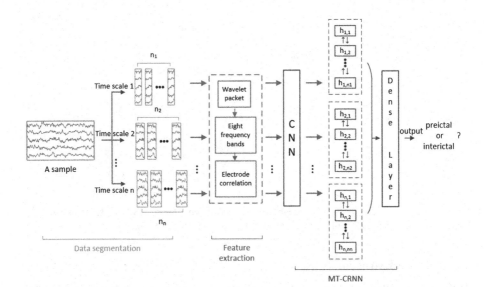

Fig. 1. A sample is split into multi-timescale segments, following by feature extraction with discrete wavelet packet and correlation coefficients among the electrodes. Then the convolutional recurrent network is used to fuse different timescale information.

2.1   Data Segmentation

Since the EEG signal is collected over a long period of time, it usually needs to be segmented before analysis. To divide the data into training set and testing set for the network inputs, preprocessing is done as described in Fig. 2. Firstly, the EEG data are split into non-overlapping segments with window length of 10 min and is denoted as S1. All S1 are labeled as 0 or 1 for preictal or interictal respectively. As the preictal data are scarce, we randomly select a subset of interictal data to match them. Then, the selected segments are divided into training set and testing set. Secondly, S1 are split into smaller segments with window length of 30 s, and denoted as S2. Each of S2 acts as a sample. In this period, 20% sliding is used for segmenting the preictal data for data augmentation. Otherwise it may lead to over-fitting without enough data. Finally, in the experiments, S2 are furtherly split into 'n' parts using window length of 1, 2, 3 s respectively without overlapping in both classes, and denoted as S3. The feature matrixes are calculated based on S3, and each of S3 is seen as the input of a time-step in recurrent neural network. In this step, different sliding windows are applied on S2 to obtain different scale segments. In addition, 1 h of data between 5 min before the current seizure and at least 4 h away from the last seizure is defined as preictal. And the data between two seizures at least 4 h away from each one is defined as interictal. There is no further pre-processing applied such as noise/artifact rejection techniques.

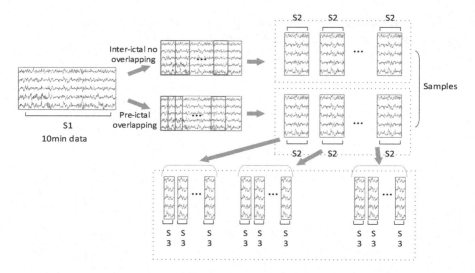

Fig. 2. The data segmentation flowchart. S1 is data of ten minutes. S2 are samples used for training the network. Different timescales data are got from S2 to S3 step.

2.2   Feature Extraction

EEG is non-stationary signal. It is important to extract features from raw data for making some analysis. In the task of classifying preictal and interictal data, the quality of features directly affect the classifier performance. Researchers usually extract features within the time domain, frequency domain, time-frequency domain, and nonlinear mechanics and graph theory is also used for EEG feature extraction [6, 11, 15, 20]. In time domain, the mean value, variance, energy, skewness, kurtosis etc. are frequently used. EEG signals can generally reflect difference in frequency domain better than time domain. So EEG is usually analyzed in frequency domain, and Fourier transform is usually used for extracting frequency information. However, there are still some limitations in frequency analysis, especially for non-stationary signals. Therefore, time-frequency analysis came into being, such as short time Fourier transform and wavelet transform. Moreover, a lot of studies have shown that EEG signals have nonlinear dynamic characteristics, thus, approximate entropy, sample entropy, and Lyapunov exponent, etc. are often employed for EEG analysis. By calculating local or global measures in a graph constructed from the correlation or the coherence matrix among electrodes to represent brain network features, graph theory is also used to analyze EEG data.

In this paper, EEG segments are processed by 3-level discrete wavelet packet to obtain wavelet coefficients in eight frequency bands. Wavelet packet transform are suitable for non-stationary signals. Compared with wavelet transform, wavelet packet transform can obtain more detail information. Considering that different patient epileptic-form signal may relate to different frequency bands, we reconstruct data in each band by correlative wavelet coefficients. Some studies have shown that the correlation values among electrodes will change with seizure onset approaching. Therefore the correlation coefficients are calculated as features. As the correlation matrix is symmetric, only upper triangle is selected to eliminate redundant data. Wavelet packet decompose is used to calculate $\{d_k^{j+1,2n}\}$ and $\{d_k^{j+1,2n+1}\}$ by $\{d_l^{j,n}\}$:

$$\begin{cases} d_k^{j+1,2n} = \sum_l h_{2l-k} d_l^{j,n} \\ d_k^{j+1,2n+1} = \sum_l g_{2l-k} d_l^{j,n} \end{cases} \tag{1}$$

$\{d_k^{j+1,2n}\}$, $\{d_k^{j+1,2n+1}\}$ are wavelet packet coefficients; h_{2l-k}, g_{2l-k} are low pass and high pass filters. And wavelet packet reconstruction is used to calculate $\{d_l^{j,n}\}$ by $\{d_k^{j+1,2n}\}$ and $\{d_k^{j+1,2n+1}\}$:

$$d_l^{j,n} = \sum_k h_{l-2k} d_k^{j+1,2n} + \sum_k g_{l-2k} d_k^{j+1,2n+1} \tag{2}$$

The correlation coefficient of two variables is defined by:

$$\rho_{X,Y} = \frac{cov(X,Y)}{\sigma_X \sigma_Y} = \frac{E[(X-\mu_X)(Y-\mu_Y)]}{\sigma_X \sigma_Y} \tag{3}$$

$cov(*,*)$ is the covariance of the variables, and σ_* denotes the standard deviation. We randomly slice 5 s data of three cases in preictal and interictal respectively and visualize the correlation coefficient among the electrodes of them as shown in Fig. 3.

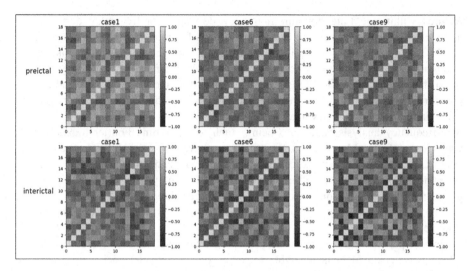

Fig. 3. The first row shows the heatmap of preictal data, and the second row is interictal data. There are three cases, case1, case6 and case9. Both the axes of each figure represent electrodes.

2.3   Multi-Timescale Convolutional Recurrent Neural Network

We propose a multi-timescale seizure prediction method based on convolutional recurrent neural network. The flowchart of the proposed method is shown in Fig. 1. After the segment processing mentioned in Sect. 2.1, we obtain a dataset defined as $\mathbf{D} = \{(X_1, Y_1), (X_2, Y_2), \ldots, (X_n, Y_n)\}, X_i \in R^{C*N}$. Where X_i is data, corresponding to S2 in Fig. 2. Y_i is label. C is the number of electrodes equals to 18, and N is the sampling points. Subsequently the features are extracted from S3 mentioned in Fig. 2 with method mentioned in Sect. 2.2. Furthermore, the mother wavelet used in Sect. 2.2 is Daubechies 4 (db4) which have been used in [20]. After that we get a matrix $M_{ij} \in R^{8*18*18}$, 8 is the number of bands and 18 is the number of electrodes. To eliminate redundant data, we only take half values of the M_{ij} and denoted as M'_{ij} acting as the input of the convolutional network:

$$u_{ij} = f\left(M'_{ij}, \theta^f\right) \tag{4}$$

$f(*)$ represents convolutional network, and θ^f are network parameters. The structure of the convolutional network refers to VGG-Net. But less layers and kernels are used to defend against overfitting. More details of the CNN structure see Table 1. In layer 4, the kernel whose size is 1 * 1 is for dimension reduction. As EEG signals are time-dependent, it is significant to consider its inherent relationship in time domain. Therefore, a bi-directional recurrent neural network (Bi-RNN), specifically the bi-direction Gated Recurrent Unit (Bi-GRU) which is a modified version of the Long Short-Term Memory Unit is used to learn the relationship in time series. Bi-RNN is generally used in natural language processing (NLP) field [29–31]. The function is given as:

$$v_i = g(\boldsymbol{u}_{i1}, \boldsymbol{u}_{i2}, \ldots, \boldsymbol{u}_{iT}, \boldsymbol{\theta}^g) \tag{5}$$

$g(*)$ represents the Bi-GRU, and $\boldsymbol{u}_{ij}$ is every time step input obtained from the CNN corresponding to S3. T denotes the number of S3 segments contained in S2. $\boldsymbol{\theta}^g$ are parameters of the recurrent network. Considering that EEG may contain different information in different timescales, so we investigate a multi-timescale approach as shown in Fig. 1. S2 are split into a sequence of S3 with three different length of sliding windows. At last, the three timescales information are fused together to make a decision:

$$d_i = h([v_i^1 \cdot v_i^2 \cdot v_i^3], \boldsymbol{\theta}^h) \tag{6}$$

$h(*)$ denotes the dense layer, and $\boldsymbol{\theta}^h$ are dense layer parameters. v_i^1, v_i^2, v_i^3 are three timescales data correlation to 1 s, 2 s, 3 s respectively. They are concatenated together as the dense layer's input.

Table 1. Structure of the CNN, conv#*#-#, the first one of the # represents the filter height, the second one represents the filter width, and the last one represents the number of the filters, maxpool is a max pooling layer.

Layer1	Layer2	Layer3	Layer4
conv1*3-16	conv1*3-32	conv1*3-64	conv1*1-16
conv1*3-16	conv1*3-32	conv1*3-64	
maxpool	maxpool	conv1*3-64	
		maxpool	

3   Dataset

CHB-MIT dataset [25, 26] is used for the investigation. The dataset contain scalp EEG (sEEG) data collected from 23 epilepsy patients (5 males, ages 3–22; and 17 females, ages 1.5–19; and one without subject information), which grouped into 24 cases (Case chb21 and case chb01 are from the same person.). Subjects are monitored for up to several days. The sample rate is 256 Hz. Most files contain 23 electrodes signals. The international 10–20 system of EEG electrode positions is used for the recordings. Most of the files contain data of one hour. The dataset has a total of 983 h data. The recorded seizure events are totally 198 times. As the experiments are done for patient-independent situations, we need to select the common channels in all of the subjects, there are 18 channels: FP1-F7, F7-T7, T7-P7, P7-O1, FP1-F3, F3-C3, C3-P3, P3-O1, FP2-F4, F4-C4, C4-P4, P4-O2, FP2-F8, F8-T8, T8-P8, P8-O2, FZ-CZ and CZ-PZ.

4   Experiments and Results

In order to make full use of the data and reduce the fluctuation of the experimental results, we use 10-fold cross-validation. The dataset is divided into ten subsets and in each experiment, one subset is used for testing and the rest is for training. Then the mean values of ten experiment results is calculated as the final result. The batch normalization layer is used after the convolutional layer to accelerate convergence of the network. The total training epoch is 30 and the momentum optimizer is used for training. The model is implemented with the tensorflow framework.

4.1   Evaluation Metrics

Three metrics are used to evaluate the proposed model, accuracy, sensitivity, specificity:

$$\begin{cases} Accuracy = \frac{TP+TN}{TP+FN+FP+TN} \\ Sensitivity = \frac{TP}{TP+FN} \\ Specificity = \frac{TN}{TN+FP} \end{cases} \tag{7}$$

True positive (TP) is defined by the number of segments that are correctly classified to be the preictal period. False positive (FP) is defined by the number of segments that are incorrectly classified to be the preictal period. True negatives (TN) is defined by the number of segments that are correctly classified to be the interictal period. False negatives (FN) is defined by the number of segments that are incorrectly classified to be the interictal period. Accuracy rate reflect the generalization ability of a model. Sensitivity reflect the correct prediction rate. And specificity can implicitly reflect the false alarming rate.

4.2   Results

To validate the effectiveness of our proposed method, we make several groups of comparison experiments as follows.

Single Band versus Multi-bands. Table 2 gives the results from the correlation coefficient of multi-frequency-bands as the input of the network compared with the correlation values in raw data. Both experiments are based on the convolutional recurrent network. The multi-bands method achieves a superior result with 93.4% accuracy, 89.5% sensitivity and 97.2% specificity compared with the single band method. The results may attribute to different pathogenesis among people. By dividing the raw data into different frequency bands may highlight the key bands correlated to epilepsy, therefore, the rest experiments are based on multi-bands.

Table 2. 30_2s_Single represents experimenting on raw data, and 30_2s_Multi represents on multi-frequency-bands. 2s indicates the experiment is based on segments of 2 s.

	Accuracy	Sensitivity	Specificity
30_2s_Single	0.859	0.762	0.954
30_2s_Multi	**0.934**	**0.895**	**0.972**

Recurrent Network Only versus Convolutional Recurrent Network. The first row of Table 3 is obtained from the network which uses Bi-GRU architecture only and the other is obtained from the integrated framework we proposed. The integrated framework gets a better result on each of the evaluation metric. It confirms that processing the data with convolutional network to reduce data dimension and extract features before the recurrent network is effective for the task. So the following experiments are based on the convolutional recurrent network to investigate the network performance with different segment lengths.

Table 3. BiGRU-only represents the experiment is done based on Bi-GRU, CNN-BiGRU represents the convolutional recurrent neural network.

	Accuracy	Sensitivity	Specificity
BiGRU-only	0.893	0.871	0.913
CNN-BiGRU	**0.934**	**0.895**	**0.972**

Single Timescale versus Multi-timescale. We furtherly make experiments based on three different lengths of segments, 1 s, 2 s and 3 s. There is no obvious variance among them as shown in Table 4. And the last row in Table 4 shows the results of combination with the three scales segments. It is seen that the multi-timescale method is superior to any of the other three single scale methods in all of the three evaluation metrics. The experiments results are consistent with the point of view that EEG in different scales reflect neuron activity in distinct scope. So we can get more information from multi-timescale and make a good decision.

Table 4. 30_1s denotes the segment length is 1 s. 30_2s is 2 s. 30_3s is 3 s. And multi-timescale represents combined with three lengths of the segments.

	Accuracy	Sensitivity	Specificity
30_1s	0.933	0.894	0.971
30_2s	0.934	0.895	0.972
30_3s	0.932	0.893	0.971
Multi-timescale	**0.948**	**0.917**	**0.977**

Some related works in recent years are listed in Table 5. The comparison focuses on studies that had been also evaluated with the CHB-MIT EEG database, since it is the frequently used public database consisting of long term recordings. And the features used in each method are presented in the table. The methods proposed in [9, 16, 28] are for patient-specific which is simpler than patient-independent. In fact, this is the first time that the convolutional and recurrent neural network are combined for seizure prediction with the multi-timescale information. Compared with [27], which uses longer preictal-duration while more complex features, we make better results in both of the evaluation metrics. The methods in [9, 28] also make good results, but require some complicated pre-processing steps before classification. [16] only uses CNN for seizure prediction, presenting lower levels of performance compared with our work.

Table 5. Comparison with previous studies.

Study	Dataset	# of cases	Total cases	Features	SEN (%)	SPEC (%)	Preictal duration (min)
Cui et al. [9]	CHB-MIT	9	24	Bag of wave	88.24	–	60
Zhang and Parhi [28]	CHB-MIT	17	24	Absolute/relative spectral power	98.68	–	60
Truong et al. [16]	CHB-MIT	13	24	STFT spectral images	81.2	–	30
Tsiouris et al. [27]	CHB_MIT	24	24	A large variety of local and global features	85.75	85.75	120
Our work	CHB_MIT	24	24	Pearson correlation coefficient of multi-band	91.7	97.7	60

5 Conclusion

In this paper, a multi-timescale seizure prediction method based on convolutional recurrent network is proposed by discriminating the preictal data from interictal data. Firstly, the extracted features are processed by a CNN to obtain high-level feature vectors and then, they are fed into an RNN to learn the synchronization patterns of the input. Considering the difference in age, gender and epilepsy type among different patients, we divide the raw EEG data into eight frequency bands to highlight the key frequency bands. Since EEG signals are always segmented for a study, it is easy to lose information at a single time scale. What's more, epilepsy signals are mutational, so we capture small-scale mutations and large-scale trends with multi-timescale signals. In the future, we can study the effect of more time scales for the task, adding a vote mechanism to make an alarm when reaching the threshold pre-set, through which to improve the stability of the prediction model.

Acknowledgements. This research is partially sponsored by National Natural Science Foundation of China (No. 61672070 ,61572004), the Project of Beijing Municipal Education Commission (No. KZ201910005008,KM201911232003), the Research Fund from Beijing Innovation Center for Future Chips (No. KYJJ2018004).

References

1. Kannathal, N., Choo, M.L., Acharya, U.R., et al.: Entropies for detection of epilepsy in EEG. Comput. Methods Programs Biomed. **80**(3), 187–194 (2005)
2. Altunay, S., Telatar, Z., Erogul, O.: Epileptic EEG detection using the linear prediction error energy. Expert Syst. Appl. **37**(8), 5661–5665 (2010)
3. Lehnertz, K., Mormann, F., Kreuz, T., et al.: Seizure prediction by nonlinear EEG analysis. IEEE Eng. Med. Biol. Mag. **22**(1), 57–63 (2003)

4. Mormann, F., Andrzejak, R.G., Elger, C.E., et al.: Seizure prediction: the long and winding road. Brain **130**(2), 314–333 (2006)
5. Alotaiby, T.N., Alshebeili, S.A., Alshawi, T., et al.: EEG seizure detection and prediction algorithms: a survey. EURASIP J. Adv. Sig. Process. **2014**(1), 183 (2014)
6. Ahammad, N., Fathima, T., Joseph, P.: Detection of epileptic seizure event and onset using EEG. BioMed Res. Int. **2014** (2014)
7. Cho, D., Min, B., Kim, J., et al.: EEG-based prediction of epileptic seizures using phase synchronization elicited from noise-assisted multivariate empirical mode decomposition. IEEE Trans. Neural Syst. Rehabil. Eng. **25**(8), 1309–1318 (2017)
8. Kitano, L.A.S., Sousa, M.A.A., Santos, S.D., Pires, R., Thome-Souza, S., Campo, A.B.: Epileptic seizure prediction from EEG signals using unsupervised learning and a polling-based decision process. In: Kůrková, V., Manolopoulos, Y., Hammer, B., Iliadis, L., Maglogiannis, I. (eds.) ICANN 2018. LNCS, vol. 11140, pp. 117–126. Springer, Cham (2018). https://doi.org/10.1007/978-3-030-01421-6_12
9. Cui, S., Duan, L., Qiao, Y., et al.: Learning EEG synchronization patterns for epileptic seizure prediction using bag-of-wave features. J. Ambient Intell. Hum. Comput., 1–16 (2018)
10. Park, Y., Luo, L., Parhi, K.K., et al.: Seizure prediction with spectral power of EEG using cost-sensitive support vector machines. Epilepsia **52**(10), 1761–1770 (2011)
11. Xiang, J., Li, C., Li, H., et al.: The detection of epileptic seizure signals based on fuzzy entropy. J. Neurosci. Methods **243**, 18–25 (2015)
12. He, K., Zhang, X., Ren, S., et al.: Deep residual learning for image recognition. In: Proceedings of the IEEE Conference on Computer Vision and Pattern Recognition, pp. 770–778 (2016)
13. Vaswani, A., Shazeer, N., Parmar, N., et al.: Attention is all you need. In: Advances in Neural Information Processing Systems, pp. 5998–6008 (2017)
14. Girshick, R.: Fast R-CNN. In: Proceedings of the IEEE International Conference on Computer Vision, pp. 1440–1448 (2015)
15. Thodoroff, P., Pineau, J., Lim, A.: Learning robust features using deep learning for automatic seizure detection. In: Machine Learning for Healthcare Conference, pp. 178–190 (2016)
16. Truong, N.D., Nguyen, A.D., Kuhlmann, L., et al.: A generalised seizure prediction with convolutional neural networks for intracranial and scalp electroencephalogram data analysis. arXiv preprint arXiv:1707.01976 (2017)
17. Mirowski, P., Madhavan, D., LeCun, Y., et al.: Classification of patterns of EEG synchronization for seizure prediction. Clin. Neurophysiol. **120**(11), 1927–1940 (2009)
18. Ma, X., Qiu, S., Zhang, Y., Lian, X., He, H.: Predicting epileptic seizures from intracranial EEG using LSTM-based multi-task learning. In: Lai, J.-H., et al. (eds.) PRCV 2018. LNCS, vol. 11257, pp. 157–167. Springer, Cham (2018). https://doi.org/10.1007/978-3-030-03335-4_14
19. Xun, G., Jia, X., Zhang, A.: Detecting epileptic seizures with electroencephalogram via a context-learning model. BMC Med. Inform. Decis. Mak. **16**(2), 70 (2016)
20. Tsiouris, K.M., Pezoulas, V.C., Zervakis, M., et al.: A Long Short-Term Memory deep learning network for the prediction of epileptic seizures using EEG signals. Comput. Biol. Med. **99**, 24–37 (2018)
21. Acharya, U.R., Oh, S.L., Hagiwara, Y., et al.: Deep convolutional neural network for the automated detection and diagnosis of seizure using EEG signals. Comput. Biol. Med. **100**, 270–278 (2018)
22. Hussein, R., Palangi, H., Ward, R., et al.: Epileptic seizure detection: a deep learning approach. arXiv preprint arXiv:1803.09848 (2018)

23. Bashivan, P., Rish, I., Yeasin, M., et al.: Learning representations from EEG with deep recurrent-convolutional neural networks. arXiv preprint arXiv:1511.06448 (2015)
24. Fei, K., Wang, W., Yang, Q., et al.: Chaos feature study in fractional Fourier domain for preictal prediction of epileptic seizure. Neurocomputing **249**, 290–298 (2017)
25. CHB-mit scalp EEG database, Physionet.org. https://www.physionet.org/pn6/chbmit
26. Goldberger, A.L., et al.: PhysioBank, PhysioToolkit, and PhysioNet: components of a new research resource for complex physiologic signals. Circulation **101**(23), e215–e220 (2000)
27. Tsiouris, K.M., Pezoulas, V.C., Koutsouris, D.D., et al.: Discrimination of preictal and interictal brain states from long-term EEG data. In: 2017 IEEE 30th International Symposium on Computer-Based Medical Systems (CBMS), pp. 318–323. IEEE (2017)
28. Zhang, Z., Parhi, K.K.: Low-complexity seizure prediction from iEEG/sEEG using spectral power and ratios of spectral power. IEEE Trans. Biomed. Circuits Syst. **10**(3), 693–706 (2016)
29. Lin, Z., Feng, M., Santos, C.N., et al.: A structured self-attentive sentence embedding. arXiv preprint arXiv:1703.03130 (2017)
30. Devlin, J., Chang, M.W., Lee, K., et al.: BERT: pre-training of deep bidirectional transformers for language understanding. arXiv preprint arXiv:1810.04805 (2018)
31. Xing, C., Wu, Y., Wu, W., et al.: Hierarchical recurrent attention network for response generation. In: Thirty-Second AAAI Conference on Artificial Intelligence (2018)

L2R-QA: An Open-Domain Question Answering Framework

Tieke He[1], Yu Li[1], Zhipeng Zou[1], and Qing Wu[2(✉)]

[1] State Key Laboratory for Novel Software Technology, Nanjing University,
Nanjing 210093, People's Republic of China
hetieke@gmail.com
[2] School of Economics, Nanjing 210093, People's Republic of China
wuqing@nju.edu.cn

Abstract. Open-domain question answering has always being a challenging task. It involves information retrieval, natural language processing, machine learning, and so on. In this work, we try to explore some comparable methods in improving the precision of open-domain question answering. In detail, we bring in the topic model in the phase of document retrieval, in the hope of exploiting more hidden semantic information of a document. Also, we incorporate the learning to rank model into the LSTM to train more available features for the ranking of candidate paragraphs. Specifically, we combine the results from both LSTM and learning to rank model, which lead to a more precise understanding of questions, as well as the paragraphs. We conduct an extensive set of experiments to evaluate the efficacy of our proposed framework, which proves to be superior.

Keywords: Question answering · Learning to rank · Topic model

1 Introduction

Since the 1960's, a large number of Question answering systems has been developed [1]. It addresses different types of questions on different domains from different data sources. We focus on the answer of factoid open-domain questions adopting Wikipedia as the knowledge source. Open-domain means the domain of questions is not limited, and its development was limited because of the absence of a viable knowledgebase. The emergence of Wikipedia, a simultaneously evolving collection of information on diverse topics, provides an opportunity to bridge the gap between the open-domain questions and knowledge base. According to Jurczyk's analysis [9], Wikipedia performs well on different types of questions. It is a suitable knowledge base for what, how and who questions. Especially for what- questions, where the coverage of answers can reach 60%. And for the other two types of questions, Wikipedia also performs better than other existing knowledge bases. Besides, Wikipedia has diverse types of data which maintains high coverage on numerical, personal, and objective facts.

© Springer Nature Switzerland AG 2019
Z. Cui et al. (Eds.): IScIDE 2019, LNCS 11936, pp. 151–162, 2019.
https://doi.org/10.1007/978-3-030-36204-1_12

Using the large-scale collection of documents brings the challenge of open-domain question along with machine reading at scale (MRS). Traditional reading comprehension supposes that the input text contains the answer, while it is not realistic for open domain QA. We need to narrow down the range of answers, i.e., selecting the most relevant documents first and then using reading comprehension to process. Besides, limiting to the single knowledge source force the model to be very precise because the answer may only appear once.

To alleviate all these problems, we propose a new open-domain question answering framework based on Wikipedia text. We mainly follow the structure as Zhao et al. [4] proposed to solve the problem of machine reading at scale. However, in their framework, they use the TF-IDF for the representation of documents, which may lose much semantic information of the documents, leading to inaccuracy when computing the similarities between documents. In seeing this, we try to enhance that by adopting some advanced models that better represent these documents, i.e., Latent Semantic Indexing (LSI) for example, and then we calculate the similarity between them to get the top 5 relevant documents according to given questions. Also, we consider more features during the reading comprehension phase, especially, our framework adds up the learning to rank (LTR) as features to better encode both the paragraph and question, in order to find the final answer more precisely. An extensive set of experiments are designed and conducted to testify the efficacy of our proposed framework, and the results demonstrate the superiority of it.

The main contributions of our work are as follows:

- We apply LDA on document representation in the task of large-scale document retrieval and compare with the accuracy and efficiency while using the tfidf model.
- We use the attention mechanism in selecting answers from candidates by using learning to rank (LTR) model to pay more attention to answers included in more relevant documents.

The rest of the paper is organized as follows: In Sect. 2, we introduce the related work. Section 3 presents our framework for open-domain question answering, the methods and techniques we used are depicted in details in this section. Section 4 describes four datasets used in our work. The details of our experiments and the evaluation part are in Sect. 5. Comparison with others' work is also presented in this section. We conclude our work In Sect. 6, along with future work.

2   Related Work

2.1   Question Answering Systems (QASs)

Question answering systems (QASs) was born to solve the "Last-Mile problem" of the search engines. Instead of presenting an ordered list of related documents or corresponding web links, the QASs generate the answers of the question asked in natural language directly, which save the time of browsing the documents to find the answer.

Based on the type of data source, QASs can be classified as knowledge-base question answering (KB-QA) and text-based question answering. For KB-QA, whether using traditional methods, such as semantic parsing [2], information extraction [15], vector modeling [3], or using deep learning methods to improve it, such as adapting Convolutional Neural Network (CNN) on the semantic analysis method [16] and vector modeling method [7], using Long Short-Term Memory (LSTM) and Convolutional Neural Network (CNN) to classify entity relationship [14] can not solve the fixed schemes and incompleteness. The incompleteness makes the system unable to use these knowledge bases. Hence the corresponding answer cannot be given. Besides, the reason for incompleteness including real lacking raw data and not mined data cannot be determined, which adds difficulties for optimization and improvement. On the other hand, triples used to represent questions cannot faithfully represent the semantic structure of the question, especially for more complicated question or its concept is rather vague. However, answering from the raw text as text-based QA avoid these problems, which motivate the proposal of our method.

2.2   QA with Wikipedia

Wikipedia is a collaborative, continuously growing, semi-structured knowledge source which is usually applied to QA as the knowledge base. Jennifer leverages the characteristics of Wikipedia to implement type independent candidate generation method for QA, which shows that Wikipedia metadata can be used to extract candidate answers from results returned by searching without inputting ontology. Pum-Mo Ryu makes full use of Wikipedia by using different types of data to answer different kinds of question [5]. They suggest that infoboxes, category structure, and definitions have their unique strength in answering factoid questions, list questions and descriptive questions perspectively. And the combination of answers from different modules using different types of semi-structured knowledge source has been improved compared with the traditional use of Wikipedia [6,13]. Wikipedia is also used as an auxiliary data source. Sergio presents a novel architecture to solve the Cross-Lingual Question Answering task with the use of multilingual relations encoded in Wikipedia when processing Name-Entities in queries [8].

In our work, we treat Wikipedia as a collection of documents and use LDA and LSI to filter the relevant documents of question from more than 5 million items. And the result is the input of the machine comprehension of text to find the final answer.

3   Datasets

3.1   Wikipedia

We use the processed Wikipedia provided by Danqi et al. (2017) as our knowledge base to find answers in the full-scale experiment. The document dump includes the latest documents of diverse topics [13]. Based on the 2016-12-21 dump of English Wikipedia, they used the WikiExtractor script to extract and clean text for machine comprehension. The output document files of the extractor are consist of a serious of articles which are represented by XML element. And each element has two attributes: (1) id, which is the identification of the article (2) URL, which is the link to the original Wikipedia page, including content that contains pure text, one sentence per line. Note that the first sentence is the title of the article.

3.2   SQuAD

The Stanford Question Answering Dataset (SQuAD) dataset is a large-scale reading comprehension dataset generated by crowdsourcing. Rajpurkar et al. [12] used Project Nayuki's Wikipedia internal PageRank to get top 10000 articles from Wikipedia and sampled 536 articles randomly as the resource, and then filtered to get 23,215 paragraphs which cover a wide range of topics. For each paragraph, up to 5 question and answering tasked to be given based on the content. Furthermore, the crowd workers are tasked to select the shortest span based on the question along with the paragraph. Hence, each example in SQuAD consist of articles and each article consists of title and paragraphs along with human-generated questions and corresponding answers in the type of span usually. Besides, they used two metrics in evaluation:

1. Exact match (EM), which is used to measure the percentage of matching where the prediction matches one of the ground truth exactly.
2. (Macro-average) F1 score, which is used to measure the average overlap between them.

In our work, we use SQuAD which contains 87k examples in training and evaluating periods of selecting an answer from candidate paragraphs returned by document retrieval to complete the reading comprehension task. And the development SQuAD containing 10k examples is used only in evaluating period of open-domain question answering, which means giving the whole Wikipedia as the resource to find the answer.

4   Framework

In our framework, we use information retrieval, topic model, bidirectional long short-term memory network (BiLSTM) and feature engineering et al. to implement answering questions based on large-scale documents. Figure 1 gives an illustration of our system.

4.1   Information Retrieval and Topic Model

In our work, we use LSI and LDA to replace TF-IDF used in Chen's DrQA [4]. Which improves efficiency and accuracy.

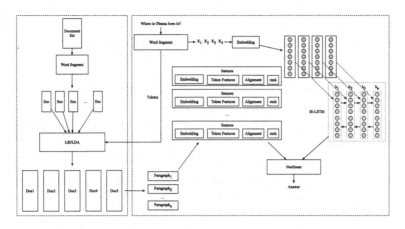

Fig. 1. The proposed framework L2R-QA

In the module of document retrieval, we use LDA to model the documents extracted from the Wikipedia and the given question. The integrity process flow is as follows.

- Use CoreNLP toolkit [10] to tokenize, name entity tags, generate lemma and part-of-speech.
- Construct a bag-of-words model and index of the document.
- Compute TF-IDF of the document.
- Model the documents with LSI.

When retrieving relevant documents of given question, we first model the question with LSI model generated, and then return the top 5 most relevant documents based on the similarity between question and documents.

4.2 Paragraph Features Extraction

In our work, we use paragraph as the smallest units of data for machine comprehension, and apply BiLSTM to process the features of paragraph to learn the model. We describe the details of our methods as follows.

We divide the documents returned by document retriever into paragraphs and denotes it as Eq. 1

$$A = \{p_1, p_2, \cdots, p_m\} \tag{1}$$

And tokenize the paragraphs as Eq. 2

$$p_i = \{t_1, t_2, \cdots, t_n\} \tag{2}$$

For each token t_i, we use five methods to extract features, including four features has been used in DrQA and additional one added by us. Hence, we introduce the first four features briefly and the added one in details.

– Word embedding: We used the fine-tuned word embedding based on the 300-dimentional Glove word embedding, which consider the keywords like what, when, who, that are crucial in QA. And denote it as Eq. 3

$$v_{embedding}(p_i) = E(p_i) \tag{3}$$

– Exact match: 3 binary features are used to indicate whether p_i can match any word in question q in original, lower-case or lemma form respectively. And denote it as Eq. 4

$$v_{exact-match} = M(p_i \in q_i) \tag{4}$$

– Token features: Include three manual factors include part of speech (POS), name entity recognition (NER) and normalized term frequency (TF), which is denoted as Eq. 5

$$v_{token-feature}(p_i) = (POS(p_i), NER(p_i), TF(p_i)) \tag{5}$$

– Aligned question embedding: Consider the doft alignment between the words with similar but not same word, Chen adds aligned question embedding. Denote as Eq. 6

$$v_{align}(p_i) = \sum_j a_{i,j} E(q_j) \tag{6}$$

where

$$a_{i,j} = \frac{exp(\alpha(E(p_i)) \cdot \alpha(E(q_j)))}{\sum_{j'} exp(\alpha(E(p_i)) \cdot \alpha(E(q_{j'})))} \tag{7}$$

Thus, we extracted all the features to represent the topic, semantic and the semantic structure of the paragraph, which is denoted as P_i. Hence, the representation of article can be denoted as $A = \sum_1^m p_i$. We also choose BiLSTM to learn the presentation of paragraph, and update the value of hidden weights step by step.

$$q = \mathbf{BiLSTM}(\{q_1, q_2, \cdots, q_l\}) \tag{8}$$

4.3 Question Semantic Modeling

Question is usually a short sentence which has simple semantic structure and few implicit semantic. Recurrent Neural Network can perform well on modeling the semantic of the question, and BiLSTM can improve the performance. In last section of our system, we tokenize question into sequences and denote it as $q = \{q_1, q_2, \cdots, q_k\}$. We suppose that $q = nonlinear(\{q_1, q_2, \cdots, q_k\})$, where q is the semantic vector of question, and q_i denotes the vector of ith word in question. We apply another BiLSTM on question embedding as Eq. 9:

$$q = \mathbf{BiLSTM}(\{q_1, q_2, \cdots, q_l\}) \tag{9}$$

Recurrent neural network (RNN) can contact the context to express the semantics of text and can perform well in modeling text. While LSTM pays more attention on the context related to the current word, i.e. the importance of words differs in different sentences, which is also known as keyword effect. So we need to set different weights on different words to accurately represent the semantic features of the text as $q = \sum_k w_k q_k$. For w_k, we can get Eq. 9, where u is the shared parameters learned by BiLSTM.

$$w_k = nonlinear(u \cdot q_k) \tag{10}$$

4.4   Candidate Answers Selection

Candidate answer selection is to find the final answer from the paragraph retrieved by the system in the first period. The framework of selecting answer is shown in Fig. 1. Generally speaking, we combine the features of questions and paragraph tokens and use deep learning algorithms to learn the weights. After feature extracting and parameters learning of paragraphs, sequences and questions, the system can represent the semantic of them very well. We use the feature vector of tokens and questions as input and train another two BiLSTMs to get the start token and end token of answer independently. Then the content between the start and end token is supposed to be the final answer. To capture the similarity between question q and paragraph p when each token being start and end, we compute the probabilities using nonlinear terms as

$$P_s(m) = nonlinear(t_m W_s q) \tag{11}$$

$$P_e(n) = nonlinear(t_n W_e q) \tag{12}$$

The original method is to maximize the $P_s(m) \times P_e(n)$ t get the start and end tokens, while we consider the effect of the paragraph using the results of learning to rank (LTR). The answer is supposed to be contained in the paragraph, so take paragraph as the smallest unit, the paragraph which has a higher score in LTR is more likely to contain the correct answer. In the candidate answer selection, we use LTR as part of weight when computing the probabilities of start and end tokens. In other words, the tokens contained in the top paragraph in LTR are rewarded to increase the possibility of choosing the final answer. And the reward is a_i, and the probability to be maximized changed to Eq. 13

$$P_{i,m,n} = \max(a_i P_s(m) P_e(n)) \tag{13}$$

In summary, we use information retrieval and current neural network to complete the machine learning at scale task. Firstly, we tokenize the documents of Wikipedia and question to get the valid sequences. Then use the BoW method [17] to encode the terms and TF-IDF to calculate the importance of terms. With the statistics of TF-IDF, we apply LDA/LSI to model the document features, i.e, using Dirichlet Probability Model/Singular Value Decomposition to compute the similarity between documents and given questions. The top 5 relevant documents are returned as the input of reading comprehension.

In the machine reading period, we divide the document into paragraphs for processing. With the same method used to tokenize in document retrieval period, and construct the features of terms manually. For each term in the paragraph, it has the feature of word embedding vector. At the same time, we add NER and POS tagging process to get the features. These features can specifically show the type of words and the location of words, which can provide a basis for the neural network to understand the semantic structure. Besides, we add LTR to improve the importance and attention of paragraphs as shown in Fig. 2. For the generated feature vectors, we use BiLSTM network to learn the rules corresponding to the segmentation vectors as the final feature of the text. As for the question, we apply another BiLSTM on tokenized text to get the semantic vector q.

In answer selection period, we use two independent neural networks to find the starting and ending positions of the answer and use a nonlinear model to calculate the probability that each word in the candidate answer becomes the beginning and the end. Additionally, we also considered the influence of LTR's paragraph ranking on the answer.

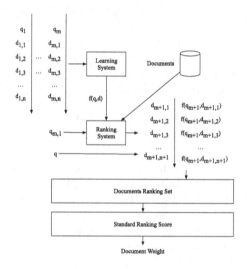

Fig. 2. Learning to Rank in our work

5   Experiments

We introduce our experiments from three aspects: (1) document retrieval based on LSI and LDA (2) reading comprehension of paragraphs and (3) open-domain question answering system with the combination of (1) and (2).

5.1   Document Retrieval Based on LSI

First, we use gensim [11] to learn a topic distribution model from the SQLite-formatted document library[1]. Based on this model, we evaluated our document

---

[1] docs.db provided by Wikipedia's open source Document Dump here.

retrieval module on all the datasets which contain SQuAD and expanded Curat-edTREC, WebQuestions, WikiMovies. Table 2 shows the examples of wiki documents found by our retriever according to the given question. And Table 4 shows the latent semantic found by our Lsi model from the aspect of weight in question and weight in Wikipedia documents respectively, while also taking the question in Table 2. As for the evaluation, we use the hit probability of top 5 document, i.e. the ratio of top 5 documents that contains the correct answer.

Table 1. Documents retrieval results

Question	Wiki Documents
Who is the president of the United States?	List of Presidents of New York University
	Vice President of Madagascar
	Vice President of Madagascar
	List of United States Presidential firsts
	...others

Table 2 shows the latent semantic found by our LSI model from the aspect of weight in question and weight in Wikipedia documents respectively, while also taking the question in Table 1 as an example. As shown in table, there are not only words in question, but some synonyms.

Table 2. Latent semantic found by LSI model.

Score type	Word	Score
Weight in question	President	0.949
	America	0.033
Weight in Wikipedia documents	President	0.949
	President	+0.959
	Prime	+0.003
	Presidential	+0.001
	Chairman	+0.001

5.2   Document Reader

Our reading model was trained and evaluated on SQuAD data set. We use the Stanford CoreNLP toolkit to tokenize, name entity recognize and encode the paragraphs and given questions. The obtained word vectors are sent to a three-layer BiLSTM neural network model. In order to avoid overfitting, we set the dropout rate of 0.3, i.e. 30% neural network units are randomly discarded in each batch of training.

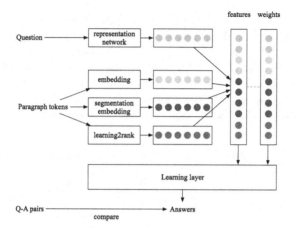

Fig. 3. The **picker** of candidate answers

We evaluated our module of machine comprehension on the SQuAD test set. We compare the results with the results of the Document Reader module in DrQA on the SQUAD test set. As can be seen from the comparison data, our module performed slightly better, indicating that the CTR feature we added during the training period had a positive effect. Although the effect is not very significant, it is not bad. The framework of selecting answer is shown in Fig. 3. Generally speaking, we combine the features of questions and paragraph tokens and use deep learning algorithms to learn the weights. They are encoded as vectors, and we merge all vectors. Then there is a weight vector to learn in the learning layer, which will be updated in the layer.

5.3   Open-Domain Question Answering System

Open-domain question answering system combines the tasks of document retrieval and reading comprehension. According to the input question, the document retrieval part will select five documents most relevant to the problem in the collection of Wikipedia documents and return it to the reading part for reading comprehension. As for reading comprehension period, documents are first divided into paragraphs. Then we choose the most relevant paragraph and locate the answer in it. Both the paragraph and the predictable answer will be returned as a result. We evaluate our system on all four datasets with Wikipedia as the single knowledge source.

According to the results of Chen's experiments, we can see the EM indicator has dropped significantly, from **69.5** to **70.0** respectively, which also appears in our experiments. In the experiment of machine comprehension, the data we give is a paragraph that is clearly related to the problem, that is, the answer is contained in the paragraph. While for the whole system, we give the entire collection of Wikipedia documents. The retriever module returns five documents that are most relevant to the question, but the correct answer may not be in these five documents, which leads to a decline in performance. That is the reason why we are interested in the improvement of retrieval module and make changes on it. Also, there is still room for improvement in this part, and we still need to work harder to move further in improving the relevance of the documents returned by the retrieval module to the problem.

We also compare our system with DrQA on four datasets, which indicates that both using LSI in the retrieving period and add learning to rank in reading comprehension period has a positive effect. Later, we may consider a more granular granularity. For example, the retrieval model returns paragraphs rather than documents, so that the reading model analyzes tasks based on paragraphs and gives predictions, or apply Learning to Rank method on sentences rather than paragraphs. We will continue to work in this area, and hope to get better results.

6   Conclusion

In this work, we propose an enhanced framework for the open-domain question answering task, in which, we introduce some topic model during the document retrieval phase, leading to a better representation of the documents, i.e., more hidden semantic information is exploited. Moreover, we incorporate the learning to rank model into the LSTM to train more available features for the ranking of candidate paragraphs. Specifically, we combine the results from both LSTM and learning to rank model, which lead to a more precise understanding of questions, as well as the paragraphs. The result of the empirical experiment demonstrates the efficacy of our proposed framework. We would like to test the framework with more advanced topic models and ranking methods in the future, as well as more knowledge bases.

Acknowledgement. The work is supported in part by the National Key Research and Development Program of China (2016YFC0800805) and the National Natural Science Foundation of China (61772014).

References

1. Androutsopoulos, I., Ritchie, G.D., Thanisch, P.: Natural language interfaces to databases-an introduction. Nat. Lang. Eng. **1**(1), 29–81 (1995)
2. Berant, J., Chou, A., Frostig, R., Liang, P.: Semantic parsing on freebase from question-answer pairs. In: Proceedings of EMNLP (2013)
3. Bordes, A., Chopra, S., Weston, J.: Question answering with subgraph embeddings. Comput. Sci. (2014)
4. Chen, D., Fisch, A., Weston, J., Bordes, A.: Reading Wikipedia to answer open-domain questions, pp. 1870–1879 (2017)
5. Chu-Carroll, J., Fan, J.: Leveraging Wikipedia characteristics for search and candidate generation in question answering. In: AAAI Conference on Artificial Intelligence, pp. 872–877 (2011)

6. Denoyer, L., Gallinari, P.: The Wikipedia XML corpus. In: Fuhr, N., Lalmas, M., Trotman, A. (eds.) INEX 2006. LNCS, vol. 4518, pp. 12–19. Springer, Heidelberg (2007). https://doi.org/10.1007/978-3-540-73888-6_2

7. Dong, L., Wei, F., Zhou, M., Xu, K.: Question answering over freebase with multi-column convolutional neural networks. In: Meeting of the Association for Computational Linguistics and the International Joint Conference on Natural Language Processing, pp. 260–269 (2015)

8. Ferrández, S., Toral, A., Ferrández, Ó., Ferrández, A., Munoz, R.: Exploiting Wikipedia and EuroWordNet to solve cross-lingual question answering. Inf. Sci. **179**(20), 3473–3488 (2009)

9. Jurczyk, T., Deshmane, A., Choi, J.D.: Analysis of Wikipedia-based corpora for question answering (2018)

10. Manning, C.D., Surdeanu, M., Bauer, J., Finkel, J., Bethard, S.J., Mcclosky, D.: The Stanford CoreNLP natural language processing toolkit. In: Meeting of the Association for Computational Linguistics: System Demonstrations (2014)

11. Řuřek, R., Sojka, P.: Gensim–statistical semantics in python (2011)

12. Rajpurkar, P., Zhang, J., Lopyrev, K., Liang, P.: Squad: 100,000+ questions for machine comprehension of text, pp. 2383–2392 (2016)

13. Vrandečić, D., Krötzsch, M.: Wikidata: a free collaborative knowledgebase. Commun. ACM **57**(10), 78–85 (2014)

14. Yan, X., Mou, L., Li, G., Chen, Y., Peng, H., Jin, Z.: Classifying relations via long short term memory networks along shortest dependency path. Comput. Sci. **42**(1), 56–61 (2015)

15. Yao, X., Durme, B.V.: Information extraction over structured data: question answering with freebase. In: Meeting of the Association for Computational Linguistics, pp. 956–966 (2014)

16. Yih, W.T., Chang, M.W., He, X., Gao, J.: Semantic parsing via staged query graph generation: question answering with knowledge base. In: Meeting of the Association for Computational Linguistics and the International Joint Conference on Natural Language Processing, pp. 1321–1331 (2015)

17. Zhao, R., Mao, K.: Fuzzy bag-of-words model for document representation. IEEE Trans. Fuzzy Syst. **26**, 794–804 (2017)

Attention Relational Network for Few-Shot Learning

Jia Shuai, JiaMing Chen, and Meng Yang[✉]

School of Data and Computer Science, Sun Yat-Sen University, Guangzhou, China
shuaij@mail2.sysu.edu.cn, chenjm26@mail2.sysu.edu.cn,
yangm6@mail.sysu.edu.cn

Abstract. Few-shot learning aims to learn a model which can quickly generalize with only a small number of labeled samples per class. The situation we consider is how to use the information of the test set to generate the better prototype representation of the training set. In this paper, based on attention mechanism we propose a flexible and efficient framework for few-shot feature fusion, called Attention Relational Network (ARN) which is a three-branch structure of embedding module, weight module and matching module. Specifically, with attention mechanism, the proposed ARN can model adaptively the constribution weights of sample features from embedding module and then generate the prototype representations by weighted fusion of the sample features. Finally, the matching module identify target sample by calculating the matching scores. We evaluated this method on the MiniImageNet and Omniglot dataset, and the experiment proved that our method is very attractive.

Keywords: Few-shot learning · Prototypical representation · Attention mechanism · Feature fusion

1 Introduction

As a general trend of machine learning, deep learning has been widely applied in various fields and shows obvious advantages and stimulating a tremendous upsurge of interest in deep neural network. However, the deep learning model met with serious challenges in practical applications, which is difficult to solve the problem with scarcity of labeled data. In comparison, humans are very good at identifying objects with few direct observation [14,16]. This gap between human and machine learning provides a fertile ground for the development of few-shot learning [3,12,19].

Few-shot learning identifies new categories that have not been seen during training through little labeled samples. In recent years, methods for solving few-shot learning can be roughly divided into three categories. The first category is based on fine-tuning, such as MAML [5] and the few-shot parameter optimization strategy [15]. The methods based on fine-tuning first pre-train a underlying

© Springer Nature Switzerland AG 2019
Z. Cui et al. (Eds.): IScIDE 2019, LNCS 11936, pp. 163–174, 2019.
https://doi.org/10.1007/978-3-030-36204-1_13

network on a large-scale dataset, such as Imagenet and then retrain it on a specific small-scale data set to learn the task-specific representations. During training, the parameters of the basic network part are frozen and only the domain-specific network parameters are trained. However, these approaches suffer from extra computational burden for fine-tuning on the target problem. The second category is based on metric [19,22,24]. This methods model a distribution distance, with which the target images can be classified by a simple linear classifier. However, the effect of metric-based method depends on the selection of distance measurement, which necessarily involves expert knowledge. The third category is based on meta-learning. Meta-learning-based methods are usually trained with an auxiliary meta-learning phase, in which the model automatically learns some transferable meta-knowledges, such as a great initial conditions for few-shot classifiers [5,15], or a shared metric [24]. The transferable meta-knowledges help automatically learning how to solve different types of problems and improving the flexibility and generalization of existing algorithms. The relational network [23] learns a nonlinear deep distance measurement through meta-learning, which can effectively reduce the influence of the improper selection of distance measurement. The relational network adopts a simple addition method to implement feature fusion, that is, all sample images share the same weight of fusion. However, for different classification problems, the contribution weight of each sample should be different since the similarity between sample images and target image is different.

Attention Mechanism [4] is derived from the study of human vision. In cognitive science, humans selectively focus on some parts which are more important and ignore the parts that are irrelevant to task. In computer vision, the attention mechanism was introduced for visual information processing, including object recognition [18], image generation [9], image caption [26] and image classification [6]. The attention module is usually used as an additional neural network that can select certain parts of the input and assign different weights to different parts of the input, which help the model learn better intermediate features.

Due to the advantage of attention mechanism and the promising performance of relational network, we propose a novel model of Attention Relational Network (ARN) for few-shot learning, by learning contribution weights for feature fusion and a nonlinear distance metric method for similarity comparison. Rather than simply averaging or summing the feature in the usual method of feature fusion [22,23], the proposed Attention Relational Network, based on the idea of the attention mechanism [11,25], can adaptively learn the individual contribution weights to obtain a better feature representation. Specifically, by using the feature information of the target image, weight value is assigned to each sample image of a sample class, and the weighted summation is used to generate better prototype representation of the class. The target image is then matched with the prototype representation of the sample class and we use the learned nonlinear distance metric method to classify the target image. Our evaluation of benchmarks on the MiniImageNet and Omniglot dataset shows that it has superior performance and outperforms the recent state-of-the-art few-shot learning methods.

2   Related Work

In this section, we first introduce existing methods for few-shot learning. Due to the wide literature on few-shot learning techniques, we focus mainly on the most related works, Relational Network. In the following, we will refer to the related researches about attention mechanism.

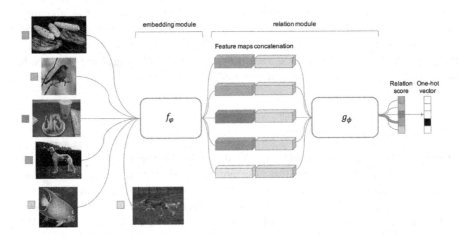

Fig. 1. Relation network [23] architecture for a 5-way 1-shot problem with 1-query example.

2.1   Relational Network

In the few-shot classification problem, the key challenge is the over-fitting to the limited data, which leads to a large gap between the training classification result and the general result. In order to reduce the impact of overfitting, a metric-based meta-learning method, Relational Network (RN), have been proposed in [23]. The RN is based on Prototypical Networks [22], and the improvement is simple yet effective.

The prototype network adopts the Euclidean distance measurement method in classification. However, as a linear method, the Euclidean distance limits prototype network to learn the general feature representations. Sung et al. [23] pointed out that people recognize images by comparing the features between images and images and then proposed a two-branch relationship network based on this idea. As shown in Fig. 1, the left part of Relational network is the embedded unit that is used to extract the feature of image, and the right part is the relation unit, which is used to calculate the similarity of the two images.

Previous few-shot learning approaches are pre-specified with metrics, such as Euclidean distance [22] or Cosine distance, which are inflexible and can not explore the complex structures of deep features. In addition, the learning part of previous few-shot learning work is mainly embodied in feature embedding, but the relational network learns the feature embedding and nonlinear metric (similarity function). The learned similarity metric has more flexibility than the artificially selected metric, and can capture the similarity between features more precise.

Although promising performance of relational network is reported, there are still some issues. When acquiring the prototype representation of the sample class, the relational network adopts a simple addition method, that is, all sample images are weighted and summed with the same weights. However, for different classification problems, the contribution weights on different samples for generating the prototype representation are different. Instead of assigning same weights on all sample, our ARN can learn weights individually for different samples for feature fusion.

2.2   Attention Mechanism

In general, the attention mechanism is able to learn the weight distribution of data and select the task-specific information. Different attention mechanisms have been proposed, with learning weights in spatial [11,27] and channel [7, 25]. For example, SENet [11] performs the attention generation on the channel of the feature map and then multiplies it by the original feature map. Woo et al. [25] proposed CBAM: the attention is applied to both channel and spatial dimensions. Attention can not only tell the network model which information is interesting, but also enhance the representation of specific regions.

Our propesed ARN performs the attention mechanism on feature fusion. Distribution weights are assigned to sample images of the same sample class to obtained the optimized prototype representation of the sample class by employing the feature information of target image.

Different pooling technologies, such as avg pooling and max pooling, have been adopted in attention methods. In [25], through comparative experiments it's confirmed that adding max pooling has a positive effect. In our proposed work, we adopt the method of superimposing the max pooling and the avg pooling. After mapping the feature vectors by using the embedding function, the avg pooling and the max pooling are performed, respectively.

3   Attention Relational Network

We solve the few-shot image classification by learning distribution weights for feature fusion and a nonlinear distance metric method for similarity comparison. Our method, based on the attention mechanism, proposes an Attention Relational Network (ARN). By using the feature information of the target image, a weight value is assigned to each sample image of a sample class, and the weighted summation is used to generate a better prototype representation of the class. The target image is then matched with the prototype representations of sample classes and sent to matching module to classify target image.

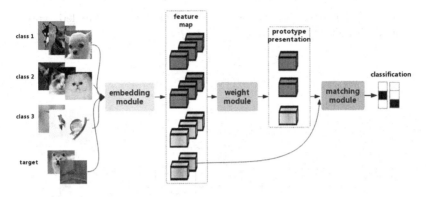

Fig. 2. Structure of attention relational network: embedding modules + weight modules + matching modules

Our attention relational network consists of three modules: embedding module, weight module and matching module, as shown in the Fig. 2. The embedded module extract the feature of the image, and then the feature of the sample image is merged by the weight module to generate prototype representation of each sample class. The matching module is used to compare the target image with the prototype representation, and produce a matching score representing the similarity. Note that the weight assigned to sample image of sample class in our work depends on the relationship between sample class and target image, and the weight is changeable based on variable classification problem.

3.1 Problem Description

We consider the task of few-shot learning classifier. We have three data sets: training set, sample set, test set. The training set belongs to one label space, while the sample set and the test set belong to another different label space. The strategy we adopted was based on episodes, which trained the model on the training set to obtain transferable knowledge. Then learn in the sample set and classify the test set. It should be noted that when training the model with the training set, the training set also simulates the settings in testing, and splits into the structure similar to the sample set and the test set [22,24]. To distinguish, we call the support set and the query set here.

During training, an episode consists of two parts: support set and query set. When a random selection of N classes from the support set and each of the classes randomly selects K labeled samples, this few-shot problem is called $N-way\ K-shot$. Support set S and class S_1 in S:

$$S = [S_1, S_2, \cdots, S_N]^K \tag{1}$$

$$S_1 = [s_1, s_2, \cdots, s_K] \tag{2}$$

Among the N categories, except for the selected $K * N$ labeled samples, n query sets are randomly selected. Query set Q:

$$Q = [Q_1, Q_2, \cdots, Q_n] \tag{3}$$

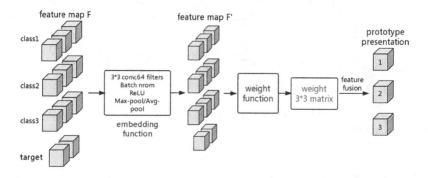

Fig. 3. Structure of weight module for a 3-way 3-shot problem with 2-query sample

3.2   Model of Attention Relational Network

The structure of the weight module is show in Fig. 3. The weight module consists of an embedded function and a weight function. The embedding function performs dimensionality reduction on the feature data, and the weight function calculates the weight value of all sample images by using the reduced-dimensional data.

Weight Calculation. When training, the sample classes in the support set and the query set are first sent to the embedding module to generate feature map of the images F_s and F_q, and then the feature maps of the support set and the query set are sent together into the weight module. The embedded function obtain the feature map F'_s and F'_q, and then calculates the distance D of the feature map F'_s and F'_q by the distance metric function (here we take sample class S_1 for example):

$$D = Euclidean\_dis\,(S_1, Q) = \left[d^1, d^2, \cdots, d^k\right] \in R^{k*n} \tag{4}$$

$$d^i = \left[d^i_1, d^i_2, d^i_3, \cdots, d^i_n\right] (i = 1, 2, 3, \cdots, k) \tag{5}$$

where D represents the distance of support class S_1 to all query image, d^i represents the distance of class i to all query image, k represents the number of samples of class S_1, n represents the number of query images. The distance D is converted into weight ∂.

$$W_i = \frac{\sum_j d^i_j}{n}, \quad j = 1, 2, 3, \cdots, n \tag{6}$$

$$\partial_i = -log\_softmax\left(W_i\right) = -log\left(\frac{exp\left(W_i\right)}{\sum_j exp\left(W_j\right)}\right) \tag{7}$$

$$\partial = [\partial_1, \partial_2, \cdots, \partial_k] \tag{8}$$

Feature Fusion. The feature map of the support set obtained by the embedding module is weighted and summed according to the weight ∂ obtained by the weight module, there we obtain the prototype representation of support class S_1 and the prototype representations of all classes in the support set S.

$$S_1 = \sum_i^k \partial_i * s_i \tag{9}$$

$$S = [S_1, S_2, \cdots, S_N] \tag{10}$$

Objective Function. We connect the prototype representation of support class with the feature vector of query image and sent to the matching module which generates a similarity score $r_{i,j}$ between 0–1.

$$\begin{aligned} P_S &= P_\alpha\left(F_\beta\left(S_i\right), F_\beta\left(Q_j\right)\right), i = 1, 2, \cdots, N \\ r_{i,j} &= G_\phi\left(C\left(P_S, F_\beta\left(Q_j\right)\right)\right), j = 1, 2, \cdots, n \end{aligned} \tag{11}$$

where, N represents the number of sample classes, F_β represents embedded module, P_α represents weight module, P_S represents prototype representation of support class. C represents cat function, G_ϕ represents matching module. $r_{i,j}$ represents match score. We use mean square error (MSE) loss (Eq. 12) to train our model, regressing the matching score $r_{i,j}$ to the ground truth: matched pairs have similarity 1 and the mismatched pair have similarity 0.

$$\alpha, \beta, \phi \leftarrow \arg\min_{\alpha,\beta,\phi} \sum_{i=1}^{C} \sum_{j=1}^{n} \left(r_{i,j} - 1\left(y_i = y_j\right)\right)^2 \tag{12}$$

where y_i represents the label class of S_i, y_j represents the predicted class of Q_j.

3.3 Network Module and Optimization

Embedded Module: The embedded module consists of four convolutional layers. The first two convolutional layers are: a 64-filters 3 * 3 kernel convolution layer, a BatchNorm, a ReLU, and a max pooling layer. The last two layers have no max pooling layer, but other structure is the same.

Weight Module: The weight module includes two convolution layers and a calculation weight function. The two convolution layers have two structures, one is a 64-filter 3 * 3 kernel convolution layer, a batchNorm, a ReLU and a max pooling layer, and the other is a 64-filter 3 * 3 convolution kernel, a batchNorm, a ReLU and an avg pooling layer, the two structures are superimposed. The

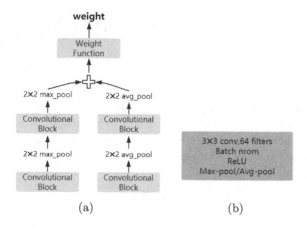

Fig. 4. (a) The network architecture of weight module. (b) The component of convolutional layer.

architecture of weight module and the component of convolutional layer is show in Fig. 4.

Matching Module: Following relational network [23], the matching module consists of two convolutional layers and two fully connected layers. The two convolutional layers are: 64-filter 3*3 kernel convolution, a batchNorm, a ReLU and a max pooling layer. The two fully connected layers are 8D and 1D, respectively.

Optimization: The last two blocks of embedded module do not contain a 2 * 2 max pooling layer while the first two do. We do so because we need the output feature maps for further convolutional layers in the weight module and matching module. In weight module, we use both avg pooled and max pooling features simultaneously, which greatly improves representation power of networks than using each independently. The output layer of matching module is Sigmoid in order to generate matching scores in a reasonable range for all versions of our network architecture.

4   Experiment

In this section, we first introduce the benchmark datasets used in our evaluation experiments and the experimental setup is followed. We evaluate our approach on two related tasks: 5-way 5-shot and 20-way 5-shot classification on Omniglot and 5-way 5-shot on miniImagenet. Then we compare our approach with state-of-the-art few-shot learning method and discuss the experimental results.

For few-shot learning experiments, we uses Adam [2] with initial learning rate 10^{-3}, annealed by half for every 100,000 episodes in all experiments. We train our model on miniImageNet with 500,000 episodes and on Omniglot with 1,000,000 episodes. All our models use no additional dataset and are end-to-end trained from scratch. To evaluate ARN, we conduct the comparisons against recent state-of-the-art baselines for few-shot image recognition, including neural statistician

[3], MANN [20], Siamese Nets with Memory [12], Convolutional Siamese Nets [19], Matching Nets [24], MAML [5], Meta-Learner LSTM [15] and Relational Nets [23].

4.1   Experiment on MiniImageNet Dataset

The miniImageNet dataset contains 100 classes, each class with 600 samples and a total of 60,000 color images. We split the dataset into 64, 16 and 20 classes for training, validation and testing, respectively. The 16 validation classes are just to monitor the performance of the model.

For evaluation, we conduct 5-way 5-shot classification experiments based on the standard settings following most existing few-shot learning methods. In addition to 5 sample images, each class in each training episode has 10 query images. The size of the input image is $84 * 84$.

Following Prototype [22] and Relational Nets [23], we analyzed 10 test target images in each episode for evaluation in 5-shot scene, and calculated the few-shot classification accuracy by averaging 600 randomly generated episodes from the test set. The experimental results on miniImageNet dataset are displayed in Table 1.

Table 1. Test result in miniImageNet dataset, All accuracy results are averaged over 600 test episodes

Model	Fine-Tune	5-way 5-shot
Matching nets [24]	N	$55.31 \pm 0.73\%$
Meta learn LSTM [15]	N	$60.60 \pm 0.71\%$
MAML [5]	Y	$63.11 \pm 0.92\%$
Relation nets [23]	N	$65.32 \pm 0.70\%$
Proposed model	N	$\mathbf{65.44 \pm 0.73\%}$

As shown in Table 1, Our model achieved the best results in the 5-way 5-shot scene. Compared to the relational network, our model has a 0.1% improvement in average classification accuracy.

4.2   Experiment on Omniglot Dataset

The data set Omniglot contains 1,623 classes from 50 different alphabets, each class containing 20 samples from different peoples. Following [1, 20], we change classes by rotating the existing data by 90, 180, 270°. And we used 1200 original or rotated categories for training, and the remaining 423 original or rotated categories were used for testing.

For each class in each training episode, in addition to K sample images, the 5-way 5-shot question contains 15 query images, and the 20-way 5-shot question contains 5 query images. The size of the input image is $28 * 28$.

For Omniglot, following Prototypical [22] and Relational Nets [23], we assess the performance of the proposed ARN by calculating the few-shot classification accuracy on Omniglot dataset by averaging more than 1000 randomly generated episodes from the test set. The experimental results on Omniglot dataset are shown in Table 2.

Table 2. Test result in Omniglot dataset, results are accuracies averaged over 1000 test episodes

Model	Fine-Tune	5-way 5-shot	20-way 5-shot
MANN [20]	N	94.9%	-
Convolutional siamese nets [19]	Y	98.4%	97.0%
Matching nets [24]	N	98.9%	98.5%
Siamese nets with memory [12]	N	99.6%	98.6%
Neural statistician [3]	N	99.5%	98.1%
Prototypical nets [22]	N	99.7%	98.9%
MAML [5]	Y	$99.9 \pm 0.1\%$	$98.9 \pm 0.2\%$
Relation net [23]	N	$99.8 \pm 0.1\%$	$99.1 \pm 0.1\%$
Proposed model	N	$\mathbf{99.92 \pm 0.03\%}$	$\mathbf{99.41 \pm 0.06\%}$

As shown in Table 2, in the 5-way 5-shot and 20-way 5-shot classification of the Omniglot dataset, We achieved state-of-the-art performance under all experiments setting with higher averaged accuracies and lower standard deviations, except 5-way 5-shot where the average precision of our model is the same as MAML method. Although many of the other methods in the table have significantly more complex architectures or need to fine-tune the target problem, the attention relational network does not.

4.3   Analysis

Compared with the traditional single-branch or double-branch model structure of the few-shot learning method, the three-branch structure proposed in our method can adaptively learn the individual weights of samples to exploit efficiently the essential information of limited data for feature fusion. Moreover, the weight module in ARN is an extra small neural network that can be ported and embedded easily in any popular network architecture. Due to the samll number of network parameters in our ARN, the low model complexity prevents our model from the overfitting to the limited data. The efficient collaboration among there modules, i.e. embedding module, weight module and matching module, make our ARN effective and achieve the best performance in the few-shot learning experiments.

5   Conclusion

In this paper, we solve the few-shot image classification by designing a flexible and efficient framework for few-shot feature fusion, called Attention Relational

Network (ARN) based on attention mechanism. The proposed ARN is a three-branch structure, which consist of embedding module, weight module and matching module. with attention mechanism, the proposed ARN can model adaptively the constribution weights of sample features from embedding module and then generate the prototype representations by weighted fusion of the sample features. The efficient collaboration among there modules can significantly improve the performance of model. The experimental results on the MiniImageNet and Omniglot dataset shows the superior performance of our method.

Acknowledge. This work is partially supported by the National Natural Science Foundation of China (Grant no. 61772568), the Guangzhou Science and Technology Program (Grant no. 201804010288), and the Fundamental Research Funds for the Central Universities (Grant no. 18lgzd15).

References

1. Ba, J., Mnih, V.K.: Multiple object recognition with visual attention (2015)
2. Chen, L.C., Yang, Y., Wang, J., Xu, W., Yuille, A.L.: Attention to scale: scale-aware semantic image segmentation, pp. 3640–3649. IEEE (2016)
3. Edwards, H., Storkey, A.: Towards a neural statistician (2017)
4. Feng, W., Tax, D.M.J.: Survey on the attention based RNN model and its applications in computer vision (2016)
5. Finn, C., Abbeel, P., Levine, S.: Model-agnostic meta-learning for fast adaptation of deep networks (2017)
6. Fu, J., Zheng, H., Mei, T.: Look closer to see better: recurrent attention convolutional neural network for fine-grained image recognition, pp. 4476–4484, July 2017. https://doi.org/10.1109/CVPR.2017.476
7. Fu, J., Liu, J., Tian, H., Fang, Z., Lu, H.: Dual attention network for scene segmentation (2018)
8. Garcia, V., Bruna, J.: Few-shot learning with graph neural networks (2017)
9. Gregor, K., Danihelka, I., Graves, A., Rezende, D.J., Wierstra, D.: Draw: a recurrent neural network for image generation. Comput. Sci., 1462–1471 (2015)
10. He, K., Zhang, X., Ren, S., Sun, J.: Deep residual learning for image recognition. In: 2016 IEEE Conference on Computer Vision and Pattern Recognition (CVPR), pp. 770–778 (2016)
11. Hu, J., Shen, L., Albanie, S., Sun, G., Wu, E.: Squeeze-and-excitation networks (2017)
12. Kaiser, Ł., Nachum, O., Roy, A., Bengio, S.: Learning to remember rare events (2016)
13. Krizhevsky, A., Sutskever, I., Hinton, G.: Imagenet classification with deep convolutional neural networks (2017)
14. Lake, B., Salakhutdinov, R., Gross, J., Tenenbaum, J.: One shot learning of simple visual concepts. In: Proceedings of the 33rd Annual Conference of the Cognitive Science Society, January 2011
15. Larochelle, S.R.H.: Optimization as a model for few-shot learning (2017)
16. Li, F.F., Rob, F., Pietro, P.: One-shot learning of object categories. IEEE Trans. Pattern Anal. Mach. Intell. **28**(4), 594–611 (2006)
17. Mishra, N., Rohaninejad, M., Xi, C., Abbeel, P.: Meta-learning with temporal convolutions (2017)

18. Mnih, V., Heess, N., Graves, A., Kavukcuoglu, K.: Recurrent models of visual attention, vol. 3 (2014)
19. Salakhutdinov, G.K.Z.: Siamese neural networks for one-shot image recognition. In: International Conference on Machine Learning (2015)
20. Santoro, A., Bartunov, S., Botvinick, M., Wierstra, D., Lillicrap, T.: Meta-learning with memory-augmented neural networks. In: Balcan, M.F., Weinberger, K.Q. (eds.) Proceedings of the 33rd International Conference on Machine Learning, Proceedings of Machine Learning Research, 20–22 June 2016, vol. 48, pp. 1842–1850. PMLR, New York (2016)
21. Simonyan, K., Zisserman, A.: Very deep convolutional networks for large-scale image recognition. arXiv preprint arXiv:1409.1556 (2014)
22. Snell, J., Swersky, K., Zemel, R.S.: Prototypical networks for few-shot learning (2017)
23. Sung, F., Yang, Y., Zhang, L., Xiang, T., Torr, P.H.S., Hospedales, T.M.: Learning to compare: relation network for few-shot learning (2018)
24. Vinyals, O., Blundell, C., Lillicrap, T., Kavukcuoglu, K., Wierstra, D.: Matching networks for one shot learning (2016)
25. Woo, S., Park, J., Lee, J.Y., Kweon, I.S.: CBAM: convolutional block attention module (2018)
26. Xu, K., et al.: Show, attend and tell: neural image caption generation with visual attention. Comput. Sci., 2048–2057 (2015)
27. Yang, D., Yuan, C., Bing, L., Zhao, L., Hu, W.: Interaction-aware spatio-temporal pyramid attention networks for action classification (2018)

Syntactic Analysis of Power Grid Emergency Pre-plans Based on Transfer Learning

He Shi[1,2], Qun Yang[1,2(✉)], Bo Wang[3,4,5], Shaohan Liu[1,2],
and Kai Zhou[1,2]

[1] College of Computer Science and Technology,
Nanjing University of Aeronautics and Astronautics, Nanjing 211106, China
657332737@qq.com, Qun.Yang@nuaa.edu.cn
[2] Collaborative Innovation Center of Novel Software Technology
and Industrialization, Nanjing 211106, China
[3] NARI Group Corporation (State Grid Electric Power Research Institute),
Nanjing 210016, China
[4] NARI Technology Co., Ltd., Nanjing 210016, China
[5] State Key Laboratory of Smart Gird Protection and Control,
Nanjing 210016, China

Abstract. To deal with the emergency pre-plans saved by the power grid dispatch department, so that the dispatcher can quickly retrieve and match similar accidents in the pre-plans, then they can learn from the experience of previous relevant situations, it is necessary to extract the information of the pre-plans and extract its key information. Therefore, deep learning method with strong generalization ability and learning ability and continuous improvement of model can be adopted. However, this method usually requires a large amount of data, but the existing labeling data in the power grid field is limited and the manual method for data labeling is a huge workload. Therefore, in the case of insufficient data, this paper aims to solve how to use deep learning method for effective information extraction? This paper modifies the ULMFiT model and uses it to carry out word vector training, adopting transfer learning method to introduce annotating datasets in the open field and combining with the data in the field of power grid to training model. In this way, the semantic relation of power grid domain is introduced into the syntactic analysis of the pre-plans, and we can further complete the information extraction. Experimental verification is carried out in this paper, the results show that, in the case of insufficient corpus or small amount of annotated data, this method can solve the problem of part of speech analysis errors, it can also improve the accuracy of syntactic analysis, and the experimental verifies the effectiveness of this method.

Keywords: Transfer learning · Word vector · Syntactic analysis

Supported by State Key Laboratory of Smart Grid Protection and Control.
Fault plan understanding and decision support based on natural language processing.

Z. Cui et al. (Eds.): IScIDE 2019, LNCS 11936, pp. 175–186, 2019.
https://doi.org/10.1007/978-3-030-36204-1_14

1 Introduction

1.1 Background

In the past time, the power grid dispatching department conducted in-depth analysis of power grid faults and made the handling plans for faults, prescribing the stability requirements, post-fault modes and disposal points of fault handling. These handling plans were written in texts and were also called the emergency pre-plans. When a fault occurs, the dispatcher can get the handling plan by searching the emergency pre-plans. However, with the expansion of power grid scale, the faults are becoming more and more complex. Meanwhile, there are more and more emergency pre-plans, which makes it difficult for the dispatcher to search the emergency pre-plans.

When an emergency fault occurs, it is particularly important for the dispatcher to learn from the experience of previous relevant situations. Obviously, if the relevant accident can be matched in the emergency pre-plans quickly, then the handling suggestions can be retrieved, and the dispatcher can judge the situation and handle the fault quickly and accurately.

To realize the quick searching in the pre-plans, it is necessary to extract information from the existing emergency pre-plans, achieving the key information, and transforming the unstructured plan into the structured form. In this paper, we propose to use the deep learning method to carry out the syntactic analysis of the emergency pre-plans. However, there are problems as followings:

(1) In the field of power grid, there is no large-scale high-quality annotated datasets for model training. The emergency pre-plans is a manual compilation of solutions to possible faults. Its compilation lacks strict specification and the quantity is limited, so it's difficult to establish large-scale available annotated datasets. In the syntactic analysis of the emergency pre-plans, insufficient data will lead to poor training effect and poor learning ability of the model, thus the more accurate analysis cannot be completed.

(2) The manual method for data annotation is a huge workload. Syntactic analysis of the emergency pre-plans requires labeling the sentence structure, however, due to the text from manual collation, syntactic structure is complex, and the part of speech analysis is prone to errors, so precisely divide the sentence structure is a difficult problem. If we use manual methods for data annotation, on the one hand, the quality of labeled samples is required to be high and the labeling time is long, on the other hand, it is difficult for labelling personnel to complete the work without sufficient domain knowledge, and only those who have relevant professional knowledge can be competent. So manually labeling samples is too difficult and costly [1, 2, 36–38].

To solve the above problems, this paper modifies the ULMFiT model, and uses it to train the word vectors, transferring the open domain word vectors to the grid field. After that, the annotated datasets in the open domain, small data in the power grid domain and the word vector after transfer are combined together for the training of the syntactic analysis model. In the case of small amount of data and difficulty in labeling, we use this method to solve the problem of errors in part of speech analysis. At the same time, the semantic relation of power grid domain is introduced into the training of syntactic analysis model. In this way, we can further improve the accuracy of part of speech analysis and the effect of syntactic analysis.

1.2 Related Work

In recent years, to achieve the computer assists decision making, it is necessary to express the logic and structure of the emergency pre-plans into a computer-aided form, hence the research on the digitalization of emergency pre-plans is increasing [3, 4]. For example, [5–7] use deep learning method to mine the power grid information, analyzing the text of the emergency pre-plans and extracting the important information needed by the dispatcher. The dispatcher can then judges the fault and handle the fault.

However, the standardization of data and the amount of data need to be improved. In the case of non-standard data or insufficient data, it should be solved that how to extract information from the emergency pre-plans. Generally, people use adversarial learning to improve the robustness and generalization ability of the model [8, 9], and use transfer learning to solve the errors caused by datasets dilution [10–12]. In the field of information extraction, when faced with problems, such as low precision of small sample data extraction, people will use the method of domain adaptation and transfer learning to solve the problem [13, 14].

At present, transfer learning has aroused wide interest [15–26]. Transfer learning has the following advantages: (1) It transfers the mature knowledge from one field to another new field. It can continuously expand the new knowledge by using the existing knowledge. (2) It can be used to train the model. It is suitable for the target field and can effectively improve the task effect of the target field. (3) It can solve the basic problem of insufficient training data [21]. Therefore, this paper applies transfer learning to the syntactic analysis of the emergency pre-plans.

In terms of word vector training, the traditional model adopts backward propagation algorithm and randomly initialize the model parameters [27–30], such as word2vec [31]. However, the above method starts training from scratch, which requires a large number of data sets and takes a lot of time to converge. Therefore, scholars have not considered the random initialization of the model. The model pre-training method has emerged. It has been improved on many tasks, and has received extensive attention from scholars [22, 23, 32], such as ELMo model, BERT model, ULMFiT model. These models demonstrate the importance of pre-training. In addition, the ULMFiT model utilizes universal field pre-training and novel fine-tuning techniques to prevent over-fitting and achieve better performance on small data sets. Therefore, this paper uses the ULMFiT model to carry out word vector training of the emergency pre-plans.

2 Syntax Analysis Method of the Emergency Pre-plans Based on Transfer Learning

First we introduce the ULMFiT model [23]. ULMFiT model has many advantages. For instance, it is suitable for the domains with a small amount of data and does not require too many files or labels in the domains. In addition, the training consists of two stages. Each stage uses the same network structure. No custom preprocessing is required.

This paper modifies the ULMFiT model and combines with the characteristics of the emergency pre-plans to improve the effect of syntactic analysis. We complete the transferring of power grid domain word vector and introduce the word vector into the training of syntactic analysis model. By this way, the syntactic analysis model can correctly analyze the data of power grid which contains many domain words.

2.1   Model Introduction

2.1.1   ULMFiT Model

The training process of ULMFiT model includes two stages which are called pre-training and fine-tuning. The first stage pretrains a model in a large open domain datasets and second stage fine-tunes the model with the target domain data. The above two stages use the same network architecture. We describe the components of the neural network architecture. It has three layer long short-term memory neural network and layers are fully connected to each other. The input layer uses one-hot coding form. The model multiplies the input by the weight of each layer and averages it after activation by the activation function. The output is computed through the softmax layer. According to the backpropagation algorithm, we can adjust the weight matrix. Through continuous training, we can obtain more accurate model parameters, thereby, we can also train the word vector with contextual semantic relationship. The ULMFiT model is shown in the following Fig. 1:

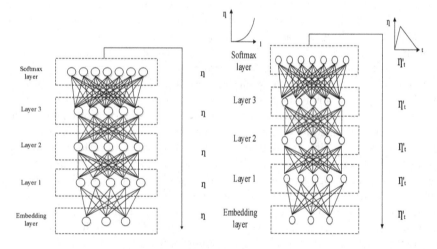

Fig. 1. LM pre-training (left), LM fine-tuning (right)

The model uses two strategies, called discriminative fine-tuning and slanted triangle learning rates. The first strategy uses the regular stochastic gradient descent (SGD) training algorithm to modify model parameters and learning rate at each level, the formula is as follows:

$$\theta_t = \theta_{t-1} - \eta \cdot \nabla_\theta J(\theta)$$

Where η represents the learning rate, θ_t represents model parameter θ under time t step, $\nabla_\theta J(\theta)$ represents the gradient of the model objective function. The second strategy which first linearly increases the learning rate and then the linearly decays it. Network training is based on epoch, each epoch has a correction for the learning rate. In this way, the change pattern of the learning rate will present a "triangle".

2.1.2   Modified ULMFiT Model

In this section, we introduce our modifications to the model. We remove the softmax layer of the LM fine-tuning model and add a new softmax layer for the grid domain datasets, initializing the softmax layer. We need to consider the characteristics of the grid domain words influence on training model. As a result, the initialization parameters use the softmax layer parameters of the LM pre-training model. It is trained by the grid domain datasets. However, the parameters of other layers are provided by the hidden layer of the neural network. It is trained by the open domain datasets. The output nodes of the network's Layer3 layer correspond to the cell of the new softmax layer, and the total number of layers in the model is unchanged. The model is shown as follows (Fig. 2):

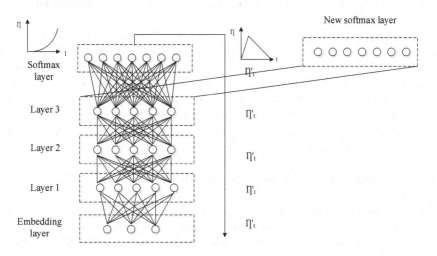

Fig. 2. Modified ULMFiT model

When modifying the model parameters and the learning rate of each layer, we consider the characteristics of the emergency pre-plans and modify the algorithm of the model as follows:

We define a function g: X → Y, where X is the set of the emergency pre-plans word vectors and Y is the set of all possible word vectors. The set of all possible word vectors for a given word x_i is defined as Y (x_i) and the transferred word vector for a word is y_i.

We first define a structured margin loss $\nabla(y_i, y)$ for getting a word vector y for a given correct word vector. The discrepancy between word vectors is called the smoothing of the source domain to the target domain.

$$\nabla(y_i, y) = \mu \cdot \frac{y_i + y}{2}$$

We set $\mu = 0.1$ in all experiments. For a given set of instances (x_i, y_i), we search for the function g with the smallest expected loss.

$$g(x) = \min(LMpretrain(\theta, x, y))$$

We redefine the gradient of the model objective function: $\nabla(y_i, y)_\theta \cdot J(\theta)$
Gradient vector of the objective function:

$$\nabla(y_i, y)_\theta J(\theta) = \left[\frac{\partial J(\theta)}{\theta_1}, \ldots, \frac{\partial J(\theta)}{\theta_n}\right]$$

2.2   Training Methods Based on Transfer Learning

2.2.1   Word Vector Training Method of the Emergency Pre-plans Based on Transfer Learning

This paper uses the modified model for training. The following is a detailed training process: The first step is that we use open field datasets and the grid field datasets to train LM pre-training model respectively, and then we use the modified LM fine-tuning model to fine-tuning, inputting the grid domain data into the network. When each training sample into the network, only the softmax layer is updated. However, the hidden layer of the neural network trained by the open domain datasets provides a shared hidden layer for the power grid domain. In the training, the two strategies are also adopted to complete the open domain data transfer to the grid field.

2.2.2   Training Method of Syntax Analysis Model Based on Transfer Learning

The training process of the syntactic analysis model includes the following steps: The datasets marked in the open field is used for the pre-training of syntactic analysis model. We put it into the probabilistic context-free grammar (PCFG). The candidate syntax tree of sentences is generated after training. The model parameters of pre-training are retained and then the grid domain datasets is input into the trained PCFG. Foremost, the word vector trained by the modified ULMFiT model and the generated candidate tree are used as the input of the recurrent neural network to train the syntactic analysis model, obtaining the syntactic analyzer. The above mentioned recurrent neural network adopts a semantic-based network (SU-RNN) [33]. It combines the word vector to syntactic analysis. When the network calculates the probability of each node in a syntax tree, it relies on the part of speech of its child nodes, and introduces the semantic relationship between words into the model training, selecting the best syntax tree from the candidate trees.

2.3 Syntactic Analysis

2.3.1 Syntactic Analysis

The following is introduced the process of syntactic analysis: First, we need to pre-process the emergency pre-plans and convert it into plain text. The plain text is used to carry out word segmentation. Then we use the maximum probability path algorithm to find the maximum segmentation combination based on word frequency, and combine with the established plan special dictionary, using dictionary-based word segmentation method to further achieve correct word segmentation. Finally we put the processed sentence into the syntactic analysis model and the model will output a syntax tree for the sentence. This syntax tree has a clearer hierarchical structure than the deep learning syntax tree. Through this syntax tree, the relationship between the parts of a sentence can be obtained. It can reflects the semantic modification relationship among the components of the sentence, and the long-distance collocation information can be obtained, regardless of the physical location of the sentence component. By analyzing the relationship between the sentence components, useful information will be extracted for the dispatcher.

2.3.2 Sample Results

Take "尖叫屋棚剩余单台主变潮流越限，转移 95%"(The remaining single main transformer current is over the limit) for example.

Figure 3 is the result of training without the transfer learning method. Since the influence of grid domain semantics on model training is not considered, the noun "主变"(main transformer) is analyzed into a verb and the verb "越限"(over the limit) is analyzed into a noun in the syntax tree. If the above analysis results are used, the action of "主变"(main transformer) will be extracted and the key information of "越限"(over the limit) will be ignored. This method causes an error in the information extraction.

Figure 4 describes the result of training with the method in this paper. It can be seen that when we consider the semantic relationship of the grid domain for syntactic analysis, the "主变"(main transformer) and "越限"(over the limit) are correctly analyzed. This result indicates that we have captured useful semantic and syntactic information.

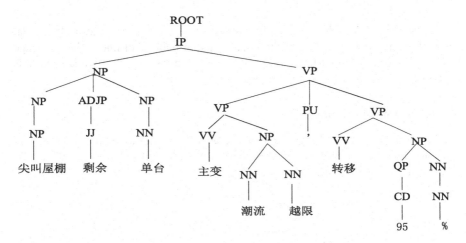

Fig. 3. The syntax tree obtained without the use of transfer learning for syntactic analysis

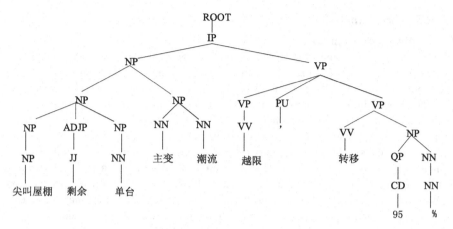

Fig. 4. The syntax tree obtained by use of transfer learning for syntactic analysis

3 Experiment and Results

3.1 Experiment Setting

The datasets used in this experiment is shown in Table 1: CTB8.0 datasets, which is used as the pre-training datasets of the model. 4196 emergency pre-plans saved by a power company are taken as the target domain datasets.

Table 1. Datasets

Classification	The datasets	The amount of data
Pre-training datasets	CTB8.0 datasets	76489
Target domain datasets	The emergency pre-plans	4196

The model includes three hidden layers with 1024 nodes in each layer. The initial value of learning rate is 0.01 and the initial iteration number of fine-tuning language model is 20. To avoid over-fitting, the number of iterations is gradually increased. It is appropriately adjusted to find the best value.

3.2 Experimental Results

We first compare the results of the syntactic analysis model using the word vectors generated by the four models, such as word2vec, ELMo, BERT, and ULMFit. The experimental results are shown in Table 2.

Table 2. Data comparison

Method	Model	LP/(%)	LR/(%)	F1/(%)
word2vec	CBOW	76.5	74.58	75.54
ELMo	Bi-LSTM	78.4	75.44	76.92
BERT	Transformer	78.9	76.57	77.74
ULMFit	Bi-LSTM	79.5	77.64	78.57

From Table 2, we can see that the ULMFiT model works best. Therefore, this paper chooses the ULMFiT model for the word vectors transfer learning. The purpose of this paper is to evaluate the performance of the modified ULMFiT model. The ULMFiT model without improvement and the method without transfer learning are selected for comparative experiments. We use the above methods in the emergency pre-plans data for experiments. We report models' performance on datasets in Table 3.

Table 3. Data comparison

Method	Weather or not to use transfer learning	LP/(%)	LR/(%)	F1/(%)
PCFG [34]	No	74.6	72.66	73.63
SSN [35]	No	76.3	74.94	75.62
CVG(RNN) [33]	No	75.7	74.69	75.20
CVG(SU-RNN) [33]	No	76.2	74.58	75.39
ULMFiT model	Yes	79.1	77.23	78.17
Modified ULMFiT model	Yes	79.8	78.56	79.18

In this paper, the index F1 is used as the evaluation result of the model. It is the harmonic average of recall rate and accuracy rate, where LP represents the accuracy rate and LR represents the recall rate.

It can be seen from Table 3 that the index F1 of the model obtained without using the transfer learning are: 73.63%, 75.62%, 75.20%, 75.39%. The adoption of the transfer learning algorithm has a greater improvement than not using it. The value of F1 increases to 78.17%, while the modified model mentioned in this paper F1 value is further improved to reach 79.18%.

It can be seen from Table 4 that the accuracy of part of speech analysis is also improved. The accuracy obtained without using transfer learning is respectively 77.46%, 80.62%, 79.75%, 80.58%. The accuracy of transfer learning method is increased increases to 85.69%, while the accuracy of the modified model in this paper is up to 88.76%.

In Tables 3 and 4, it is found that the model F1 value obtained without transfer learning, which is slightly lower than that of transfer learning. But the accuracy of part-of-speech analysis is more obvious. Since the sufficient data in the source area, the training effect of the syntactic analysis model is better. However the application of the model to the power grid field will cause errors in part-of-speech analysis.

Table 4. Comparison of accuracy of syntactic analysis

Method	Correct rate of part of speech
PCFG [34]	77.46%
SSN [35]	80.62%
CVG(RNN) [33]	79.75%
CVG(SU-RNN) [33]	80.58%
ULMFiT model	85.69%
Modified ULMFiT model	88.76%

In the case of less corpus or small amount of annotated data, the experimental results show that we can use transfer learning method to solve the problem of errors in part of speech analysis. It is worth noting that the modified model can further improve the syntactic analysis accuracy.

4   Conclusion

When using deep learning method in the field of power grid, there will be many problems in syntactic analysis, such as the mark corpus is insufficient and manual labeling is difficult. Our main contribution is to show that we adopt transfer learning method and modify the ULMFiT model to solve these problems. We use this method to solve the tasks in the target domain by learning knowledge from the source domain. Simultaneously we use this model to train the word vector for existing high-quality non-target domain public data and transfer the vector to the power grid field. In addition, the semantic relation of power grid domain is introduced into the pre-plans processing.

In the test of 4196 emergency pre-plans, the experiment shows that this method can effectively improve the accuracy of lexical analysis in the professional field. It is better than the traditional method and can effectively improves the effect of syntactic analysis.

References

1. Lu, Q., Liu, X.Y.: Knowledge map q&a semantic matching model based on transfer learning. Comput. Appl. **335**(7), 22–28 (2018)
2. Long, M., Cao, Y., Wang, J.: Learning Transferable Features with Deep Adaptation Networks. arXiv:1408.0872 (2015)
3. Liu, D., Chen, Y., Shen, C.: Study on digital method of power emergency plan. Autom. Power Syst. **33**(21) (2009)
4. Wu, W.C., Zhang, B.M., Cao, F.C.: Design and key technology of grid emergency command technical support system. Power Syst. Autom. **15**, 1–6 (2008)

5. Zhao, X.S., Xie, B.M., Zhang, H.W.: A power grid fault diagnosis method based on deep learning algorithm. Henan Sci. Technol. **23**, 53–54 (2016)
6. Jiang, Q., Shen, L., Zhang, W., He, X.: Research on fault diagnosis method based on deep learning. Comput. Simul. **35**(7), 409–413 (2018)
7. Qiu, J., Wang, H.F., Ying, J.L.: Text information mining technology and its application in the life cycle state evaluation of circuit breakers. Power Syst. Autom. **40**(6), 107–112 (2016)
8. Zhang, Y.K., Zhang, P.Y., Yan, Y.H.: Language model data enhancement technology based on antagonistic training strategy. Acta Automatica Sinica **44**(05), 126–135 (2018)
9. Zhang, Q.L., Du, J.C., Xu, R.F.: Research on irony recognition based on antagonistic learning. J. Peking Univ. (nat. Sci. Edn.) **55**(01), 32–39 (2019)
10. Chen, C., Shen, F., Yan, R.Q.: Bearing fault diagnosis based on improved LSSVM migration learning method. J. Instrum. (2017)
11. Gu, T.Y., Guo, J.S., Li, Z.X.: Fault probability prediction of airborne equipment based on interpolation-fit-transfer learning algorithm. Syst. Eng. Electron. Technol. **40**(1), 114–118 (2018)
12. Ren, J., Hu, X.F., Li, N.: Transfer learning prediction algorithm based on SDA and SVR hybrid model. Comput. Sci. **45**(1) (2018)
13. Miwa, M., Sætre, R., Miyao, Y.: A rich feature vector for protein-protein interaction extraction from multiple corpora. In: Proceedings of the 2009 Conference on Empirical Methods in Natural Language Processing. Association for Computational Linguistics (2009)
14. Li, L.S., Guo, R., Huang, D.G.: Extraction of protein interaction based on transfer learning. J. Chin. Inf. Process. **30**(2), 160–167 (2016)
15. Oquab, M., Bottou, L., Laptev, I.: Learning and transferring mid-level image representations using convolutional neural networks. In: Computer Vision and Pattern Recognition. IEEE (2014)
16. Razavian, A.S., Azizpour, H., Sullivan, J.: CNN features off-the-shelf: an astounding baseline for recognition. In: Computer Vision & Pattern Recognition Workshops (2014)
17. Yosinski, J., Clune, J., Bengio, Y.: How transferable are features in deep neural networks? Eprint Arxiv **27**, 3320–3328 (2014)
18. Wu, Y., Ji, Q.: Constrained deep transfer feature learning and its applications. In: Computer Vision and Pattern Recognition, pp. 5101–5109. IEEE (2016)
19. Tzeng, E., Hoffman, J., Zhang, N.: Deep domain confusion: maximizing for domain invariance. Comput. Sci. (2014)
20. Wang, N.: Conceptual relationship extraction of basic education geography based on transfer learning. Wuhan University of Technology (2017)
21. Pan, C.W.: Research on vector optimization method for pre-training Chinese words in transfer learning. Beijing Jiaotong University (2018)
22. Peters, M.E., et al.: Deep contextualized word representations. NAACL (2018)
23. Howard, J., Ruder, S.: Universal Language Model Fine-tuning for Text Classification (2018)
24. Zhuang, F.Z.: Research progress in transfer learning. J. Softw. **26**(1), 26–39 (2015)
25. Pan, S.J., Yang, Q.: A survey on transfer learning. IEEE Trans. Knowl. Data Eng. **22**(10), 1345–1359 (2010)
26. Tan, C., Sun, F., Kong, T.: A Survey on Deep Transfer Learning (2018)
27. Zhang, J., Qu, D., Li, Z.: Language model of cyclic neural network based on word vector features. Pattern Recogn. Artif. Intell. **28**(4), 000299–305 (2015)
28. Hinton, G.E.: Learning distributed representations of concepts. In: Proceedings of the Eighth Annual Conference of the Cognitive Science Society, pp. 1–12. Erlbaum, Hillsdale (1986)
29. Mikolov, T., Chen, K., Corrado, G.: Efficient estimation of word representations in vector space. Comput. Sci. (2013)

30. Mikolov, T., Sutskever, I., Kai, C.: Distributed representations of words and phrases and their compositionality. Adv. Neural. Inf. Process. Syst. **26**, 3111–3119 (2013)
31. Rong, X.: word2vec Parameter Learning Explained. Comput. Sci. (2014)
32. Devlin, J., Chang, M.W., Lee, K.: BERT: Pre-training of Deep Bidirectional Transformers for Language Understanding (2018)
33. Socher, R., Bauer, J., Manning, C.D., Ng, A.Y.: Parsing with compositional vector grammars. In: Proceedings of 51st Annual Meeting of the Association for Conputational Linguistics (Volume 1: Long papers), pp. 455–465 (2013)
34. Klein, D.: Accurate unlexicalized parsing. In: Proceedings of the 41st Meeting of the Association for Computational Linguistics, Sapporo, Japan (2003)
35. Henderson, J.: Discriminative training of a neural network statistical parser. In: Proceedings of the 42nd Annual Meeting of the Association for Computational Linguistics, Barcelona, Spain. DBLP (2004)
36. Yao, Y., et al.: Towards automatic construction of diverse, high-quality image dataset. IEEE Trans. Knowl. Data Eng. (2019). https://doi.org/10.1109/tkde.2019.2903036
37. Yao, Y., Shen, F., Zhang, J., Liu, L., Tang, Z., Shao, L.: Extracting privileged information for enhancing classifier learning. IEEE Trans. Image Process. **28**(1), 436–450 (2019)
38. Yao, Y., Zhang, J., Shen, F., Hua, X., Xu, J., Tang, Z.: Exploiting web images for dataset construction: a domain robust approach. IEEE Trans. Multimedia **19**(8), 1771–1784 (2017)

Improved CTC-Attention Based End-to-End Speech Recognition on Air Traffic Control

Kai Zhou[1,2], Qun Yang[1,2(✉)], XiuSong Sun[1,2], ShaoHan Liu[1,2], and JinJun Lu[3]

[1] College of Computer Science and Technology, Nanjing University of Aeronautics and Astronautics, Nanjing 210016, Jiangsu, China
qun.yang@nuaa.edu.cn
[2] Collaborative Innovation Center of Novel Software Technology and Industrialization, Nanjing, China
[3] State Key Laboratory of Smart Grid Protection and Control, Nanjing, China

Abstract. Recently, many end-to-end speech recognition systems have been proposed aim to directly transcribes speech to text without any predefined alignments. In this paper, we improved the architecture of joint CTC-attention based encoder-decoder model for Mandarin speech recognition on Air Traffic Control speech recognition task. Our improved system include a Vggblstm based encoder, an attention LSTM based decoder decoded with CTC mechanism and a LSTM based ATC language model. In addition, several tricks are used for effective model training, including L2 regularization, attention smoothing and frame skipping. In this paper, we compare our improved model with other three popular end-to-end systems on ATC corpus. Result shows that our improved CTC-attention model outperforms CTC, attention and original CTC-attention model without any tricks and language model. Taken these tricks together we finally achieve a character error rate (CER) of 13.15% and a sentence error rate (SER) of 33.43% on the ATC dataset. While together with a LSTM language model, CER and SER reach 11.01% and 22.75%, respectively.

Keywords: Automatic speech recognition · End-to-end speech recognition · CTC mechanism · Attention model

1 Introduction

Air traffic control (ATC) means airport ground controllers using communication and navigation technology to monitor and control aircraft flight activities to ensure flight safety and orderly flight. So far, the communication mode of ATC still stays on manual monitoring and radio communication. However, with the popularity of air transportation, flights are increasing, which leads to great pressure on air traffic controllers and more potential safety hazards at the same time. Therefore, it is very urgent to realize intelligent ATC. Before that, speech recognition as the most convenient way of human-computer interaction is crucial to intelligent ATC. However, there are still many challenges to apply automatic speech recognition systems to this field, such as strong background noise, fast speaking, special pronunciation of words and. In this paper, we

© Springer Nature Switzerland AG 2019
Z. Cui et al. (Eds.): IScIDE 2019, LNCS 11936, pp. 187–196, 2019.
https://doi.org/10.1007/978-3-030-36204-1_15

present our contribution to apply a hybrid CTC-attention end-to-end system in ATC task, including improving the encoder architecture with CNN network to address noise, integrating CTC mechanism into decoder to give model more robustness and using some training tricks for effective model training.

Traditional large vocabulary continuous speech recognition (LVCSR) systems usually consist of several elaborate modules including a GMM/DNN-HMM based acoustic model, a pronunciation lexicon and an external word-level language model [1, 2]. However, completing such a system is a very complex task requiring professional knowledge. For this reason, end-to-end system gradually become a research hotspot. Generally, these end-to-end speech recognition methods can be categorized into two main approaches: Connectionist Temporal Classification (CTC) [3–6] and attention based encoder-decoder [7–11].

Both approaches are aim to transcribe variable-length speech input to text and both have its defect. CTC introduce intermediate label 'blank' to address the problem of variable length input output sequence. But it makes a strong assumption that the labels are conditionally independent of each other. The attention approach which learns a mapping from acoustic frame to character sequences does not use any conditional independence assumption, it has often shown to improve CER than CTC when no external language model is used [11]. However, it also has two issues. Firstly, the alignment estimated in the attention mechanism is easily corrupted due to the noise. Secondly, the model is hard to learn from scratch due to the misalignment on longer input sequences. So a CTC-attention based encoder-decoder method has been proposed to overcome the above misalignment issues [12]. The key idea of CTC-attention method to improve performance and accelerate learning is using a joint CTC-attention model within the multi-task learning framework. Their proposed method improves the performance by rectifying the alignment problem using the CTC loss function based on the forward-backward algorithm. Along with improving performance, their framework significantly speeds up learning with fast convergence.

In this paper, we improved the encoder of CTC-attention system by adding six deep CNN layers before four bidirectional LSTM layers in order to improve the translation invariance of the model in time-frequency domain and enhance model robustness. In addition, CTC is used as an aligner in decoder to make up for the weakness of attention model that lack of left-to-right constraints. Finally, we also present several tricks for effective model training, including L2 regularization, Gaussian weight noise and frame skipping. We evaluate our model on the ATC tasks, and show that our system out-performs all the CTC, attention and original CTC-attention based system in CER and SER.

The next section reviews models from CTC to attention and original joint CTC-attention model. Section 3 describes the architecture of the improved CTC-Attention based encoder-decoder model. Section 4 compares CTC-attention to other ASR systems on the ATC speech corpus. Section 5 shows experiment results and discusses some key differences between our model and other end-to-end systems, and the paper concludes with Sect. 6.

2 Original CTC-Attention Encoder-Decoder

2.1 Connectionist Temporal Classification (CTC)

The main idea of CTC [3] is using intermediate label representation $\pi = (\pi_1, \ldots, \pi_T)$, allowing repetitions of labels and occurrences of a blank label $(-)$, which represents the special emission without labels, i.e., $\pi_t \in \{1, \ldots, K\} \cup \{-\}$. CTC trains the model to maximize $p(y|x)$ according to:

$$p(y|x) = \sum\nolimits_{\pi \in \Phi(y')} P(\pi|x) \tag{1}$$

where $\Phi(y')$ denotes the probability distribution over all possible label sequences, y' is a modified label sequence of y, which is made by inserting the blank symbols between each label and the beginning and the end for allowing blanks in the output (i.e., y = (c, a, t), y' = (−, c, −, a, −, t, −)).

CTC is generally applied on top of Recurrent Neural Networks (RNN). Each RNN output unit is interpreted as the probability of observing the corresponding label at particular time. The probability of label sequence $P(\pi|x)$ is modeled as being conditionally independent by the product of the network outputs:

$$P(\pi|x) \approx \prod\nolimits_{t=1}^{T} P(\pi_t|x) = \prod\nolimits_{t=1}^{T} q_t(\pi_t) \tag{2}$$

where $q_t(\pi_t)$ denotes the softmax activation of π_t label in RNN output layer q at time t. The CTC loss to be minimized is defined as the negative log likelihood of the ground truth character sequence y^*, i.e.

$$L_{CTC} \approx -lnP(y^*|x) \tag{3}$$

The probability distribution $p(y|x)$ can be computed efficiently using the forward-backward algorithm as

$$p(y|x) = \sum\nolimits_{u=1}^{|y'|} \frac{\alpha_t(u)\beta_t(u)}{q_t(y'_u)} \tag{4}$$

where $\alpha_t(u)$ is the forward variable, representing the total probability of all possible prefixes $(y'_{1:u})$ that end with the u-th label, and $\beta_t(u)$ is the backward variable of all possible suffixes $(y'_{u:U})$ that start with the u-th label. The network can then be trained with standard backpropagation by taking the derivative of the loss function with respect to $q_t(k)$ for any k label including the blank.

Since CTC does not explicitly model inter-label dependencies based on the conditional independence assumption in Eq. (2), there are limits to model character-level language information.

2.2   Attention Based Encoder-Decoder

The attention based encoder-decoder architecture is composed of an encoder, which transforms the low level acoustic inputs into high level representations, and a decoder which produces linguistic outputs (i.e., characters or words) from the encoded representations. The challenge is the input and output sequences have variable (also different) lengths, and usually alignments between them are unavailable. Given the input sequence $x = (x_1, \ldots, x_T)$ of length T, and the output sequence $y = (y_1, \ldots, y_U)$ of length U, with each y_u being a V-dimensional one-hot vector, the encoder maps the input x into a high level representation $h = (h_1, \ldots, h_{T'})$, which can be shorter than the input $(T' \leq T)$ with time-scale downsampling. The decoder consists of an aligner and attention mapping from h to y.

The model emits each label distribution at u conditioning on previous labels according to the following recursive equations:

$$p(y|x) = \prod_u p(y_u|h, y_{1:u-1}) \tag{5}$$

$$h = Encoder(x) \tag{6}$$

$$y_u \sim AttentionDecoder(h, y_{1:u-1}) \tag{7}$$

The loss function of the attention model is computed from Eq. (5) as:

$$L_{attention} \approx -lnP(y^*|x) \tag{8}$$

$$= -\sum_u lnP(y_u^*|x, y_{1:u-1}^*) \tag{9}$$

where $y_{1:u-1}^*$ is the ground truth of the previous characters.

The decoder generates next label y_u as:

$$y_u \sim Generate(c_u, s_{u-1}) \tag{10}$$

where c_u is calculated by integrating all the inputs h based on their attention weight vectors and s_{u-1} is the decoder state.

In practice, the model has a great weakness on noisy speech data. The attention model is easily affected by noises, and generates misalignments because the model does not have any constraint that guides the alignments be monotonic as in CTC.

2.3   Original CTC-Attention Encoder-Decoder

The idea of original CTC-attention model is using a CTC objective function as an auxiliary task to train the attention model encoder within the multitask learning (MTL) framework. The objective is represented as follows by using both attention model in Eq. (8) and CTC in Eq. (3):

$$L_{MTL} = \lambda L_{CTC} + (1 - \lambda)L_{Attention} \qquad (11)$$

with a tunable parameter λ: $0 \leq \lambda \leq 1$.

The original CTC-attention model consists of a BLSTM based encoder and an attention based decoder. CTC mechanism is used as an aligner to keep high-level feature h and output y as the same length. It dose strengthen the anti-noise ability of the model, but it is not enough for the noise intensity of ATC task. In order to solve the problem of fast speaking speed in ATC tasks, the decoder of the model needs to be strengthened accordingly. What's more, we need to prepare a language model for the special words pronunciation and grammatical features of ATC task.

3   Improved CTC-Attention Encoder-Decoder

Figure 1 shows us the architecture of improved CTC-attention encoder-decoder. Like the original one, it consists of an encoder and a decoder. The difference is that encoder adds deep CNN structure while decoder incorporates CTC mechanism to assist decoding.

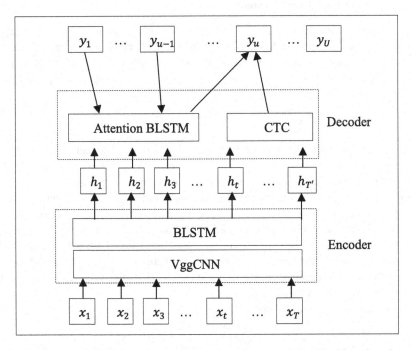

Fig. 1. The improved CTC-Attention model consists of two modules: Vggblstm based encoder and CTC-attention based decoder. Vggblstm based encoder is implemented by stacking six VggCNN layers and multiple BLSTM layers. CTC-Attention based decoder receive the output deep features from Vggblstm encoder and compute objective functions.

3.1   VggCNN and Bidirectional LSTM Based Encoder

The six layers we add in encoder are four 2D convolutional neural networks and two max-pooling layers. We use the initial layers of the VGG net architecture detailed introduced in [13]. Table 1 gives a detailed description of these 6 CNN layers.

Table 1. Architecture of the six layer VggCNN

Convolution2D	In:3	Out:64	Filter:3×3
Convolution2D	In:3	Out:64	Filter:3×3
Maxpool2D	Patch:3×3		Stride:2×2
Convolution2D	In:64	Out:128	Filter:3×3
Convolution2D	In:128	Out:128	Filter:3×3
Maxpool2D	Patch:3×3		Stride:2×2

The initial three input channels are composed of the spectral features, delta, and delta features as suggested in [20].

3.2   CTC-Attention Based Decoder

In this section, we intend to use CTC mechanism to assist attention based decoder in decoding fast and noisy speech audio. We take the CTC probabilities into account to find a better aligned hypothesis to the input speech [20]. Given speech input x, the attention based decoder will find the character sequence $\hat{C}$ with the highest probability.

$$\hat{C} = \arg\max_{C \in u^*} \{\lambda \log P_{CTC}(C|x)\} + (1 - \lambda)\log P_{att}(C|x) \tag{12}$$

During the beam search, the number of partial hypotheses for each length is limited to a predefined number, called a beam width, to exclude hypotheses with relatively low scores, which dramatically improves the search efficiency. But combining CTC and attention scores in beam search is not feasible, because the attention decoder performs it character-synchronously while CTC does it frame-synchronously. There are two methods to incorporate CTC probabilities in the score according to [20]: rescoring method and one-pass decoding. The former method rescores each hypothesis after beam searching while the latter computes the probability of each partial hypothesis using both CTC and attention model. In this paper, we borrowed the second method to improve the attention based decoder.

The attention score can be computed as

$$\alpha_{att}(g_l) = \alpha_{att}(g_{l-1}) + \log p(c|g_{l-1}, x) \tag{13}$$

Where g_l is a partial hypothesis with length l, and c is the last character of g_l, which means $g_l = g_{l-1} + c$.

The CTC score can be computed as

$$\alpha_{CTC}(g_l) = \text{logp}(g_l, \ldots | x) \tag{14}$$

$$\text{logp}(g_l, \ldots | x) = \sum\nolimits_{v \in (u \cup \{\langle eos \rangle\}) +} P(g_l * v | x) \tag{15}$$

where v denotes all possible label sequences, and <eos> represent the end of sentence. Finally, $\alpha_{CTC}(g_l)$ and $\alpha_{att}(g_l)$ can be combined together using λ according to Eq. (12).

4 Experiments

4.1 ATC Corpus

The data set we use to evaluate our system has a total of 50 h audio called Air Traffic Control (ATC) corpus. They are recorded by airport equipment and compressed. The ATC corpus has no speaker label, so speaker adaptation is not applicable. Conventional transfer learning also has little effect due to the recording equipment channel mismatch. Other challenge including strong noise, fast speaking rate and special pronunciation of words are overcome by our improved CTC-attention end-to-end system with a LSTM based language model. We divide the data into training set development set and test set according to 8:1:1.

4.2 Training Tricks

We detail the tricks we used in the improved CTC-attention end-to-end speech recognition which are inspired by [16].

L2 Regularization. L2 regularization introduced in [17] in order to reduce the risk of over-fitting in model training.

Attention Smoothing. For ATC task long context information is very important. The key idea of attention smoothing is getting longer context information in the attention mechanism. It is detailed introduced in [16].

Frame Skipping. Frame skipping is a simple and reported effective for fast model training and decoding in [16], so we borrow this idea in the training of BLSTM in both encoder and decoder.

4.3 Experimental Setup

The improved CTC-Attention based system and other systems for comparison are experimented with the Pytorch [14] backend of the ESPNET toolkit [15]. CTC and attention-based systems are implemented by setting λ to 1 and 0, respectively. For comparison, no lexicon, language model and tricks are used in all the recognition systems. Finally, we will show the result of our system which trained with those tricks and decode with LSTM based language model.

We used 80-dimensional mel-scale filterbank coefficients with 3-dimensional pitch as input features as suggested in [18–20]. The original CTC-attention system experiment setup is similar to the one introduced in a [20]. The encoder is a 4-layer BLSTM with 256 cells in each layer and direction, and linear projection layer is followed by each BLSTM layer. The decoder is a 2-layer BLSTM with 128 cells. For our improved CTC-attention system, the encoder is a 6-layer VggCNN followed with 4-layer BLSTM as same as the one in original system. And decoder is same as original, but combines CTC scores.

4.4   LSTM Based Language Model Using Character Units

The LSTM based language model probabilities are used to predict the output label jointly with the decoder network. The network of language model used in our system is a 2-layer LSTM with 650 cells. We use character as unit to train language model. The benefits are easily processing out-of-vocabulary words and reducing the workload of data preparation. The language model is trained separately using training set transcription and texts in ATC field totally about 170 thousand sentences and 3231 characters.

5   Result

Table 2 reports a comparison between CTC, attention, original CTC-attention and improved CTC-attention in terms of CER and SER. The results show that our improved CTC-attention model significantly outperformed original CTC-attention model in CER on ATC task and also better than both CTC and attention model. Our model showed 27.1–42.5% and 17.5–38.4% relative improvements on CER and SER without any tricks or language model, respectively.

Table 2. Result of experiments in CER and SER.

System	CER (%)	SER (%)
CTC	22.87	46.93
Attention	21.45	54.34
Original CTC-attention	18.04	40.53
Improved CTC-attention	13.15	33.43
With tricks	12.32	31.50
With tricks and LSTM-LM	**11.01**	**22.75**

When we use those tricks in model training, we find that it is helpful not only in the performance of system but also in accelerating training process. Figure 2 shows the CER curves of two training process: with or without those tricks. Obviously, these tricks help the algorithm train faster and improve the performance of the model. We also find that the CTC-attention model often gives poor generalization to new data without regularizations.

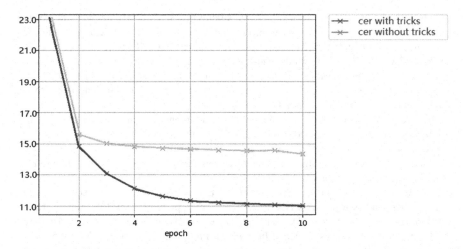

Fig. 2. The CER curves of our improved CTC-attention model with and without training tricks.

Finally, the best result is produced by our improved CTC-attention model trained with L2 regularization, attention smoothing and frame skipping and decode with the LSTM-LM describe in Sect. 4.4. It achieves a CER of 11.01% and a SER of 22.75% on the ATC test set.

6 Conclusion

In this paper, we improved the CTC-attention based end-to-end speech recognition to apply in ATC corpus. Firstly, we add VggCNN network in the encoder to improve the translation invariance of the model in time-frequency domain and enhance model robustness. Secondly, CTC mechanism was merged with attention model as the decoder. Thirdly, we also applied several tricks to improve the model training. Taken together, these tricks allow us to finally achieve a CER of 12.32% and a SER of 31.50% on the ATC dataset. While together with a LSTM language model, CER and SER reach 11.01% and 22.75%, respectively.

Acknowledgements. This work was supported by State Key Laboratory of Smart Grid Protection and Control.

References

1. Rabiner, L.R.: A tutorial on hidden Markov models and selected applications in speech recognition. Proc. IEEE **77**(2), 257–286 (1989)
2. Hinton, G., Deng, L., Yu, D., et al.: Deep neural networks for acoustic modeling in speech recognition. IEEE Sig. Process. Mag. **29**, 82–97 (2012)

3. Graves, A., Fernández, S., Gomez, F., et al.: Connectionist temporal classification: labelling unsegmented sequence data with recurrent neural networks. In: Proceedings of the 23rd International Conference on Machine Learning, pp. 369–376. ACM (2006)
4. Graves, A., Jaitly, N.: Towards end-to-end speech recognition with recurrent neural networks. In: International Conference on Machine Learning, pp. 1764–1772 (2014)
5. Hannun, A., Case, C., Casper, J., et al.: Deep speech: scaling up end-to-end speech recognition. arXiv preprint arXiv:1412.5567 (2014)
6. Miao, Y., Gowayyed, M., Metze, F.: EESEN: end-to-end speech recognition using deep RNN models and WFST-based decoding. In: 2015 IEEE Workshop on Automatic Speech Recognition and Understanding (ASRU), pp. 167–174. IEEE (2015)
7. Bahdanau, D., Cho, K., Bengio, Y.: Neural machine translation by jointly learning to align and translate. arXiv preprint arXiv:1409.0473 (2014)
8. Chorowski, J., Bahdanau, D., Cho, K., et al.: End-to-end continuous speech recognition using attention-based recurrent NN: first results. arXiv preprint arXiv:1412.1602 (2014)
9. Chorowski, J.K., Bahdanau, D., Serdyuk, D., et al.: Attention-based models for speech recognition. In: Advances in Neural Information Processing Systems, pp. 577–585 (2015)
10. Chan, W., Jaitly, N., Le, Q., et al.: Listen, attend and spell: a neural network for large vocabulary conversational speech recognition. In: 2016 IEEE International Conference on Acoustics, Speech and Signal Processing (ICASSP), pp. 4960–4964. IEEE (2016)
11. Bahdanau, D., Chorowski, J., Serdyuk, D., et al.: End-to-end attention-based large vocabulary speech recognition. In: 2016 IEEE International Conference on Acoustics, Speech and Signal Processing (ICASSP), pp. 4945–4949. IEEE (2016)
12. Kim, S., Hori, T., Watanabe, S.: Joint CTC-attention based end-to-end speech recognition using multi-task learning. In: 2017 IEEE International Conference on Acoustics, Speech and Signal Processing (ICASSP), pp. 4835–4839. IEEE (2017)
13. Simonyan, K., Zisserman, A.: Very deep convolutional networks for large-scale image recognition. arXiv preprint arXiv:1409.1556 (2014)
14. Ketkar, N.: Introduction to PyTorch. In: Deep Learning with Python, pp. 195–208. Apress, Berkeley (2017)
15. Watanabe, S., Hori, T., Karita, S., et al.: ESPnet: end-to-end speech processing toolkit. arXiv preprint arXiv:1804.00015 (2018)
16. Shan, C., Zhang, J., Wang, Y., et al.: Attention-based end-to-end speech recognition on voice search. In: 2018 IEEE International Conference on Acoustics, Speech and Signal Processing (ICASSP), pp. 4764–4768. IEEE (2018)
17. Hinton, G., Van Camp, D.: Keeping neural networks simple by minimizing the description length of the weights. In: Proceedings of the 6th Annual ACM Conference on Computational Learning Theory (1993)
18. Ghahremani, P., BabaAli, B., Povey, D., et al.: A pitch extraction algorithm tuned for automatic speech recognition. In: 2014 IEEE International Conference on Acoustics, Speech and Signal Processing (ICASSP), pp. 2494–2498. IEEE (2014)
19. Miao, Y., Gowayyed, M., Na, X., et al.: An empirical exploration of CTC acoustic models. In: 2016 IEEE International Conference on Acoustics, Speech and Signal Processing (ICASSP), pp. 2623–2627. IEEE (2016)
20. Hori, T., Watanabe, S., Zhang, Y., et al.: Advances in joint CTC-attention based end-to-end speech recognition with a deep CNN encoder and RNN-LM. arXiv preprint arXiv:1706.02737 (2017)

Revisit Lmser from a Deep Learning Perspective

Wenjin Huang, Shikui Tu$^{(\boxtimes)}$, and Lei Xu$^{(\boxtimes)}$

Department of Computer Science and Engineering,
Centre for Cognitive Machines and Computational Health (CMaCH), SEIEE School,
Shanghai Jiao Tong University, Shanghai, China
{huangwenjing,tushikui,leixu}@sjtu.edu.cn

Abstract. Proposed in 1991, Least Mean Square Error Reconstruction for self-organizing network, shortly Lmser, was a further development of the traditional auto-encoder (AE) by folding the architecture with respect to the central coding layer and thus leading to the features of Duality in Connection Weight (DCW) and Duality in Paired Neurons (DPN), as well as jointly supervised and unsupervised learning which is called Duality in Supervision Paradigm (DSP). However, its advantages were only demonstrated in a one-hidden-layer implementation due to the lack of computing resources and big data at that time. In this paper, we revisit Lmser from the perspective of deep learning, develop Lmser network based on multiple fully-connected layers, and confirm several Lmser functions with experiments on image recognition, reconstruction, association recall, and so on. Experiments demonstrate that Lmser indeed works as indicated in the original paper, and it has promising performance in various applications.

Keywords: Autoencoder · Lmser · Bidirectional deep learning

1 Introduction

Least Mean Square Error Reconstruction (Lmser) self-organizing network was first proposed in 1991 [13,14], and it is a further development of autoencoder (AE) with favorable features. Early efforts on AE can be traced back to 1980s. Three-layer networks, i.e., networks with only one hidden layer, were used to make auto-association to learn inner representations of observed signals [1,2]. In this framework, the network architecture is considered to be symmetric with the hidden layer as the central coding layer Y. The input pattern X is mapped through the encoder part to the central coding layer, while the output $\hat{X}$ is constrained to reconstruct the input via the decoder part to decode the internal representations back to the data space.

The Lmser architecture is obtained from folding AE along the central coding layer Y, and then is improved into a distributed cascading by not only constraining $X \rightarrow Y$ and $Y \rightarrow \hat{X}$ to share the same architecture, but also using the

© Springer Nature Switzerland AG 2019
Z. Cui et al. (Eds.): IScIDE 2019, LNCS 11936, pp. 197–208, 2019.
https://doi.org/10.1007/978-3-030-36204-1_16

same neurons for the two layers symmetrically paired between the encoder and decoder with respect to the central coding layer, and using the same connection weights for the bidirectional links along the directions of $X \to Y$ and $Y \to \hat{X}$. One neuron takes dual roles in encoder and decoder, and this nature is shortly called Duality in Paired Neurons (DPN), which can be regarded as adding short-cut connections between the paired neurons. Using the same connection weights for the bidirectional links is referred to $A_j = W_j^T$, where W_j is the weight matrix for layer j in the direction $X \to Y$ and A_j is the weight matrix at the corresponding layer in the direction $Y \to \hat{X}$. The matrix equality $A_j = W_j^T$ indicates that each connection weight between a pair of neurons plays a dual role for both directions, and thus this nature is shortly called Duality in Connection Weight (DCW). DCW enables Lmser to approximate identity mapping for each layer simply through $W_j A_j = A_j^T A_j \approx I$ which holds exactly for an orthogonal matrix A_j. Therefore, AE implements a direct approximation of inverse of $X \to Y$ by $Y \to \hat{X}$ in a simple cycle, whereas Lmser improves the direct cascading into a distributed cascading by DPN and DCW.

Due to the above architectural features, Lmser works in two phases, i.e., perception phase and learning phase. In perception phase, the signal propagation from two directions $X \to Y$ and $Y \to \hat{X}$ constitute a dynamic process which will approach equilibrium in a short term [14]. If the reconstruction $\hat{X}$ is not close to the input X, then Lmser enters the learning phase to update the connection weights to reduce the discrepancy between X and $\hat{X}$. Moreover, part of the central (or top) layer Y can be used for label prediction Y_L and thus supervised learning at the top by labeled data and unsupervised learning at the bottom by unlabeled data are made jointly. This nature is shortly called Duality in Supervision Paradigm (DSP). More advances about Lmser are referred to a recent review in [15].

As discussed in [13], Lmser potentially has many functions. However, due to the lack of powerful computing facility and big data at the time of 1990's, Lmser was implemented by computer simulations with only one hidden layer. It was shown that a neuron in Lmser net behaved similar to a feature detector in the cortical field during learning [13]. In recent years, some features of Lmser were also adopted in the literature. For example, stacked restricted Boltzmann machines (RBMs) [5] constrained the weight parameters to be symmetrically shared by the encoder and decoder network, while neurons in U-Net [11] and deep RED-Net [10] shared values by skip connections from the layers in the encoder to the layers in the decoder. However, whether Lmser indeed works on deep network structures and whether it is effective for those potential functions as indicated in [13], are still not systematically explored.

In this paper, we revisit Lmser by implementing it on a multi-layer network, and confirm that it indeed works as indicated in [13,14]. Since the dynamic process in the perception phase of multi-layer Lmser net can not be exactly implemented in practice, we present an effective implementation to approximate the dynamic process by updating neurons in a layer-by-layer way. Meanwhile, instead of using original Lmser learning rule to calculate gradient when updating parameters, we compute the gradients via back propagation, which is easy to implement and works well in practice. Experiments confirm several potential functions of Lmser and demonstrate its promising performance in various applications. Our contributions are summarized as follows:

- We revisit Lmser net by implementing it on multiple layers of neural networks. Our implementation can train the deep Lmser networks stably and effectively.
- Experiments are conducted on image reconstruction, generation, associative recall, and classification, in comparisons with AE as a baseline. The results not only confirm that Lmser works as indicated in the original paper, but also demonstrate its promising performance in various applications.

2   Related Work

2.1   Networks with Symmetrically Weighted Connections

A stack of two-layer restricted Boltzmann machines (RBMs) with symmetrically weighted connections was used in [6] to pretrain an AE that has multiple hidden layers in a one-by-one-layer way. AE consisting of multiple layers may have a slow convergence speed when optimizing its weights because the gradients vanishes as it propagates from the last layer of decoder to early layers in encoder. The pretrain can help to remedy this problem. The stacked RBMs worked well in dimensionality reduction.

2.2   Networks with Symmetrically Skip Connections

One recent typical example network architecture with symmetrically skip connections is the U-Net [11]. It consists of a contracting path as encoder and an expansive path as decoder, and both paths form a U-shaped architecture. The feature map from each of the layer of the contracting path was copied and concatenated with the symmetrically corresponding layer in the expansive path. Such skip connections can be regarded as a type of sharing the encoder neuron values with the correspondingly paired decoder neuron. Experiments in [11] demonstrated that U-Net worked very well for biomedical image segmentation. Such skip connections were also adopted in deep RED-Net [10]. Different from U-Net, it directly adds old feature map with present top-down signal. Feature map must be in the same size, so it only add skip connections for specific layers. It has been shown in [10] that deep RED-Net worked well in super-resolution image restoration. Moreover, with the appearance of ResNet [4] and DenseNet [7], deep neural networks with skip-connections become very popular and showed impressive performance in various applications.

3   A Brief Review of Lmser

The Lmser self-organizing net was proposed in [13,14] based on the principle of Least Mean Square Error Reconstruction (LMSER) of an input pattern. An example of Lmser architecture is demonstrated in Fig. 1(a). When Lmser net consists of multiple layers, each neuron z_k in the k-th layer receives both bottom-up signal y_k from the lower layer and top-down signal u_k from the upper layer, and is activated by their summation.

The Lmser net works in two phases, i.e., perception and learning. In the perception phase, the input pattern X triggers the dynamic process by passing the signals up from the bottom layer, while simultaneously the signals in the upper layers will be passed down to the lower layers. It has been proved that the process will converge into an equilibrium state [13]. The top-down signal to the input layer is regarded as reconstruction of the input. In the learning phase, the parameters are updated by minimizing the mean square error between the input and reconstruction. Given by Eq. (5a) in [14], the loss function for Lmser learning is:

$$J = \frac{1}{2}E(\|\overrightarrow{x} - W_1^T \overrightarrow{z_1}\|^2) \tag{1}$$

where $E(\cdot)$ denotes the expectation, W_1 is the weights of the first layer, $z_1 = s(y_1 + u_1)$ is the activity of the first layer neurons which receive both the bottom-up signals y_1 and the top-down signals u_1. As given by Eqs. (6a)–(8b) in [14], the gradients to update the network parameters are restated below:

$$\text{for} \quad k = 1, \quad \varepsilon_{i0} = x_i - u_{i0}, \varepsilon_{i1} = y_{i1} - y_{i1}^r, \quad \frac{\partial J}{\partial w_{pq1}} = \varepsilon_{q0} z_{p1} + s'_{p1} \varepsilon_{p1} x_q$$

$$\text{for} \quad k \geq 2, \quad \varepsilon_{ik} = \sum_{j=1}^{n_{(k-1)}} \varepsilon_{j(k-1)} w_{ijk}$$

$$\frac{\partial J}{\partial w_{pqk}} \approx s'_{q(k-1)} \varepsilon_{q(k-1)} z_{pk} + s'_{pk} \varepsilon_{ik} z_{q(k-1)} \tag{2}$$

4 Methods

4.1 Revisit Lmser and Implement It on Multiple Fully-Connected Layers

In order to further study the features and functions of Lmser, we need to implement it on multiple fully-connected layers. There is one technical challenge that it is hard to exactly implement the dynamic process in the perception phase and it may be time consuming to reach the convergent state. To overcome this problem, we propose an effective and simple way to approximate the dynamic process.

As shown in Fig. 1(b), we update the neurons in a layer-by-layer way. At the very beginning, a signal vector x is placed at the input layer into the network, and then it will trigger the bottom-up signal propagation through the layers one by one. In the first bottom-up pass, there is no input signal for the top layer, so at the i-th layer, the top-down signals from $(i + 1)$-th layer to the i-th layer are

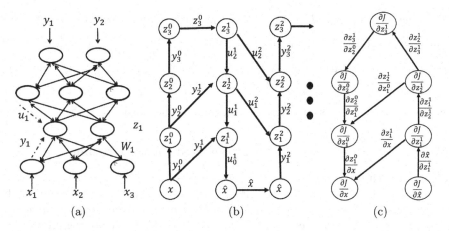

Fig. 1. (a) The architecture of multi-layer Lmser, where connection is bidirectional and symmetric, u_k is the top-down signal, y_k is the bottom-up signal, and each neuron is output as a sigmoid activation, i.e., $z_k = s(u_k + y_k)$, where $u_k = W_{k+1}^T z_{k+1}$, $y_k = W_k z_{k-1}$. (b) The reflection implementation, where z_k^t denotes the value of z_k at time t. (c) The calculation of gradients during back propagation.

initialized to be zero. After the first bottom-up pass, all neurons are given activity values. Then, the first top-down pass propagates the signals backwards and the neuron activation at the i-th layer is calculated as a sigmoid of the summation of u_i and y_i. In analogy to light reflections, we call such one bottom-up and one top-down pass as one reflection. In this way, bottom-up signal and top-down signal can be updated alternatively until they become stable. In practice, we use the Rectified Linear Units (ReLU) as the neuron activating function, and we find that instead of reaching the stable state of the updating process, one reflection followed by learning phase works well for the whole Lmser learning.

Another difference in our implementation from the original Lmser net is the learning rule. In the learning phase, for the calculation of gradient, instead of directly using Eq. (2), we compute it via an approximate back propagation. Such implementation enables us to use the available computational platform and back propagation library efficiently, and has been shown to work well in practice.

$$\frac{\partial J}{\partial W_k} = \frac{\partial J}{\partial z_k^1} \frac{\partial z_k^1}{\partial W_k} + \frac{\partial J}{\partial z_{k-1}^1} \frac{\partial z_{k-1}^1}{\partial W_k}$$

$$\frac{\partial z_k^1}{\partial W_k} = \frac{\partial z_k^1}{\partial z_{k-1}^0} \frac{\partial z_{k-1}^0}{\partial W_k} + \frac{\partial z_k^1}{\partial z_{k+1}^1} \frac{\partial z_{k+1}^1}{\partial W_k}, \quad \frac{\partial z_{k-1}^1}{\partial W_k} = \frac{\partial z_{k-1}^1}{\partial z_{k-2}^0} \frac{\partial z_{k-2}^0}{\partial W_k} + \frac{\partial z_{k-1}^1}{\partial z_k^1} \frac{\partial z_k^1}{\partial W_k}$$

$$\frac{\partial J}{\partial W_k} = \left(\frac{\partial J}{\partial z_k^1} + \frac{\partial J}{\partial z_{k-1}^1} \frac{\partial z_{k-1}^1}{\partial z_k^1} \right) \frac{\partial z_k^1}{\partial z_{k+1}^1} \frac{\partial z_{k+1}^1}{\partial W_k} \tag{3}$$

When we set reflection $T = 1$, as shown in Eq. (3), we can decompose $\frac{\partial J}{\partial W_k}$ to sum of multiplications of factors in the form of $\frac{\partial J}{\partial z_k^t}$, $\frac{\partial z_{k-1}^1}{\partial z_k^1}$, and $\frac{\partial z_l^t}{\partial W_k}$, where $\frac{\partial z_l^t}{\partial W_k}$ can be further decomposed or easily computed. Finally, $\frac{\partial J}{\partial W_k}$ is composed of factors in the form of $\frac{\partial J}{\partial z_k^t}$, $\frac{\partial z_{k-1}^1}{\partial z_k^1}$, $\frac{\partial z_k^1}{\partial z_{k-1}^0}$, and $\frac{\partial z_k^0}{\partial z_{k-1}^0}$, which can be calculated effectively in the process of back propagation, as shown in Fig. 1(c).

4.2   Jointly Supervised and Unsupervised Lmser Learning

In Lmser, each input pattern X can be recognized by a label Y_L output at the top layer, which plays the same role as a classifier. Thus, for the input with labels, Lmser can be implemented jointly both in an unsupervised manner to reduce the reconstruction error at the bottom layer and in a supervised way by minimizing the discrepancy between the predicted label Y_L and the true label. The reconstruction error will push the network to do self-organizing, which helps network to learn structure information of input and facilitates concept abstracting and formation at the top domain. The self-organizing learning directed by reconstruction error plays the role of regularization, which help network to prevent over fitting and make classification be more robust such as defense against adversarial attacks.

In perception phase, the implementation of dynamic process is the same as in Sect. 4.1, but with two parts computed at the top layer, i.e., one for predicted labels and the other as input to the decoder for top-down reconstruction. In learning phase, the loss function includes two terms for reconstruction and classification separately:

$$J = \frac{1}{2}E(\|\overrightarrow{x} - W_1^T\overrightarrow{z_1}\|^2) + L(f(x), y) \tag{4}$$

where the additional term $L(f(x), y)$ measures the error between the Lmser predicted $f(x)$ and the given label y.

5   Experiments

In this section, we demonstrate the effectiveness and strengths of the deep Lmser learning by some promising results on image recognition, reconstruction, generation, and associative recall. We use Lmser(un) to denote the Lmser network trained only in the unsupervised manner, use Lmser(un-n) to denote the Lmser(un) by removing DPN, use Lmser(sup) to denote the Lmser network trained jointly in the supervised and unsupervised manner.

5.1   Datasets and Experimental Settings

We evaluate Lmser on two benchmark datasests: MNIST [9] and Fashion-MNIST [12].

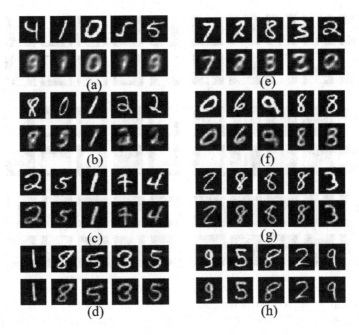

Fig. 2. Examples of reconstructed images on MNIST. (a) AE, $\tau = 500$; (b) Lmser(un-n), $\tau = 500$; (c) Lmser(un), $\tau = 500$; (d) Lmser(sup), $\tau = 500$; (e) AE, $\tau = 5000$; (f) Lmser(un-n), $\tau = 5000$; (g) Lmser(un), $\tau = 5000$; (h) Lmser(sup), $\tau = 5000$. (τ: the number of training iterations.)

- The MNIST contains $60,000$ training images and $10,000$ testing images. The images are handwritten digits from 250 people. Each picture in the data set consists of 28×28 pixels, each of which is represented by a gray value.
- Fashion-MNIST (F-MNIST) is a MNIST-like dataset which shares the same image size and structure of training and testing splits. F-MNIST is served as a replacement for the original MNIST dataset for benchmarking machine learning algorithms.

For MNIST and F-MNIST, we use 4 layers with the numbers of neurons $[10, 100, 300, 784]$ for Lmser based on fully-connected layers. We train every network model with Adam (Adaptive moment estimation) optimiser method [8]. In the training process, the batch size is set to be 50 and the adaptive learning rate is set to be 0.01 with decay rate 0.9999.

5.2   Reconstruction

We evaluate the reconstruction performance on MNIST dataset and F-MNIST dataset. Lmser is a development of AE, thus for comparisons, AE is used as a baseline. Detailed reconstruction errors are summarized in Table 1. Examples of the reconstructed results are given in Fig. 2 for different number of training iterations.

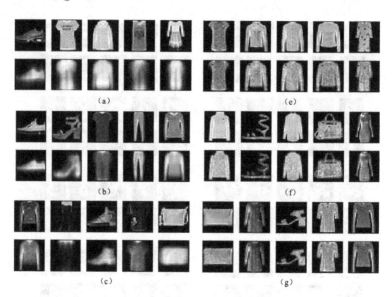

Fig. 3. Examples of reconstructed images on Fashion-MNIST. (a–c) AE, $\tau = 1000, 5000, 10000$; (e–g) Lmser(un), $\tau = 1000, 5000, 10000$. (τ: the number of training iterations)

Table 1. Reconstruction error on MNIST, where τ denotes the number of training iterations.

Model	$\tau = 500$	$\tau = 5000$	$\tau = 20000$
AE	0.071	0.036	0.020
Lmser(un-n)	0.030	0.021	0.016
Lmser(un)	0.0067	0.0018	0.0006
Lmser(sup)	0.0078	0.0025	0.0007

It can be observed that Lmser converges faster with smaller reconstruction errors than AE, and Lmser(un) is better than Lmser(un-n), which suggests that shortcuts between paired neurons play a significant role in reconstruction performance. Lmser(sup) is slightly worse than Lmser(un), but still much better than Lmser(un-n) and AE, indicating that there is a trade-off between the classification and reconstruction. Moreover, we also test the reconstruction performance on F-MNIST. Examples are shown in Fig. 3. Obviously, Lmser converges faster and is better than AE, which is consistent to the observations from Fig. 2.

5.3 Recognition

In this section, we investigate the classification performance of Lmser(sup), which is trained jointly in supervised manner and unsupervised manner. The classification accuracies of Lmser(sup) and a fully-connected feedforward network (FCN)

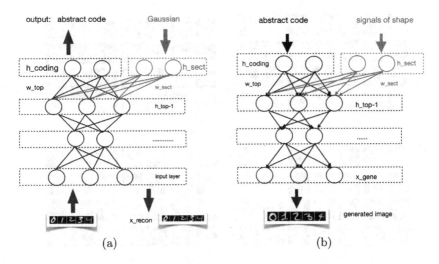

Fig. 4. Structure of supervised Lmser for image generation: (a) training stage; (b) generation stage

are comparably high, both over 98%. When there exist adversarial attacks at intensity level 0.3 from the Fast Gradient Sign Method (FGSM) [3], a well-known adversarial attack method, the classification accuracy of FCN drops down to 2.68%, while Lmser(sup) still gets 31.16%.

In this section, we investigate the performance of deep Lmser learning in image generation by manipulating the latent code to get different styles of handwritten digit numbers and clothes. In addition to the categorical coding units, we add two more hidden coding units h_{sect}^1 and h_{sect}^2 to the top layer of Lmser(sup) to get different styles of handwritten digits, as shown in Fig. 4. We assume that the two additional coding units are independent and follow Gaussian distributions. In practice, when we train the model Lmser(sup), the means of the two additional coding units are computed by encoder. Before they are fed into the decoder, we perturb them by a zero-mean Gaussian. When generating digital numbers, as shown in Fig. 4(b), the decoder will generate digital numbers according to not only the label by the categorical coding units but also randomly sampled style codes by the two additional hidden coding units. The style codes may also be assigned at specific values. Figure 5(e) show that when assigning the coding unit with different values, the digital number is gradually changing in different styles.

5.4   Generation

For the input patterns without labels, no labels can be used to guide the separation of categorical information and non-categorical styles. Lmser is still able to form a self-organized top coding domain, which preserves the neighbourhood relations and topological similarities. By manipulating one coding unit in the top

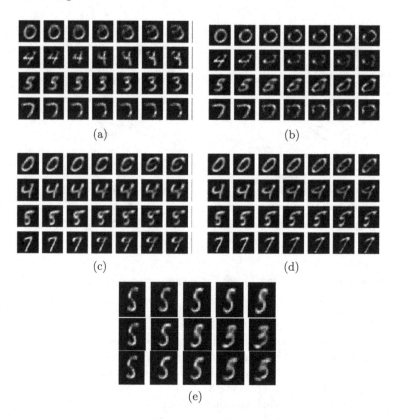

Fig. 5. Generated digital numbers by Lmser(un). There are 10 codes in the latent space, where Z_i represent the i-th coding unit. (a) manipulating Z_3; (b) manipulating Z_0; (c) manipulating Z_5; (d) manipulating Z_9. (e) Generated digital number by Lmser(sup-n) with changing two coding units. Top: manipulating h^1_{sect}; middle: manipulating h^2_{sect}; bottom: manipulating both the two units. It seems that the two units can control different shape changing.

layer with others fixed, we can observe pattern changing smoothly as the coding value varies, which indicates that the coding regions are well self-organized and clustered. With such property, we are able to synthesize images in a controllable way or in a creative way for novel synthesis and reasoning. Figure 5(c) shows that when specifying the coding unit with values from 0 to 10, the digital number is gradually changed to have a circle created over the head, while in Fig. 5(d) another coding unit seems to control the tilt degree of generated images.

5.5   Association

For the incomplete input, Lmser is able to recover the missing part by associative memory from the observed part. This function is related to tasks such as associative recall of recover the whole image with some key parts blocked.

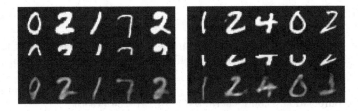

Fig. 6. Examples of associative recall from partial mnist images. (top) the ground-truth images; (middle) partial images with half blocked; (bottom) associative recall by Lmser(un-n).

Specifically, the observed part of the image is fed into the Lmser network, and triggers the bottom-up signals passing to the top layers. Then, the activated neurons passing the top-down signals back to the bottom layer to give a complete image, which actually recovers unobserved part based on what has been learned by the network layers. The sharing connection weights by both directions of the links between the consecutive layers enables Lmser to catch invertible structures under input patterns for restoring these structures under partial input.

We train the Lmser net on the MNIST dataset, and then feed the masked images into the model for the output of reconstructed complete images. In practice, we have found that shortcuts between paired neurons is helpful in reconstruction, but not so beneficial in associative memory, because it can also pass hole to the low layers of decoder when it pass the detailed information from encoder to decoder. Therefore, we only present the results by the Lmser net without using the paired neurons for both encoder and decoder in Fig. 6. Examples of the half blocked images are given in the middle row in Fig. 6, and the corresponding reconstructed complete digits from the associative memory by Lmser are shown in the bottom row, which are similar to the ground-truth images in the top row. The results demonstrate that Lmser is promising for this task of associative recall from partial input.

6   Conclusion

In this paper, we have revisited the Lmser network for the practical implementation of multiple layers, and confirmed that several of its potential functions indeed work effectively via experiments on image recognition, reconstruction, associative memory, and generation. Experiments demonstrate that Lmser not only works as preliminarily discussed in the original paper, but also is promising in various applications. It deserves further investigations on Lmser in the future for its improvements and comparisons with state-of-the-art methods in many real applications.

Acknowledgement. This work was supported by the Zhi-Yuan Chair Professorship Start-up Grant (WF220103010), and Startup Fund (WF220403029) for Youngman Research, from Shanghai Jiao Tong University.

References

1. Ballard, D.H.: Modular learning in neural networks. In: Proceedings of the Sixth National Conference on Artificial Intelligence, AAAI 1987, vol. 1, pp. 279–284 (1987)
2. Bourlard, H., Kamp, Y.: Auto-association by multilayer perceptrons and singular value decomposition. Biol. Cybern. **59**(4–5), 291–294 (1988)
3. Goodfellow, I., Shlens, J., Szegedy, C.: Explaining and harnessing adversarial examples. In: International Conference on Learning Representations (2015). http://arxiv.org/abs/1412.6572
4. He, K., Zhang, X., Ren, S., Sun, J.: Deep residual learning for image recognition. In: Proceedings of the IEEE Conference on Computer Vision and Pattern Recognition, pp. 770–778 (2016)
5. Hinton, G.E., Osindero, S., Teh, Y.W.: A fast learning algorithm for deep belief nets. Neural Comput. **18**(7), 1527–1554 (2006)
6. Hinton, G.E., Salakhutdinov, R.R.: Reducing the dimensionality of data with neural networks. Science **313**(5786), 504–507 (2006)
7. Huang, G., Liu, Z., Weinberger, K.Q.: Densely connected convolutional networks. CoRR abs/1608.06993 (2016). http://arxiv.org/abs/1608.06993
8. Kingma, D.P., Ba, J.: Adam: a method for stochastic optimization. CoRR abs/1412.6980 (2014). http://arxiv.org/abs/1412.6980
9. LeCun, Y., Cortes, C., Burges, C.: MNIST handwritten digit database, vol. 2. AT&T Labs (2010). http://yann.lecun.com/exdb/mnist/
10. Mao, X., Shen, C., Yang, Y.B.: Image restoration using very deep convolutional encoder-decoder networks with symmetric skip connections. In: Advances in Neural Information Processing Systems, pp. 2802–2810 (2016)
11. Ronneberger, O., Fischer, P., Brox, T.: U-Net: convolutional networks for biomedical image segmentation. In: Navab, N., Hornegger, J., Wells, W.M., Frangi, A.F. (eds.) MICCAI 2015. LNCS, vol. 9351, pp. 234–241. Springer, Cham (2015). https://doi.org/10.1007/978-3-319-24574-4_28
12. Xiao, H., Rasul, K., Vollgraf, R.: Fashion-MNIST: a novel image dataset for benchmarking machine learning algorithms. CoRR abs/1708.07747 (2017)
13. Xu, L.: Least MSE reconstruction for self-organization: (i)&(ii). In: Proceedings of 1991 International Joint Conference on Neural Networks, pp. 2363–2373 (1991)
14. Xu, L.: Least mean square error reconstruction principle for self-organizing neural-nets. Neural Netw. **6**(5), 627–648 (1993)
15. Xu, L.: An overview and perspectives on bidirectional intelligence: Lmser duality, double IA harmony, and causal computation. IEEE/CAA J. Autom. Sin. **6**(4), 865–893 (2019). https://doi.org/10.1109/JAS.2019.1911603

A New Network Traffic Identification Base on Deep Factorization Machine

Zhenxing Xu[1], Junyi Zhang[1], Daoqiang Zhang[1(✉)], and Hanyu Wei[2(✉)]

[1] College of Computer Science and Technology,
MIIT Key Laboratory of Pattern Analysis and Machine Intelligence,
Nanjing University of Aeronautics and Astronautics, Nanjing 211106, China
dqzhang@nuaa.edu.cn
[2] Network Technology Laboratory, Huawei Technologies Co Ltd.,
Nanjing, China
weihanyu@huawei.com

Abstract. Effective network traffic identification has important significance for network monitoring and management, network planning and user behavior analysis. In order to select and extract the most effective attribute as well as explore the inherent correlation between the attributes of network traffic. We proposed a new network traffic identification method based on deep factorization machine (DeepFM) which can classify and do correlation analysis simultaneously. Specifically, we first embed the feature vector into a joint space using a low-rank matrix, then followed by a factorization machine (FM) which handle the low-order feature crosses and a neural network which can learn the high-order feature crosses, finally the low-order feature crosses and high-order feature crosses are fused and give the classified result. We validate our method on Moore dataset which is widely used in network traffic research. Our results demonstrate that DeepFM model not only have a strong ability of network traffic identification but also can reveal some inherent correlation between the attributes.

Keywords: Traffic identification · Neural networks · Factorization machine

1 Introduce

Nowadays, computer network technology develops rapidly, the number of Internet users continues growing. Also, there comes out more and more new network application and network services, which results in a sharp increase in the amount of network traffic and the complexity of network communication protocol. Meanwhile, some network applications occupy a large amount of bandwidth [1], which greatly reduces the performance of the network. Low security applications have information leakage and other security risks. Moreover, some malicious applications endanger the Internet intentionally. Jointly lead to great challenges on Internet management.

Network traffic identification has important application value in network monitoring and management, network planning, user behavior analysis and so on. Effective

Z. Cui et al. (Eds.): IScIDE 2019, LNCS 11936, pp. 209–218, 2019.
https://doi.org/10.1007/978-3-030-36204-1_17

identification of network traffic can not only reduce network security risks, optimize network configuration, but also analyze user behavior according to traffic characteristics to improve service quality.

Traditional network traffic identification methods mainly use port number-based method, deep packet inspection method and host behavior-based method, but these methods have some limitations [2–4]. Especially with the growing of new applications, as well as the extensive use of dynamic port numbers technology and traffic encryption technology, the accuracy of these network traffic identification methods decrease significantly. In recent years, the network traffic identification methods based on statistical characteristics of network traffic has received extensive attention. In network communication, the traffic characteristics produced by different kinds of network applications are quite different. For example, traffic in a VoIP application can be characterized by a small number of bytes over a long period of time, while traffic in a Web application can be characterized by a large number of bytes over a short period of time. The method based on statistical features does not care about the local characteristics of network traffic, nor does it need to disassemble data packets and byte matching. Instead, it takes its statistical characteristics as discrimination, and then combines machine learning algorithm to train classification model to achieve effective identification of network traffic with relatively low computational complexity. Research shows that this method can obtain relatively high recognition accuracy. Widely used machine learning algorithms include SVM, Decision Tree, Bayesian, K-means and Neural Network.

The statistical characteristics of network traffic are usually high-dimensional feature vectors. Among many high-dimensional network traffic characteristics, how to select and extract the most effective features for network traffic identification and whether there are different forms of feature vector for different network applications identification are all problems to be solved. Tong et al. [5] selected and combined network traffic characteristics manually, and then used C4.5 decision tree algorithm to classify eight applications. Williams et al. [6] trained and compared classification accuracy of 5 kinds of machine learning algorithms using NLANR dataset in which 22 kinds of characteristics are manually pre-selected. Kornycky et al. [7] created a WLAN dataset containing multiple traffic types, and manually extracted 63 traffic characteristics for training 6 different machine learning algorithms. PCA is a method of analyzing high-dimensional data, which can produce a new set of basis. When the data is projected on the new basis, it can effectively measure differences between the data. Cao et al. [8] use PCA to extract features of network traffic and reduce the dimension of original traffic characteristics to reduce the computational cost. However, PCA takes all network traffic as a whole and ignores label information, which is important classification information.

The previous work either focuses on the improvement of network traffic classification method or on the selection of a subset of network traffic attributes, the way of manual combination and selection of traffic characteristics is also adopted. Deep learning model has achieved great success recently. While deep learning can avoid manual feature engineering, it is close to a black-box model with poor interpretability, which has great demand for data, high training cost and long training time. Although great progress has been made in deep network compression in recent years, deployment and application of deep network have great limitations in the face of strong real-time and large-scale network traffic.

The network traffic identification method based on DeepFM that we d can implicitly learn network, and extract or combine features from training data, which not only avoids the trouble of manual feature extraction, but also reduces the difference of feature extraction between different machine learning algorithms, and has a very high accuracy of network traffic identification. Compared with the deep learning model, DeepFM has the advantage of high interpretability. The model trained by training data can reflect the relationship between features. Compared with the general linear model, DeepFM considers not only each feature, but also the correlation between attributes. In practice, a large number of features are correlative. If we can find out these correlative features, it is obvious that our model will be more realistic and meaningful. In addition, the sudden, long tail and uncertainty of network traffic will lead to the characteristics of network traffic in different geographical locations or different time periods have extreme variation, which will cause many missing values of attribute. Missing values often require implicit generative models to fill in or estimate. Semi-supervised learning, weakly supervised learning, multicentric fusion, and domain adaptation are used to solve it. But we point out that DeepFM implicitly solves the above problems elegantly. The specific reasons of which will be explained in the next section.

Our main contributions are summarized as follows. Firstly, the method based on DeepFM is used to solve the problem of network traffic identification for the first time, moreover it achieves high accuracy and has strong interpretability. Secondly, we find the potential use of DeepFM in the discovery of attribute correlation and low-dimensional embedding.

The rest of this article will be arranged as follows. In the second section, we will introduce the principles of FM and DeepFM models, and view the essence of these two models from a new perspective. In the third section, we will validate our views on Moore dataset, and prove that DeepFM is very suitable for network traffic identification. Finally, in the fourth section, we made a brief summary.

2 Method

The DeepFM model that we propose is shown in Fig. 1. It includes an embedding layer, a traditional FM module and a deep neural network module. FM and deep neural network module share the output of embedding layer and influence the embedding layer in the process of backward propagation.

Traditional linear models consider each feature individually, but do not consider the relationship between features and features. In fact, certain combinations of attributes are effective discriminator. In addition, it is difficult to evaluate the impact of each feature on the results directly in high-dimensional situations. Linear model is generally expressed in Eq. (1), where n denotes the number of features of the sample and x_i denotes the i-th eigenvector, $w_0 \in R, w_i \in R_n$.

$$y = w_0 + \sum_{i=1}^{n} w_i x_i \tag{1}$$

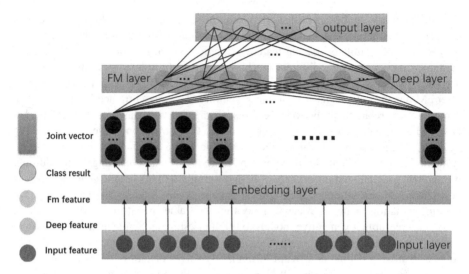

Fig. 1. Overview of the DeepFM model

It can be seen from Eq. (1) that the traditional linear model does not take into account the correlation between features and features. It is obviously meaningful to use the polynomial model to express the related features. For the case of second-order feature crosses, the combination of features x_i and x_j is represented by $x_i x_j$, and the second-order polynomial model is expressed by Eq. (2).

$$y = w_0 + \sum_{i=1}^{n} w_i x_i + \sum_{i=1}^{n} \sum_{j=i+1}^{n} w_{ij} x_i x_j \tag{2}$$

It can be seen that, compared with Eq. (1), Eq. (2) adds a part of second-order feature crosses, and the coefficients of this part are as much as $n(n-1)/2$. When the feature dimension is high, the time complexity of the model will increase greatly. Factorization machine (FM) [9] is a machine learning algorithm based on matrix decomposition proposed by Steffen Rendle. In order to solve w_{ij}, auxiliary vectors $V_i = [v_{i1}, v_{i2}, \ldots, v_{ik}]^T$ are introduced for each eigenvector x_i, the coefficient w_{ij} of feature crosses is estimated by $V_i V_j^T$, that is, $w_{ij}^* = V_i V_j^T$. So that, Eq. (2) can be expressed as Eq. (3), where $V \in R^{n*k}$, $\langle V_i, V_j^T \rangle$ denotes the inner product of the auxiliary vectors V_i and V_j, whose dimensions are k, and $k \in N^+$ is called a hyperparameter. Thus, the number of coefficients of the second-order feature crosses is reduced from $n(n-1)/2$ to $n*k$. In addition, the factorization makes the coefficients of $x_i x_h$ and $x_i x_h$ correlate with the eigenvector of x_h and are no longer independent of each other.

$$y = w_0 + \sum_{i=1}^{n} w_i x_i + \sum_{i=1}^{n} \sum_{j=i+1}^{n} <V_i, V_j> x_i x_j \tag{3}$$

From the above equation, it can be seen that the second-order correlation coefficients between attribute features are actually determined by the inner product of the auxiliary vectors corresponding to each attribute feature. We learn a k-dimensional auxiliary vector V_k for each input feature, and define their inner product as the correlation between them. It can be seen that when the auxiliary vectors corresponding to x_i and x_j are perpendicular to each other, the attributes x_i and x_j are independent of each other. When two auxiliary vectors are collinear with each other, the degree of correlation of attributes x_i and x_j is greatest. The learning process of embedding layer by gradient descent method can mine the correlation between attribute features in data. In fact, this process is to learn a linear low-dimensional embedding, which maps the original input features into a low-dimensional common auxiliary vector space, and eliminates the differences in the representation of attributes, and further consider the correlation between attributes. We point out that the key point of FM model is the embedding layer. Equation (3) shows that compared with the traditional linear model, FM model actually trains a set of embedding matrices composed of $[v_1, v_2, \ldots, v_n]$. And FM model only considers the projection relationship between two eigenvectors in a common space. The inner product part of Eq. (3) is projection in essence.

Although FM can achieve high-order feature crosses theoretically, in fact, it only uses second-order feature crosses because of the high computational cost. For different network applications, the low-order crosses and high-order crosses of network traffic statistical features may have an different impact on the final identification results. Therefore, second-order combinations of attributes are not enough. In DeepFM model, deep neural network is used to solve high-order feature crosses. Neural network has good self-learning ability, and the learning process is divided into forward propagation and back propagation [10]. Under the action of the corresponding weight and activation function, the input data passes through the hidden layer from the input layer and finally reaches the output layer to complete the forward propagation. Then the error between the output of the output layer and the expected values are calculated, and the network parameters are adjusted by back propagation. The error is reduced to the specified accuracy through the iterative learning.

We emphasize that the previous work on network traffic identification rarely involves the discovery of attribute correlation. In DeepFM model, FM and deep neural network modules share the output of the embedding layer, and influence the embedding layer through backward propagation. The embedding layer transforms the original high-dimensional sample space into the low-dimensional common auxiliary vector space, which not only eliminates the influence of the representation difference of the original input features on the training, but also enables the model to have a strong ability to extract the feature crosses. Traditional dimensional reduction methods usually learn a low-dimensional common feature space, extract only one or several attributes from the correlative attributes, and regard other attributes in feature crosses as redundant information, while those feature crosses may have strong discrimination ability. But DeepFM model can not only maximize the correlation between feature crosses and class by inner product of auxiliary vectors, but also quantify the correlation between attributes by comparing the cosine of the angle between auxiliary vectors. In addition, the low-dimensional auxiliary vector space also reduces the input width of FM and deep neural network modules, and reduces the spatial complexity and time complexity of the model.

The network traffic identification method based on deep factorization machine learn the low-order feature crosses separately by FM module firstly, then deep neural network is used to learn the high-order feature crosses. The low-order feature crosses and high-order feature crosses are combined by a parallel structure. The features of network traffic are extracted and combined implicitly, and finally the network traffic identification is achieved. As shown in Fig. 1, the DeepFM model is divided into two parts: FM and neural network. FM module extracts low-order feature crosses, and neural network module extracts high-order feature crosses. Finally, the prediction values of FM module and neural network module are combined at the output layer, and the final results are obtained.

FM module consists of three parts: input layer, FM layer and output layer. FM layer consists of addition and inner product terms, the addition term is the first-order representation of input features, and inner product term is the second-order combination representation of input features, corresponding to $\sum_{i=1}^{n} w_i x_i$ and $\sum_{i=1}^{n} \sum_{j=i+1}^{n} \langle V_i, V_j \rangle x_i x_j$ in Eq. (3), respectively. The output of FM module is marked as y_{FM}, which indicates the influence of low-order feature crosses on the final identification result.

The deep neural network module consists of three parts: input layer, hidden layer and output layer. The features of the input layer are shared with FM module. Under the operation of the corresponding weights and activation functions, the input data passes through the hidden layer from the input layer and finally reaches the output layer to obtain higher-order feature crosses. The output of the deep neural network module is marked as y_{deep}, which indicates the influence of low-order feature crosses on the final identification result.

The low-order feature crosses were learned by FM module, then neural network was used to learn the high-order feature crosses, the low-order feature crosses and high-order feature crosses were combined by a parallel structure [11]. The prediction values of FM module and deep neural network module are combined at the output layer according to Eq. (4).

$$y^* = sigmoid(y_{FM} + y_{Deep}) \tag{4}$$

3   Experiment

3.1   DataSet and Processing of Missing Values and Imbalanced Data

We adopt the authoritative Moore dataset which is authoritative in network traffic identification task. As shown in Table 1, Moore dataset contains WWW, P2P, MAIL and so on adding up to 10 kinds of application flow. Every flow in Moore dataset is described as a feature vector which consist of average interval time of packages, max bytes of packages and so on. As shown in Table 2. There are 248 features for a specified flow.

There are 248 attributes in the Moore data set, and the label is the application layer protocol. There exists a small amount of data in the training dataset which have missing attributes. For different attribute missing, we have adopted the following strategies:

1. Mark missing: The missing data accounts for about 0.04%. If we combine the techniques of semi-supervising or weakly labeling (15), we can utilize this part of data. Since its proportion is very small in the training dataset, we directly discard this part of the data.
2. Boolean attribute missing: The attributes with numbers 65, 66, 67, 68, 71, 72, 102 are Boolean attributes. The No. 102 attribute is missing for about 0.002%. We discard the missing values directly. For the rest, we convert the Boolean type to a discrete tag of 0, 1, 2 representing False, True and Unknown, respectively.
3. Numeric attribute missing: For attributes whose missing values account for over 0.01%, we use the interpolation method. Mean filling is used for the rest attributes.

As shown in Fig. 1, the Moore dataset is an extremely unbalanced dataset. Top-5 adds up to 98.79% of the total dataset while Top-1 occupies 86.91% of the total dataset. There are many ways to handle imbalances. Over-sampling, under-sampling and even sample generation can be applied. In the training process, reasonable sample weights can be set according to the proportion of categories in the training set. In the partition stage, we sample 80% from each category as the training set, leaving 20% as the test set. Finally, applying the weight to the final cross-entropy loss function.

3.2 Metrics

When evaluating the performance of the classification model, the concept of the confusion matrix is usually given according to the actual value and the predicted value of the label. Each evaluation parameter can be described by a mathematical formula [12].

Table 1. Composition of the Moore dataset

Label	Application	Flow count	Proportion
1	WWW	328091	86.910
2	BULK	11539	3.056
3	MAIL	28567	7.567
4	DATABASE	2648	0.701
5	SERVICES	2099	0.556
6	P2P	2094	0.555
7	ATTACK	1793	0.475
8	MULTIMEDIA	1152	0.305
9	INTERACTIVE	110	0.029
10	GAMES	8	0.002

The confusion matrix is shown in Table 3. Every classifier is trained on the Moore dataset, then we calculate the confusion matrix for each classifier. Based on the confusion matrix of label, Accuracy, Precision, Recall, and F1 can be calculated from the table according to formula (5)–(8). Obviously, for the k classification problems, we need to calculate k times. Finally, we use macro average and weighted average to get the global metrics.

$$A = \frac{TP + TN}{TP + FN + FP + TN} \tag{5}$$

$$P = \frac{TP}{TP + FP} \tag{6}$$

$$R = \frac{TP}{TP + FN} \tag{7}$$

$$F1 = 2 \times \frac{P \times R}{P + R} \tag{8}$$

Table 2. Some features of Moore dataset.

Label	Feature abbreviation	Feature description	Direction
1	Mean-IAT	Average interval time of packages	——
2	Max-IAT_ b a	Max interval time of packages	Server -> Client
3	Total-packets_ a b	Count of packages	Client -> Server
4	Mean-data-ip	Average number of bytes of IP packets	——
5	Max-data-ip	Maximum number of bytes of IP packets	——
6	Duration	Connection duration	——

Table 3. Confusion matrix

	PREDICTED true	Predicted false	Total
Actual true	True positive (TP)	False negative (FN)	TP + FN
Actual false	False positive (FP)	True negative (TN)	FP + TN
Total	TP + FP	FN + TN	TP + FN + FP + TN

3.3   Experimental Results and Analysis

From the Table 4. We can find that DeepFM performs better than FM, NaiveBayes, SVM and deep neural network. FM module extracts low-order feature crosses, while the neural network module extracts high-order feature crosses. In contrast, the DeepFM model is divided into two parts: FM and neural network. FM module extracts low-order feature crosses, and neural network module extracts high-order feature crosses. So, DeepFM combines the two approaches to get better result.

Table 4. Classfication metrics

Metrics	SVM	NaiveBayes	FM	Neural network	DeepFM
Weight-precision	0.92	0.93	0.98	1	1.0
Weight-recall	0.98	0.77	0.98	0.99	1.0
Weight-F1-score	0.92	0.83	0.98	1	1.0
Macro-precision	0.40	0.47	0.48	0.59	0.62
Macro-recall	0.37	0.64	0.51	0.72	0.80
Macro-F1	0.38	0.49	0.49	0.62	0.64

The remarkable fact is that all the approaches have a gap in metrics between weight-metrics and macro-metrics, the extremely imbalance dataset in which Top-5 adds up to 98.79% of the total dataset while Top-1 occupies 86.91% of the total dataset, may be the reason for this phenomenon. Feature work may pay more attention to solve this weakness.

4   Conclusion

A network traffic identification based on deep factorization machine (DeepFM) is proposed. Specifically, considering the relationship between attributes, the combinations of low-order statistical features of the network flow are extracted by the factoring machine FM module, and then apply the neural network module to learning the combination of high-order statistical features of the network flow simultaneously, implicitly extracting and combining the statistical features of the network traffic flow. Finally, the predicted value of the FM module and the deep neural network module are merged at the output layer to identify network traffic. The Moore datasets commonly used in the field of traffic identification are used for testing experiments. The results show that DeepFM-based identification method can improve the accuracy of network traffic identification.

References

1. Dong, S., Zhou, D.D., Zhou, W., et al.: Research on network traffic identification based on improved BP neural network. Appl. Math. Inf. Sci. **7**(1), 389–398 (2013)
2. Madhukar, A., Williamson, C.: A longitudinal study of P2P traffic classification. In: 14th IEEE International Symposium on Modeling, Analysis, and Simulation, pp. 179–188. IEEE, Monterey (2006)
3. Ma, J., Levchenko, K., Kreibich, C., et al.: Unexpected means of protocol inference. In: 6th ACM SIGCOMM Conference on Internet Measurement, pp. 313–326. ACM, Rio de Janeiro (2006)
4. Hurley, J., Garcia-Palacios, E., Sezer, S.: Host-based P2P flow identification and use in real-time. ACM Trans. Web **5**(2), 1–27 (2011)
5. Tong, D., Qu, Y.R., Prasanna, V.K.: Accelerating decision tree-based traffic classification on FPGA and multicore platforms. IEEE Trans. Parallel Distrib. Syst. **28**(11), 3046–3059 (2017)

6. Williams, N., Zander, S., Armitage, G.: A preliminary performance comparison of five machine learning algorithms for practical IP traffic flow classification. ACM SIGCOMM Comput. Commun. Rev. **36**(5), 5–10 (2006)

7. Kornycky, J., Abdul-Hameed, O., Kondoz, A., et al.: Radio frequency traffic classification over WLAN. IEEE/ACM Trans. Network. **25**(1), 56–68 (2016)

8. Cao, J., Fang, Z., Qu, G., et al.: An accurate traffic classification model based on support vector machines. Int. J. Netw. Manag. **27**(1), 1962 (2017)

9. Rendle, S.: Factorization machines. In: 10th IEEE International Conference on Data Mining, pp. 995–1000. IEEE, Sydney (2010)

10. Sedki, A., Ouazar, D., El Mazoudi, E.: Evolving neural network using real coded genetic algorithm for daily rainfall–runoff forecasting. Expert Syst. Appl. **36**(3), 4523–4527 (2009)

11. Guo, H., Tang, R., Ye, Y., et al.: DeepFM: a factorization-machine based neural network for CTR prediction (2017)

12. Zhu, W., Zeng, N., Wang, N.: Sensitivity, specificity, accuracy, associated confidence interval and ROC analysis with practical SAS implementations. In: NESUG Proceedings: Health Care and Life Sciences, vol. 19, p. 67 (2010)

3Q: A 3-Layer Semantic Analysis Model for Question Suite Reduction

Wei Dai[1], Siyuan Sheni[2], and Tieke Hei[2(✉)]

[1] Electric Power Research Institute, State Grid Jiangsu Electric Power Company,
Nanjing 210003, People's Republic of China
152955205920139.com
[2] State Key Laboratory for Novel Software Technology, Nanjing University,
Nanjing 210093, People's Republic of China
hetieke@gmail.com

Abstract. Question generation and question answering are attracting more and more attention recently. Existing question generation systems produce questions based on the given text. However, there is still a vast gap between these generated questions and their practical usage, which acquires more modification from human beings. In order to alleviate this dilemma, we consider reducing the volume of the question set/suite and extracting a lightweight subset while conserving as many features as possible from the original set. In this paper, we first propose a three-layer semantic analysis model, which ensembles traditional language analysis tools to perform the reduction. Then, a bunch of metrics over semantic contribution is carefully designed to balance distinct features. Finally, we introduce the concept of *Grade Level* and *Information Entropy* to evaluate our model from a multi-dimensional manner. We conduct an extensive set of experiments to test our model for question suite reduction. The results demonstrate that it can retain as much diversity as possible compared to the original large set.

Keywords: Question answering · Suite reduction · Grade level

1 Introduction

Existing Question Generation (QG) technologies can automatically generate large quantities of questions based on text from reading comprehension scenarios. Such QG techniques indicate potential contributions in the education field for the examination of understanding or construction of chat robots. However, in [4], we can see that there is still a considerable gap in the sense of readability and acceptability between machine-generated questions and human beings. For some expressions in the text, QG models are likely to produce obscure questions, which is unacceptable in reading comprehension practice. In that way, these questions need to be reorganized and modified manually before putting in use. Considering the high labor cost and time spent, we endeavor to propose a

© Springer Nature Switzerland AG 2019
Z. Cui et al. (Eds.): IScIDE 2019, LNCS 11936, pp. 219–231, 2019.
https://doi.org/10.1007/978-3-030-36204-1_18

feasible question reduction scheme to lessen the pressure of manual operation. With question suite reduction, workers will be assigned with fewer questions, and this little suite can contribute almost the same effect in examining the students.

On the other hand, question sets are common inputs for question answering (QA) models. If we can filter the question set while achieve the same test effect as the original set with a lightweight subset, we can significantly shorten the iteration training time of QA model and accelerate the researching process. In summary, high-quality streamlined sets are essential for many cases.

As an examination item of reading comprehension, it is necessary for the chosen questions to cover different grammatical structures and deal with different parts of speeches. In seeing this, we introduce a large number of traditional semantic analysis tools, including part-of-speech parsing [1], dependency syntactic parsing [8] and semantic role labelling [5]. We apply these analysis schemes synthetically and construct our own three-layer semantic analysis model.

With the help of the analysis model, a large number of statistical data can be obtained. And based on this, a semantic contribution metric for balancing and preserving features is proposed to measure the characteristic of each sentence.

Our following task is to validate and evaluate the efficacy of our work. Since there is no universal criterion for sentences' feature, we first evaluate the changes in the necessary information and then introduce the way of experimental proof to indicate the effect of selection. In the context of QA, the accuracy of the outcome of the same given training model is used to measure the degree of retention between selection sets and universe set, thus reflecting the degree of retention of features. Besides, the value of information entropy in evaluating diversity has also been noticed. By introducing the concept of information entropy in information theory, we propose our scheme of estimating the value of entropy. The change of information entropy can reflect the complexity of information, that is, the complexity of features, laterally. Finally, the classification of reading ability has attracted our attention. Grade Level formula [9] is introduced to evaluate our scheme more comprehensively.

The main contributions of our work are as follows:

- We introduce a new problem in question generation, which is question suite reduction.
- We propose a three-layer semantic analysis and processing model, i.e., the *3Q*.
- A multi-dimensional approach is proposed for the evaluation of question characteristics.

The rest of the paper is organized as follows. Section 2 discusses the related work, Sect. 3 presents our framework for the three-level analysis of sentences, Sect. 4 explains the metrics we used, Sect. 5 shows the evaluation methods we apply, Sect. 6 states our experiment and Sect. 7 concludes our work.

2   Related Work

2.1   QG and QA

There is a concept that only by asking appropriate questions can we learn how to answer. The research of problem generation is based on this. Traditionally, question generation (QG) is defined as an optimization problem, which maximizes the probability of generating the problem on the premise of a given answer. In education, QG helps students to ask questions; in the field of dialogue, as for some cold start system, it helps to start a topic or get feedback by asking questions. Problem generation is primarily realized by rules, such as text classification and fixed template generation [11,12]. With the development of the sequential model, people began to combine attention model and seq2Seq model [14] for problem generation, which brought the effect of problem generation to a new level [4]. Question generation and question answering (QA) can be viewed as a set of dual questions. The classical solution for QA is relying on information retrieval to complete the process of problem analysis, extraction and answer extraction. The latest solution is to construct a knowledge map and to query structurally through the relationship between knowledge [3].

In this study, we do not change the question generation and question answering model itself but hope to construct a better correlation between them and improve the efficiency.

At present, as a measure of evaluating complexity and diversity, information entropy is widely used in the Medical field, Biogenetic Engineering, and even Geography. In the field of data analysis, it can be used in the fuzzy model to select features with distinguishing potential [15] or verify the objectivity of the model as in [16].

In the process of searching for indicators that can cast quantity comparison on sentences, we notice the grading system of reading ability in the field of education. This project evaluates the acceptable grade level of the article directly by scoring the sentences involved in the article with a formula. The higher the score, the higher the corresponding grade level is set. This gives us great inspiration since this method has distinctive significance for human beings in learning, it can be helped in question answering model. For this reason, we introduce this indicator in this work and compared the effects of different reading level questions on the results of the QA model.

3   Question Analysis Model

Generally speaking, we demand to screen a modest scale set with great structural diversity and coverage. Based on this given proposition, we consider starting with the characteristics of the question itself. A sentence contains not only the characters like permutation and combination but also the linguistic meaning generated in conjunction and the specific organizational structure. Therefore, we propose a three-level analysis method consisting of semantics, syntax, and part of speech analysis. The main task is to conduct the syntax and part of speech analysis from the semantic analysis.

3.1   First Layer Analysis

To begin with, a point of view with high significant is that it is essential to let the machine understand human language from the perspective of language itself. From this point of view, combined with the current technological development, we start our work with the semantics of question, using semantic role labeling to analyze the sentence. Semantic role labeling is a technology viewing predicate as the core, trying to describe the relationship between different segments and the predicates. This is consistent with the logic in human language. It is a kind of shallow semantic analysis technique that annotates the arguments (semantic roles) of certain phrases in a sentence. Depending on the results of the syntactic analysis, it tags the identified phrases with argument roles such as agent, time or place. The Deep SRL model proposed by [6] is cited for dealing with this step.

On output results, semantic role labeling breaks questions into phrases by refining the core structure of sentences, parts of the meaningless phrases are ignored during this process. Here, we have to explain why the loss of this part of speech makes no sense. In people's daily application scenes of language, for example, oral communication and literature reading, all that we need is information contained in it. All the other parts are irrelevant information. What is more, questions often have more than one verb, the result of tagging is usually not unique. The combination of the most extended sum of the remaining phrases after tagging is chosen as our final decision and those questions that cannot be labeled are discarded.

Therefore, for the follow-up operations after the semantic role labeling, the so-called universe set has changed. It is necessary to define and explain these sets here. We define the original question set as U, the questions that can be tagged with semantic roles consist $U-$ and the remaining phrases after discarding meaningless phrases as S^{SRL} (Fig. 1).

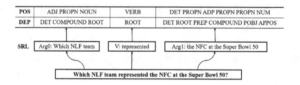

Fig. 1. Three-layer question analysis model

3.2   Second Layer Analysis

Since we get the phrases, the second layer may go further down and re-examine these phrases from the sentence level. We use the method of syntactic analysis, more specifically, the dependency syntactic parsing to do this job.

In this part, it is critical to make clear why we define semantic analysis as the first level. Because in fact, there do exact a process of syntactic analysis in semantic analysis, and why not use syntactic analysis directly? For this choice, we give the following reasons: **Difference in Objects**: These two methods are facing different objects: SRL for semantic and syntactic analysis for sentence. To see from a mainstream perspective, the semantic meaning of a sentence can be view as a higher level than syntax structure on sentence features, which deserves more attention. **Stability**: Semantic role analysis can achieve the same annotation results for the same role whether it is in active voice or passive voice. For example, given two sentences with the same semantics:

- Mike broke the window.
- The window broke.

Syntactic analysis expresses the word 'window' as the direct object of the verb in the first sentence and its subject in the second sentence. These differences in parsing impose an enormous computational burden for the research. However, for semantic role analysis, both words 'window' are labeled as the patient party (Arg1). This characteristic provides it with excellent stability for it suffers slightly from personal expression habits. **Efficiency**: Syntactic analysis usually abstracts sentences into nested structures with complex structure while the result of the semantic role analysis structure is a stable three layer mechanism. The critical point we are refusing the complex structure is, it may result in a considerable amount of computation. It makes this part of the feature challenging to represent in the scale of the sentence. In this section, the tools provided by [7] is cited.

3.3 Third Layer Analysis

After completing the processing of sentence level, we continue to conduct a more in-depth deconstruction of the question. We consider handling from the level of words, that is, lexical analysis. We mainly use Part of Speech Tagging at this part, and it is a procedure for marking the correct part of speech for each word in a word segmentation result. This part of our work is accomplished with POS tools provided by spaCy.

As an organizational structure, sentences focus mainly on the relationship between words; in other words, what role the word play in the sentence. It is natural that we view from the word side, to make sure if it is a noun or an adjective. As we all know, there is an enormous distance between people's expression habits, which is quite apparent in the questions. Some texts tend to use nouns, and some tend to use verbs. The phenomenon while the same syntax structure encountering different lexical structures is widespread. Considering the types of words, we can indirectly reflect the expression type of questions. Based on the truth that syntactic analysis only implements tagging of sentence components and will not change the basic structure of the sentence, we directly conduct the lexical analysis on the phrases of semantic role label.

4   Downsize of Questions Set

4.1   Screening Strategy

Our strategy is to evaluate questions according to the information obtained from the semantic analysis. Including the score of semantic contribution of each question, and select out questions with a high semantic contribution to form our new set S^{SCR}. What we want to explain further is why we choose question as the smallest unit. We do have considered using the semantic phrases after semantic role analysis as the smallest units to screen and to reorganize these units as new questions. Unfortunately, even if these newly created questions are grammatically reasonable, they often suffer from lacking logic and coherence, therefore being meaningless questions. Such questions are not suitable for our requirement of QG and QA testing and processing.

After evaluating all the questions, we choose the questions with the highest semantic contribution to form a new set, a basic TopN strategy. The process of this scheme has the idea of the greedy algorithm, and its implementation is straightforward. However, there are high requirements for evaluation schemes, so the design of the evaluation algorithm is essential. This method will be abbreviated as SCR in the following article.

Below, we present the primary metrics we use and give a clear explanation of the reason why we use it.

Statistical Metric: The results of the analysis models are sorted out, and the proportion of each semantic role, syntactic role and part of speech are calculated separately as $P_{i_{srl}}^{S^{srl}}, P_{i_{syn}}^{S^{srl}}, P_{i_{pos}}^{S^{srl}}$. It is worth noting that these proportions are for the remaining set S^{SRL}. At the same time, we reorganize and classify the semantic roles, construct the set of semantic roles S_i for each semantic role i, and further statistics the corresponding proportion of syntax $P_{i_{syn}}^{S_i}$ and lexicon $P_{i_{pos}}^{S_i}$ on S_i.

To implement the idea of balancing features and preserving as many features as possible, we restructured the above metrics. In this process, the above statistical indicators are transformed from the statistical frequency to the probability of occurrence approximately. Unlike the top-down analysis process to study from sentence to word step by step, our analysis process is the process of deducing sentence scores from the word.

Contribution of Syn-pos Label: Firstly, we observe the results of the current analysis. Each of the words contains three necessary information: the semantic role it belongs to, the syntactic label and the lexical label. Since both syntactic tags and lexical tags are directly attached to words, we use the combination of syntax and lexical tags as new tags to calculate the scores of each combination in the whole corpus.

Since we calculate the syntax and lexicon under each semantic role, the proportion of the same syntactic tag under different semantic roles is different. Therefore, for each occurrence of each combination, we calculate the probability of occurrence of each combination according to their semantic roles. At the same time, to achieve full coverage of the syntactic and lexical structure, we inverse the probability, that is, the smaller the probability, the higher the result. We define this score as the contribution of this syn-pos label under semantic role S^{srl}.

For all combinations, the statistics of all their appearing records are carried out, and the average score of each combination in the whole corpus can be further obtained. The calculation formula can be written as below:

$$con^{S^{srl}}_{syn-pos} = \frac{\sum_{j=1}^{n} \frac{1}{P^{S^j srl}_{syn} * P^{S^j srl}_{pos}}}{n} \tag{1}$$

where $con^{S^{srl}}_{syn-pos}$ represents the average contribution of words with the same syn-pos label in the corpus, n is the number of times this tag appears in the whole corpus.

Contribution of Question: Consider the syn-pos label as brick, and we can keep reversed the contribution of the phrase it settled.

$$con^{arg}_J = \frac{\sum_{j=1}^{m} con^{S^{srl}}_{j_{syn-pos}}}{m} \tag{2}$$

where con^{arg}_J stands for the average contribution of all the words in the semantic role phrase J, m is the number of words in this phrase.

In the last step of the statistical analysis, we turn the task into the calculation of question contribution.

$$con^{question} = \frac{\sum_{k=1}^{t} con^{arg}_k}{t} \tag{3}$$

where $con^{question}$ shows the average contribution of all the roles in the question, t is for the number of semantic roles in this sentence.

5 Evaluation

5.1 Experiment Proof

To test the effect of problem screening, we naturally introduce the QA model to simulate the actual application scene. The result QA model gets with the giving set will be analyzed as a sample for evaluation.

In terms of purpose, it can be divided into two aspects:

Consistency: To begin with, the result of the selected set should be consistent with that of the original set. If the actual processing ability of a question-answering model is like about 75 points, it can not be mistakenly elevated to 90 or underestimated to 60 after screening. For this reason, we design a Difference of Squares test scheme. That is, set the comprehensive answer prediction results of the complete set as the baseline, calculate the Square of the difference between the extracted results and the complete set.

Completeness: The set should retain the original syntactic and lexical combination features. By comparing the differences between the basic features of the data set, the test is carried out. The detailed characteristics are as follows: the proportion of lexical composition, the proportion of syntactic composition, the number of words, the average question length, the average word length, etc.

To evaluate our method more convincingly, the scheme of introducing entropy is further explained.

5.2 Computation of Information Entropy

Entropy has been widely accepted in the evaluation of complexity, diversity and other issues. Our idea is to propose a set of evaluation indicators from the perspective of entropy to measure the diversity of collected sets.

To distinguish it from the original semantic contribution, we consider the results of syntactic analysis and lexical analysis separately in the part of entropy. The independent distribution of syntax and lexicon in the universe set U can be obtained by directly cast dependency parsing and POS tagging on the questions. For each syntax and lexical label, its proportion is calculated, and the syntactic entropy $H_{syn}(x)$ (x for syntax role) and lexical entropy $H_{pos}(y)$ (y for POS tag) are obtained.

After obtaining the corresponding entropy of each syntax and lexical tag, we calculate the information entropy of the whole question. We simply calculate the entropy to avoid the interference of other factors. Therefore, for each word in the question, its entropy value is $H_{syn}(w_{syn}) + H_{pos}(w_{pos})$, and the information entropy of the sentence is the sum of the information entropy of all its words.

$$IG^{que} = \frac{\sum_{w=1}^{t} H_{syn}(w_{syn}) + H_{pos}(w_{pos})}{t} \tag{4}$$

where IG^{que} is the quantities of information of question and t is the number of words in the question.

We also consider the effect of question length on accuracy. As an answer model, the longer the question is the more information it contains in most cases, that is, higher in the theoretical accuracy. To exclude the influence of sentence length, we try to calculate the average entropy, that is, dividing the entropy of the sentence by the number of words contained in the sentence. However, the subsequent experiments show that this metric has no apparent feature effect and bring in poor discrimination, so it is abandoned.

5.3 Grade Level

The last evaluation metric we consider is the grade level feature based on the independent features of sentences. The scheme is supported by a general formula, which is different from the entropy metric we designed and the metric in the analysis tool.

$$G(S) = 0.39(\frac{n_{word}}{n_{sen}}) + 11.8(\frac{n_{syl}}{n_{word}}) - 15.59 \tag{5}$$

where n_{word}, n_{sen}, n_{syl} are numbers of all the words, sentences, syllables in the set S, $G(S)$ is the final score of the given set.

This indicator takes note of a distinctive feature from other indicators, that is, the number of syllables in a word. The syllable of a word is quite close to the number of letters in a word, but they are different from each other. It also reflects the difficulty and complexity of a word to a certain extent. This metric is widely used in the assessment of students' reading ability, the higher the score, the higher the grade. Therefore, we believe that it is also of some significance to evaluate the auto-generated problems.

6   Experiment

6.1   Data and Model

We choose Stanford Question Answering Dataset (SQuAD) [13], which is a highly used reading comprehension data set in recent years, and use the development data it provides in version 1.1 as the universe set. In this set, we have about 12,000 questions, and our job is to extract a subset with 1000 to 2000 questions from it.

For the QA model, we choose DrQA model [2]. The model was proposed in 2017, which combines semantic text retrieval and machine learning text comprehension. In recent years, it has been proved to be active on many problems.

6.2   Comparative Methods

Random: When it comes to extracting subsets in a complete set, the first thing that comes to mind is the random sampling method [10], and it can also be regarded as a baseline for comparison. According to the random principle, samples are extracted from the set U by using random numbers. All of the samplings are independent with the same probability of being pulled out. The subset S^{RAN} extracted this way (name as RAN) is usually accompanied by huge fluctuation. Therefore, for the results of random sampling, we conducted five experiments with the same scale and use the mean square difference as the result of the current scale.

Entropy: Since we trust entropy's ability in complexity evaluation, we also captured the possibility that it could be used directly as an indicator for screening. Therefore, we come up with an Information Gain selection method (name as IG) which embezzle the idea of TopN. Questions with the highest entropy value are selected to form a new subset S^{IG} based on the calculation method proposed in Sect. 5.2.

Grade Level: Grade Level is the most basic indicator scheme, starting from attributes as sentence length and number of letters in words. Because of the inherent characteristics of indicators, we still use TopN to do the screen and get S^{GL}, this method is abbreviated as GL in the following article.

6.3 Result

Firstly, we compare the set size, distinct word number, average question length and average word length of U, $U-$, S^{SCR}, S^{RAN}, S^{IG} and S^{GL}, as is shown in Table 1.

Table 1. Statistics of dataset

	U	$U-$	S^{SCR}	S^{RAN}	S^{IG}	S^{GL}
SIZE	10570	9293	1000	1000	1000	1000
DIS-WORD	12033	11033	3496	3219	4057	3687
AVE-QUE	11.443	11.396	13.155	11.321	19.049	14.288
AVE-WORD	4.437	4.432	4.491	4.422	4.454	5.226

It can be seen that there just exists little difference between U and $U-$, and the S^{RAN} set roughly maintains the characteristics of question length and word length of the original set. The other three methods all cause certain extent impact on the number of words retained, question length and word length. In contrast, the overall change of SCR is the smallest and closest to the result of RAN. Therefore, the preliminary observation shows that SCR is basically consistent with the original set.

Next, we compare the difference between the set accuracy and the original set. The exact string match (EM) method provided in DrQA is used to evaluate the accuracy of model calculation results, the universe set has got an accuracy score of about 29.489. Figure 2 shows the comparison of the results on the other four sets with the square difference of the value (the mean of the 5 times' square difference for the RAN set).

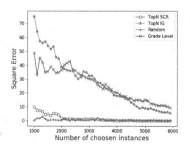

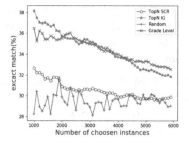

Fig. 2. Comparison of SE   Fig. 3. Comparison of EM

It is clearly that as the size of the filter set shrinks (100 questions at a time), both the SCR set and the RAN set are basically consistent with the original set.

The square deviation is less than 10 while The results obtained by IG and GL show great instability and volatility.

In the study, we encountered an interesting phenomenon in the measure of the accuracy of several methods, as shown in Fig. 3.

Although the fluctuation degree is different, SCR, IG and GL all show a positive correlation with accuracy, that is, the higher the metric score, the higher the accuracy of the model. This is not possible in RAN, and this can be applied in the QG field.

In the previous test, the results obtained by SCR are very close to RAN. So we further focus on comparing the changes in the amount of information contained in SCR.

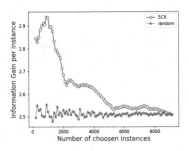

Fig. 4. Comparison of Informa Gain   **Fig. 5.** Comparison of Grade Level

Just as Figs. 4 and 5, the entropy value and Grade Level of each question is calculated respectively. Combined with Table 1, it can be seen that the quantity of information of the question in SCR is far more than that in RAN while the unit question length is not that different. Moreover, by Grade Level's definition, SCR is suitable for children of higher grades to read, that is, relatively more complex sentences. It fully reflects that SCR retains more features of the original set and has good integrity.

7 Conclusion

Generally speaking, our work originates from the idea of screening the problems generated by the QG process to reduce the following manual workload. For this reason, we design a three-layer semantic analysis model based on the characteristics of the question itself and the extensive use of different object-oriented semantic processing tools. To balance the probability of each feature appearing in the final selected set, a unique measurement method of semantic contribution metric is proposed. Finally, in order to prove the validity and rationality of our results, we further design a scheme of entropy detection based on syntactic dependency analysis and lexical analysis. The rationality and feasibility of our semantic contribution metric are proved in many aspects.

In practice, we realize that this method is not only suitable for problem selection, but also more widely used in corpus generation and other fields. It can reduce the size of training sets in various natural language processing research. Also, its positive correlation also makes it have great potential to improve the efficiency of the model. We believe that this method can extract training data through stable processing model on the one hand, and test the processing model efficiently through refined data on the other hand.

Acknowledgement. The work is supported in part by the National Key Research and Development Program of China (2016YFC0800805) and the National Natural Science Foundation of China (61772014).

References

1. Brill, E.: Transformation-based error-driven learning and natural language processing: a case study in part-of-speech tagging. Comput. Linguist. **21**(4), 543–565 (1995)
2. Chen, D., Fisch, A., Weston, J., Bordes, A.: Reading Wikipedia to answer open-domain questions. arXiv preprint arXiv:1704.00051 (2017)
3. Cui, W., Xiao, Y., Wang, H., Song, Y., Hwang, S., Wang, W.: KBQA: learning question answering over QA corpora and knowledge bases. Proc. VLDB Endow. **10**(5), 565–576 (2017)
4. Du, X., Shao, J., Cardie, C.: Learning to ask: neural question generation for reading comprehension. arXiv preprint arXiv:1705.00106 (2017)
5. Hacioglu, K.: Semantic role labeling using dependency trees. In: Proceedings of the 20th International Conference on Computational Linguistics, p. 1273 (2004)
6. He, L., Lee, K., Lewis, M., Zettlemoyer, L.: Deep semantic role labeling: what works and what's next. In: Proceedings of the 55th Annual Meeting of the Association for Computational Linguistics (Volume 1: Long Papers), vol. 1, pp. 473–483 (2017)
7. Honnibal, M., Johnson, M.: An improved non-monotonic transition system for dependency parsing. In: Proceedings of the 2015 Conference on Empirical Methods in Natural Language Processing, pp. 1373–1378 (2015)
8. Johansson, R., Nugues, P.: Dependency-based semantic role labeling of PropBank. In: Conference on Empirical Methods in Natural Language Processing (2008)
9. Kincaid, J.P., Fishburne Jr., R.P., Rogers, R.L., Chissom, B.S.: Derivation of new readability formulas (automated readability index, fog count and flesch reading ease formula) for navy enlisted personnel (1975)
10. Marshall, M.N.: Sampling for qualitative research. Fam. Pract. **13**(6), 522–526 (1996)
11. Ming, L., Calvo, R.A., Rus, V.: Automatic question generation for literature review writing support. In: Intelligent Tutoring Systems, International Conference, ITS, Pittsburgh, PA, USA, June 2010
12. Mitkov, R., et al.: Computer-aided generation of multiple-choice tests. In: Proceedings of the HLT-NAACL 2003 Workshop on Building Educational Applications Using Natural Language Processing (2003)
13. Rajpurkar, P., Zhang, J., Lopyrev, K., Liang, P.: SQuAD: 100,000+ questions for machine comprehension of text. arXiv preprint arXiv:1606.05250 (2016)
14. Vinyals, O., Le, Q.: A neural conversational model. arXiv preprint arXiv:1506.05869 (2015)

15. Zhang, X., Mei, C., Chen, D., Yang, Y.: A fuzzy rough set-based feature selection method using representative instances. Knowl.-Based Syst. **151**, 216–229 (2018)
16. Zhou, Q., Luo, J.: The study on evaluation method of urban network security in the big data era. Intell. Autom. Soft Comput. 1–6 (2017)

Data Augmentation for Deep Learning of Judgment Documents

Ge Yan[1,2], Yu Li[1,2], Shu Zhang[1,2], and Zhenyu Chen[1,2(✉)]

[1] State Key Laboratory for Novel Software Technology,
Nanjing University, Nanjing, China
`zychen@nju.edu.cn`
[2] Software Testing Engineering Laboratory of Jiangsu Province,
Nanjing, China

Abstract. With the increasing number of machine learning parameters, the requirements on data quantity are getting higher and higher to train a good model. The choice of methods and the optimization of parameters can improve the model while the quality and quantity of the data determine the upper limit of the model. However, in realistic scenarios, it is quite challenging to get a lot of tag data. Therefore, it is natural to realize data augmentation by transforming the original data. We use three methods for data augmentation on different scales of original data in solving the crime prediction problem based on the description of the cases, and find that the effects of data augmentation are different for different models and different fundamental data quantities.

Keywords: Data augmentation · Text classification · Legal intelligence · FastText

1 Introduction

With the emergence of more and more machine learning and deep learning methods, the model relies more and more on the amount of data, especially for neural network, which has many uninterpretable parameters. To get accurate parameters, the amount of required training data is growing.

Too little training data can lead to two consequences. The trained model may not be accurate, or it may be over-fitting [2], which means that the model thoroughly learns all the features of the existing data, resulting in no generalization ability for data not included in the training set. Low accuracy means that the model does not learn enough features for representing, so we can solve the problem by increasing times of learning, but to solve the over-fitting problem, we often need some more complicated methods.

There are several common methods to solve the overfitting problem. The easiest method is to divide the dataset into a training set and a verification set, we train the model on the training set, and then test it on the verification set. The test on the verification set should be conducted after each iteration until the

Z. Cui et al. (Eds.): IScIDE 2019, LNCS 11936, pp. 232–242, 2019.
https://doi.org/10.1007/978-3-030-36204-1_19

accuracy reaches a certain threshold. This strategy is also called early-stopping [11]. But the determination of the threshold can also be a problem, which can only be obtained through continual experiments in practical use.

The second method is to reduce the size of the network, namely regularization, which means adding a regularization term after the objective function when performing the objective function or the cost function optimization. We often use L1 and L2 regularization [9]. The introduction of the regularization term limits the complexity of the model, with the same principle as the drop out method in neural network, reduce the complexity through randomly deactivating neurons of a neural network [12].

Both of the above methods prevent overfitting on the model. The most straightforward solution is to increase the training data. According to a saying in the data mining field, sometimes having more data worth more than a good model. Since we make predictions about unknown data on training data by acknowledging the assumption that data is independently distributed, more data makes the estimation and prediction task of unknown data more accurate. However, due to the limitation of actual conditions, it is not realistic to manually label large quantities of data. As a result, many researchers turn their attention to the existing tag data and generate new data by conducting some calculation methods and strategies on existing data, which is called data augmentation.

Data augmentation is a method of augmenting datasets by changing training data rather than labels. It can increase both the number and the diversity of training datasets. Traditional data augmentation is generally applied to the image field [10]. Increasing the number and diversity of the image through cropping, flipping, scaling, rotating, etc. But methods applied to the image field do not fit the text field, for word order is very important information in text data. For example, 'I love you' and 'You love me' are completely different, changes of word order can result in changes of meaning of the whole sentence.

As a result, we use sentences as basic units in our experiment, perform data augmentation on text descriptions of the cases, also known as facts, through randomly changing the sentence order, randomly deleting the sentences in the text and adding sentences with the same label in other texts to amplify the case description text in legal field.

To summarize, our contributions are as follows:

- We use three ways to perform data augmentation tasks on the case description text in legal documents, increase the scale of the training set, and compare the effects of the three data augmentation methods on the same model.
- We use TextCNN and FastText to predict the law articles according to the case description in legal documents, and compare the impact of different data volumes on different models.

This paper is structured as follows. In Sect. 2, we briefly introduce the definition of data augmentation, as well as commonly used image and text augmentation methods. We explain the details of our experiment in Sect. 3, including the data augmentation methods and prediction models we use. The experiment results are analyzed in Sect. 4. Section 5 contains the conclusion, our next experiment plan and presents future prospect.

2   Related Work

2.1   Legal Judgement Prediction

Research on legal intelligence has long been studied and the judicial document is the main research object of the academia. With the openness of judicial big data, recent years have witnessed the rapid development of machine learning and deep learning methods being introduced from the image field into the text field.

With regard to legal judgement prediction, previous works mostly consider it as a task of multi-class classification. Some works aim at extracting efficient features and make full use of them. Liu and Hsieh [7] build a domain dependent word list based on PAT tree-based method and HowNet, and use the generated word list for defining relevant phrases to classify Chinese judicial documents. Some are inspired by the success of neural network methods. Kim et al. [5] use Convolutional Neural Network(CNN) with multiple filter widths in document embedding. Zhang et al. [16] apply ConvNets only on characters, and achieve state-of-the-art results. Luo et al. [8] propose an attention-based neural network method that models charge prediction and relevant article extraction tasks jointly. Zhong et al. [17] formalize the dependencies among legal judgement subtasks as a Directed Acyclic Graph(DAG) and present TOPJUDGE, a topological multi-task learning framework.

Text classification task can easily obtain a large number of tagged data, so deep learning method is easier to get better results than traditional methods. At present, the main deep learning models used in text classification include FastText, TextCNN, HAN, DPCNN, etc. And we select FastText and TextCNN in our experiment. FastText has a simple structure and is suitable for speed demanding scenarios. It adds up all the word vectors (and N-gram vectors) in the text to get the average value directly, and then uses a single-layer neural network to get the final classification results. TextCNN has a slightly more complex structure. It uses the combination of convolution and max-pooling, which is very successful in the image field, to know which keywords appear in the text and to discover the intensity distribution of similarity. The state of the art deep learning text classification models have achieved fairly good results on data over 100 thousand to 1 million and there are also some very explanatory models available.

2.2   Data Augmentation

There are two main ways of data augmentation, offline augmentation and online augmentation. In offline augmentation, all data is augmented by transforming methods before being input into the model, which can multiply the size of

datasets. Online augmentation, also called augmentation on the fly, is to transform part of the data (mini-batch) before inputting. Some machine learning
frameworks support online augmentation and can be accelerated on GPU. The
specific choice of method depends on the size of data set. When the amount of
data is small, we can use the first method to double the data directly. But it is
not possible to process all the data when the amount of data is large, because the
cost of computing will increase explosively, then we adopt the second method.

For image data, the transforming operation is very similar to the operation
we perform on the image in real life. Supposing that we do not consider any-
thing other than the boundary of the image, common processing methods are as
follows, flipping - flip the image horizontally or vertically , rotation - rotate the
image at a specified angle, scaling - scale the image inward or outward, cropping
- randomly extract a portion of the original image and scale it to the size of
the original image, translation - move the image along the x-axis or y-axis (or
both), Gaussian noise - prevent over-fitting by adding noise appropriately in the
image. The methods mentioned above are common methods of image augmentation, there are also some advanced processing methods. For example, using
conditional GANS to change the season in landscape pictures [18], generate different pictures of the same object in different scenes by fetching the texture,
atmosphere, appearance of an image and fuse it with another image.

However, methods suitable for images usually do not fit the text data, because
the word order is a significant part in text containing much information. If we
change the word order, the newly generated sample will no longer have the same
meaning as the original sample. Consequently, new text augmentation methods
such as synonym replacement are generated aiming at the characteristics of the
text. We randomly select some words in the text, then use their synonyms to
replace them and generate new data with the same labels without changing the
meaning of the text. This method also has certain problems. When constructing
the feature vector of the text, if we use the trained word vector for text representation, then the feature vector of the synonym will be similar, so the model
will treat the two texts as the same. In essence it does not achieve the purpose
of data augmentation.

A more effective way to augment text data is back translation [3] and text
tailoring. Back translation means that the original document will be converted
into text in other languages by means of an intermediate language, and then be
translated back to form a new text. In most cases, the text obtained by back
translation is different from the original text, but with the same meaning. Compared with the synonym replacing method, back translation can not only replace
the synonyms, but also has the ability to add or delete words and reorganize sentences, which is no longer the same the original text and achieves the purpose
of data augmentation. Text clipping is aimed at several particular application
scenes. Text with a certain topic will be repeatedly mentioned in multiple parts
of the text content, so the text after clipping still has the same topic label as
the original text. The tailored text can serve as augmented data and be added
to the dataset.

There are also many works concerning modifying deep learning methods to make full use of the original data. To improve the effectiveness of Convolutional Neural Networks (CNN) in the area of image recognition, sufficient training data is usually needed. Tang et al. [13] perform the robust neighborhood preserving low-rank and sparse recovery step over the original CNN features, in order to extract salient key information and remove the included noise. They use three deep networks for evaluation, i.e., VGG, Resnet, and Alexnet, and find that the extracted joint low-rank and sparse CNN features can indeed obtain the enhanced results, compared with the original CNN features. Also, in the study of OCR, the Convolutional Recurrent Neural Networks (CRNN) is very useful. However, existing deep models usually apply the down-sampling in pooling operation and drop some feature information to reduce the size, which may result in the missing of some relevant characters with low occupancy rate. Tang et al. [14] use Dense Convolutional Network (DenseNet) to replace the convolution network of the CRNN to connect and combine multiple features, and use deeper structure to extract informative features. Besides, they directly use the output of inner convolution parts to describe the label distribution of each frame to improve the process efficiency and propose a new OCR framework, termed DenseNet with Up-Sampling block joint with the connectionist temporal classification, for Chinese recognition.

3   Experiments

3.1   Datasets and Preprocessing

We use the Chinese AI and Law challenge dataset (CAIL) [15] published by the Supreme People's Court of China, including over 155 thousand cases in the training dataset and over 33 thousand cases in the testing dataset. The legal document is well-structured and contains fields like fact description, court view, parties, judgment result and other information. In order to observe the performance of data augmentation under the circumstances with different data volume, we randomly select 10 thousand, 50 thousand, 100 thousand and 150 thousand cases to form 4 datasets of different scales. Since the original data is in the form of json, we extract the 'fact' and 'charge' fields according to the labels, serialize the 'accusation' field and perform word segmentation on fact description text. We use Jieba as the word segmentation tool and remove meaningless stop words in the text according to the commonly used stop word list.

3.2   Data and Augmentation

Our task is to predict the accusations of a case giving the fact of a legal document. It can be seen as a multi-label classification problem. There are 202 accusations in total. We perform data augmentation on sentences in order not to change the meaning of the original text. As is shown in Fig. 1, there are three different ways, randomly scramble sentences in the sample, randomly delete sentences in the sample and randomly insert the sentences with the same label in other samples. All three data augmentation methods are done automatically by the machine.

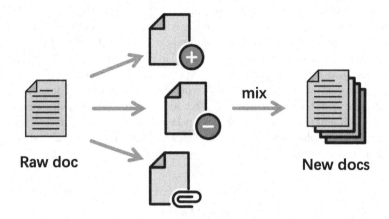

Fig. 1. Three methods of augmentation

- Random Scrambling Since we use a sentence as a unit, and the order of sentences in the case description text does not have a great influence on the meaning of the original text, so we randomly shuffle the case description text by the sentence. The new text obtained after scrambling has the same labels as the original one. In this way, we get a new dataset of the same size as the original dataset.
- Random deletion Since there are many irrelevant statements in the fact description text, deleting them will not affect the understanding of the case, so we adopted a random deletion method to delete one sentence in the original sample randomly. If the original sample only contains one sentence, then no process will be conducted and the new sample will have the same label with the original sample. We perform the same process on every sample and will get new datasets of the same scale.
- Random Insertion We find that cases with the same accusation will have many similar sentences in the case description, which lead to judgements of the same accusation. As a result, we divide data with the same accusation label into one category. We insert a random sentence from sample data with the same label into the original sample to get a new dataset. We do not process data without the same labels, thus we get a new dataset of the same scale.

Data augmentation is very effective in expanding the dataset and restraining the model from overfitting. Through three data augmentation methods, we obtain three new datasets of the same scale as the original dataset. We mix it with the original dataset and randomly scramble it into a new dataset, which is four times as large as the original dataset.

3.3   FastText

FastText is a three-layer neural network model [4]. As is shown in Fig. 2, the first layer is the input layer. After accepting the word sequence, they are converted into feature vectors, and then be mapped to the hidden layer through linear variation. Afterwards, the hidden layer maps them to labels by nonlinear variation and the probability distribution of the text on each label will be output in the end. This model is very similar to the CBoW model used in Word2vec, but Word2vec is used for getting the feature vectors of the intermediate result while FastText is aimed at the final label prediction. The processed text is directly input into the input layer of the FastText algorithm, then it will calculate the probability value of each label in the output layer. If the probability value exceeds 0.5, then the text belongs to the certain label, and finally the entire labels of the text will be integrated.

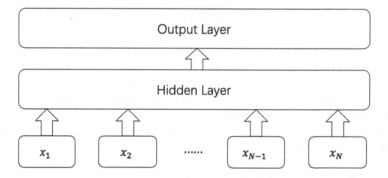

Fig. 2. The architecture of FastText. $x_1, \cdots, x_N$ are N ngram features of a sentence, they are embedded and averaged to form the hidden variable.

3.4   TextCNN

Convolutional neural network [6] consists of an input layer, an output layer, and multiple hidden layers. There are many hidden layers, including the convolution layer, the pooling layer, the fully connected layer, and the relu layer. CNN was first used in the image field, and later research found that it is also of great capability in text representation [1]. As can be seen in Fig. 3, the first layer is the embedding layer, and the input text sequence is converted into word vectors. We can make use of the pre-trained vector or construct features automatically during the training process. Next will be the convolution layer, the context information of the text is captured in the field of vision of the convolution kernel, which is also the core of the whole convolutional neural network. We use 256 convolution kernels in the size of 5 to extract a higher level of information in the text. Then we will have the max-pooling layer, also known as down-sampling, which preserves useful information while reducing data amount. Afterward will be two fully connected layers of two neurons with a total number of 128 in order to improve the generalization ability of the model. And finally, the output layer will output the labels, which is the same as the FastText model.

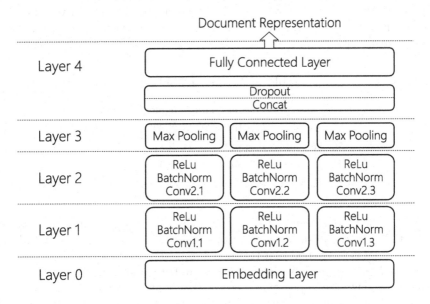

Fig. 3. The architecture of TextCNN

Parameters: Our experiment does not change the original parameters. The word vector dimension of the convolutional neural network is 64, the sequence length is 600, there are 256 convolution kernels, the convolution kernel is in the size of 5, the vocabulary is in the size of 5000, and there are 128 neurons in the fully connected layer. To prevent the overfitting problem, the dropout retention ratio is set to 0.5, the learning rate is 1e−4, the number of trainings per batch is 64 and we conduct 10 iterations.

4 Results and Discussion

Table 1 displays the F1-Score obtained by accusation prediction through the FastText method using models trained on both the original dataset and the augmented dataset. We find that the model trained with augmented data is more powerful than that trained with the original data when the data amount is small. When there are 10 thousand cases, the augmentation dataset increases F1-Score by about 10%. But the advantage will weaken gradually with the increase of the data amount. But what we need to pay attention to is that with the growth of the dataset, the augmented dataset may have a bad influence on experimental results. In the experiment with the data size of 100 thousand and 150 thousand, the model of the augmented data do not perform as well as the original data. And it is evident that the expanded data can't play the same role as the original one. For example, 200 thousand data can be obtained from 50 thousand data set, but the experimental results are not as high as 100 thousand, not to mention or even 100 thousand. Moreover, the defect of the expanded data will increase with the increase of the original data set, and will even have an adverse impact on the model.

Table 1. F1 score of FastText

	Original data	Augmented data
10 thousand	0.1910	0.3000
50 thousand	0.4134	0.4611
100 thousand	0.7890	0.7856
150 thousand	0.8347	0.8234

Table 2 shows the accuracy of obtained by accusation prediction through the TextCNN method using models trained on both the original dataset and the augmented dataset. We find that the model trained by augmented data performs better than that trained by original data, and it is not affected by the data amount. Although the improvement on the small dataset is larger than the large one, there is no significant difference in the magnitude of the data. Compared with the FastText method, it is noted that the augmented data does not have a negative impact on the experimental results. And similarly, the quality of the augmented data is not up to the original data. The effect of twice the augmented data in training model is not as good as that of the original data.

Table 2. Accuracy of TextCNN

	Original data	Augmented data
10 thousand	12.45%	14.91%
50 thousand	35.38%	39.18%
100 thousand	71.75%	72.20%
150 thousand	76.70%	78.09%

5 Conclusion

In general, data augmentation helps improve the effectiveness of the model because it provides more reference to the model for feature learning. However, it has different performance on different training methods. Take two methods used in this paper as an example, the benefits of augmented data on a small-scale dataset are quite significant for FastText, but on a large-scale dataset, there may be no improvement and even reduce the efficiency of the model. For TextCNN, data augmentation is helpful for different scales of datasets. It has a more significant improvement on a small dataset, but the difference is not significant. Moreover, there still lies a problem in the experiment. The training effect of augmenting the data from 50 thousand to 200 thousand is not as good as amplifying it to 100 thousand, which means there is a quality problem in the data we have expanded. It can not only improve the data but also mislead the model to learn wrong features.

We will try more types of data augmentation methods in our future work and research, including getting more diversified data by conducting more detailed augmenting methods on the replacing objects, and carry out more experiments and research on evaluating the data quality of the augmented data. We hope that the augmented data can function similarly as real data, and reduce the difficulty of data acquisition fundamentally.

Acknowledgment. The work is supported in part by the National Key Research and Development Program of China (2016YFC0800805) and the National Natural Science Foundation of China (61472176, 61772014).

References

1. Grefenstette, E., Blunsom, P.: A convolutional neural network for modelling sentences. In: ACL (2014)
2. Hawkins, D.M.: The problem of overfitting. J. Chem. Inf. Comput. Sci. **44**(1), 1–12 (2004)
3. Hayashi, T., et al.: Back-translation-style data augmentation for end-to-end ASR. In: 2018 IEEE Spoken Language Technology Workshop (SLT), pp. 426–433. IEEE (2018)
4. Joulin, A., Grave, E., Bojanowski, P., Douze, M., Jégou, H., Mikolov, T.: Fast-Text.zip: compressing text classification models. arXiv preprint arXiv:1612.03651 (2016)
5. Kim, Y.: Convolutional neural networks for sentence classification. arXiv preprint arXiv:1408.5882 (2014)
6. Lawrence, S., Giles, C.L., Tsoi, A.C., Back, A.D.: Face recognition: a convolutional neural-network approach. IEEE Trans. Neural Netw. **8**(1), 98–113 (1997)
7. Liu, C.-L., Hsieh, C.-D.: Exploring phrase-based classification of judicial documents for criminal charges in Chinese. In: Esposito, F., Raś, Z.W., Malerba, D., Semeraro, G. (eds.) ISMIS 2006. LNCS (LNAI), vol. 4203, pp. 681–690. Springer, Heidelberg (2006). https://doi.org/10.1007/11875604_75
8. Luo, B., Feng, Y., Xu, J., Zhang, X., Zhao, D.: Learning to predict charges for criminal cases with legal basis. arXiv preprint arXiv:1707.09168 (2017)
9. Ng, A.Y.: Feature selection, L1 vs. L2 regularization, and rotational invariance. In: Proceedings of the Twenty-First International Conference on Machine Learning, p. 78. ACM (2004)
10. Perez, L., Wang, J.: The effectiveness of data augmentation in image classification using deep learning. arXiv preprint arXiv:1712.04621 (2017)

11. Prechelt, L.: Automatic early stopping using cross validation: quantifying the criteria. Neural Netw. **11**(4), 761–767 (1998)
12. Srivastava, N., Hinton, G., Krizhevsky, A., Sutskever, I., Salakhutdinov, R.: Dropout: a simple way to prevent neural networks from overfitting. J. Mach. Learn. Res. **15**(1), 1929–1958 (2014)
13. Tang, Z., Zhang, Z., Ma, X., Qin, J., Zhao, M.: Robust neighborhood preserving low-rank sparse CNN features for classification. In: Hong, R., Cheng, W.-H., Yamasaki, T., Wang, M., Ngo, C.-W. (eds.) PCM 2018. LNCS, vol. 11164, pp. 357–369. Springer, Cham (2018). https://doi.org/10.1007/978-3-030-00776-8_33
14. Tang, Z., Jiang, W., Zhang, Z., Zhao, M., Zhang, L., Wang, M.: DenseNet with up-sampling block for recognizing texts in images. Neural Comput. Appl. 1–9
15. Xiao, C., et al.: CAIL2018: a large-scale legal dataset for judgment prediction. arXiv preprint arXiv:1807.02478 (2018)
16. Zhang, X., Zhao, J., LeCun, Y.: Character-level convolutional networks for text classification. In: Advances in Neural Information Processing Systems, pp. 649–657 (2015)
17. Zhong, H., Zhipeng, G., Tu, C., Xiao, C., Liu, Z., Sun, M.: Legal judgment prediction via topological learning. In: Proceedings of the 2018 Conference on Empirical Methods in Natural Language Processing, pp. 3540–3549 (2018)
18. Zhu, J.Y., Park, T., Isola, P., Efros, A.A.: Unpaired image-to-image translation using cycle-consistent adversarial networks. In: Proceedings of the IEEE International Conference on Computer Vision, pp. 2223–2232 (2017)

An Advanced Least Squares Twin Multi-class Classification Support Vector Machine for Few-Shot Classification

Yu Li, Zhonggeng Liu, Huadong Pan[(⊠)], Jun Yin,
and Xingming Zhang

Advanced Research Institute of Zhejiang Dahua Technology Co. Ltd.,
Hangzhou, China
pan_huadong@dahuatech.com

Abstract. In classification tasks, deep learning methods yield high performance. However, owing to lack of enough annotated data, deep learning methods often underperformed. Therefore, we propose an advance version of least squares twin multi-class classification support vector machine (ALST-KSVC) which leads to low computational complexity and comparable accuracy based on LST-KSVC for few-shot classification. In ALST-KSVC, we modified optimization problems to construct a new "1-versus-1-versus-1" structure, proposed a new decision function, and constructed smaller number of classifiers than our baseline LST-KSVC. We empirically demonstrate that the proposed method has better classification accuracy than LST-KSVC. Especially, ALST-KSVC achieves the state-of-the-art performance on MNIST, USPS, Amazon, Caltech image datasets and Iris, Teaching evaluation, Balance, Wine, Transfusion UCI datasets.

Keywords: Few-shot · Multi-classes classification · Support vector machine

1 Introduction

Deep learning models have achieved high performance on classification tasks such as image recognition which has become a fundamental challenging problem in computer vision. The goal is to obtain the category of the subject in the image. The recent research show that deep convolutional neural networks have achieved the state-of-the-art performance, including VGG [1], Inception [2], ResNet [3], and DenseNet [4]. However, in the actual task requirements, we only have few-shot of annotated datasets, which is contrary to the need for deep learning of a large number of training datasets. Therefore, this paper applies the SVM family methods [5, 6] which advantage is generalization to solve the actual situation. When we lack enough annotated data, SVM can use few-shot datasets to find support vectors, so that a classification model with better performance can be obtained.

In contrast with artificial neural network [7] which aims to reduce empirical risk, the SVM [8] implements the structural risk minimization. However, the computational complexity of quadratic programming problem (QPP) is high so that influences to

© Springer Nature Switzerland AG 2019
Z. Cui et al. (Eds.): IScIDE 2019, LNCS 11936, pp. 243–252, 2019.
https://doi.org/10.1007/978-3-030-36204-1_20

deploy. Twin support vector machines (TWSVM) [9–11] were proposed for solving two smaller-sized QPPs. In real situation, multi-class classification problem is common occurred, so SVM and TWSVM family methods resolve the problem by "1-versus-rest" and "1-versus-1" structure [12, 13]. Each constructed classifier is involved with the training data of two classes. And then the remaining patterns omitted may receive unfavorable results. A new multi-class support vector classification regression for K-class classification (K-SVCR) [14] provides better forecasting results based on "1-versus-1-versus-rest". However, Twin-KSVC [15] and least version of Twin-KSVC to reduce its time complexity in constructing decision classification are proposed.

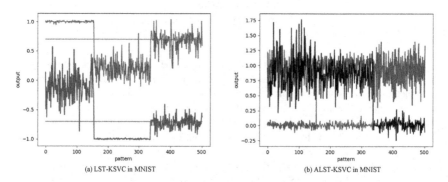

(a) LST-KSVC in MNIST   (b) ALST-KSVC in MNIST

Fig. 1. Illustration of multi-class LST-KSVC (a) and ALST-KSVC (b) in MNIST datasets. The horizontal axis represents the input pattern and the vertical axis represents the output of the model. (Color figure online)

In this paper, we follow the line of research in [16], and we propose an advanced version of LST-KSVC, called ALST-KSVC. The advanced content of the algorithm is to modify the optimization problems based a new "1-versus-1-versus-1" structure, propose a new decision function and construct smaller number of classifiers than our baseline LST-KSVC. The solution of ALST-KSVC follows directly solving three QPPs and two equality constraints like Twin-KSVC and LST-KSVC. So it not only takes less time in complexity but also achieves higher classification accuracy. Our contributions can be summarized as follows:

- we modify the quadratic programming problems of LST-KSVC to construct a new "1-versus-1-versus-1" structure and propose a new decision function to suit the structure. LST-KSVC not only need to set threshold ε, but also need to get the ensemble classifiers before voting. Compared with LST-KSVC, ALST-KSVC is more convenient.
- we propose a new multi-classes classification approach based on "1-versus-1-versus-1" structure which only constructs K ALST-KSVC classifiers. However, LST-KSVC need constructs $K(k-1)/2$ LST-KSVC classifiers.
- we conduct comprehensive experiments on MNIST, USPS, Amazon, Caltech, Iris, Teaching evaluation, Balance, Wine, Transfusion few-shot datasets, and achieve superior performance compared with LST-KSVC.

The rest of the paper is organized as follows. Section 2 reviews the related work of LST-KSVC. We introduce the proposed method which includes linear, nonlinear cases and decision function in Sect. 3. Section 4 provides the evaluation and analysis, and we make the conclusion and future work in Sect. 5.

2 Related Work

In this section, we give a brief formulated description LST-KSVC based on "1-versus-1-versus-rest" structure for multi-class classification purpose in the training set.

LST-KSVC: LST-KSVC was introduced in [16] which modify the primal problem of Twin-KSVC with the inequality constraint replaced with equality constraints, and then evaluates all the training data into a "1-versus-1-versus-rest" structure. Let matrix $A \in R^{l_1 \times m}$ represent the training data belong to "+1", $B \in R^{l_2 \times m}$ represents the training data belong to "−1" and $C \in R^{l_3 \times m}$ indicates the rest training data belong to "0". The l_1, l_2, l_3 represents the number of patterns per class, m represents the number of attributes. This method is to find two nonparallel planes that can separate the two classes of K-classes. Then the remaining patterns are mapped to the region between two nonparallel planes. These two nonparallel hyperplanes are shown below:

$$x^T w_1 + b_1 = 0 \tag{1}$$

$$x^T w_2 + b_2 = 0 \tag{2}$$

And the two nonparallel hyperplanes can be obtained by resolving the following pair of QPPs:

$$\begin{aligned}
\underset{w_1, b_1}{Min} \quad & \tfrac{1}{2}\|Aw_1 + e_1 b_1\|^2 + \tfrac{c_1}{2} y^T y + \tfrac{c_2}{2} z^T z \\
s.t \quad & -(Bw_1 + e_2 b_1) + y = e_2 \\
& -(Cw_1 + e_3 b_1) + z = e_3(1 - \varepsilon)
\end{aligned} \tag{3}$$

$$\begin{aligned}
\underset{w_2, b_2}{Min} \quad & \tfrac{1}{2}\|Bw_2 + e_2 b_2\|^2 + \tfrac{c_3}{2} y^T y + \tfrac{c_4}{2} z^T z \\
s.t \quad & (Aw_2 + e_1 b_2) + y = e_1 \\
& (Cw_2 + e_3 b_2) + z = e_3(1 - \varepsilon)
\end{aligned} \tag{4}$$

The positive parameter ε is restricted to be lower than 1 to avoid overlapping in the stage of decision. The square of 2-norm of slack variables are y and z. By substituting the equality constraints into the primal problem and setting the gradient with respect to w1, b1, w2, b2 to zeros. The solution can be shown as follows:

$$\begin{bmatrix} w_1 \\ b_1 \end{bmatrix} = - \begin{bmatrix} A^T A + c_1 B^T B + c_2 C^T C & A^T e_1 + c_1 B^T e_2 + c_2 C^T e_3 \\ e_1^T A + c_1 e_2^T B + c_2 e_3^T C & e_1^T e_1 + c_1 e_2^T e_2 + c_2 e_3^T e_3 \end{bmatrix}^{-1} \begin{bmatrix} c_1 B^T e_2 + c_2 C^T e_3(1 - \varepsilon) \\ c_1 e_2^T e_2 + c_2 e_3^T e_3(1 - \varepsilon) \end{bmatrix} \tag{5}$$

$$\begin{bmatrix} w_2 \\ b_2 \end{bmatrix} = \begin{bmatrix} B^T B + c_3 A^T A + c_4 C^T C & B^T e_2 + c_3 A^T e_1 + c_4 C^T e_3 \\ e_2^T B + c_3 e_1^T A + c_4 e_3^T C & e_2^T e_2 + c_3 e_1^T e_1 + c_4 e_3^T e_3 \end{bmatrix}^{-1} \begin{bmatrix} c_3 A^T e_1 + c_4 C^T e_3 (1 - \varepsilon) \\ c_3 e_1^T e_1 + c_4 e_3^T e_3 (1 - \varepsilon) \end{bmatrix}$$

$$(6)$$

LST-KSVC computes two low-dimensional inverse matrices, so the computational complexity is low. A new testing point x_i gets the category by the following decision function:

$$f(x_i) = \begin{cases} +1 & if \ x_i w_1 + e b_1 > -1 + \varepsilon \\ -1 & if \ x_i w_2 + e b_2 < 1 - \varepsilon \\ 0 & otherwise \end{cases} \qquad (7)$$

In the "1-versus-1-versus-rest" structure, the best existed method LST-KSVC need to constructs LST-KSVC K(k−1)/2 classifiers. Each (class(i), class(j))-LST-KSVC is trained over all the training points, where class(i) and class(j) represent labels "+1" and "−1". And then, if we have a testing point xi, a vote is given to the class(i) or class(j) based on which condition is satisfied. Finally, the result label of xi is up to the most votes. This method not only has poor classification effect, but also needs to build more classifiers as show in Fig. 1a. LST-KSVC is based on the red line of ε parameter to determine the current pattern category in three classes of MNIST data. Therefore, we change the classification structure to "1-versus-1-versus-versus-1" by modify the constraints of LST-KSVC, and maximize the distance between the current class and other class data. At the same time, a new decision function is established to extract the pattern category through the minimum of the current prediction output as shown in Fig. 1b.

3 Advanced Least Squares Twin Multi-class Classification Support Vector Machine

In this section, we will introduce the algorithm of Advanced Least Squares Twin Multi-Class Support Vector Machine (ALST-KSVC) which following the idea of LST-KSVC for image recognition. Let matrix $A \in R^{l_1 \times m}$ represent the training data belong to "1", $B \in R^{l_2 \times m}$ represents the training data belong to "2" and $C \in R^{l_3 \times m}$ represents the training data belong to "3".

3.1 Linear ALST-KSVC

We modify the optimization function of (3) and (4) in the new equality constrains as follows:

$$\begin{aligned} \underset{w_1, b_1}{Min} \quad & \tfrac{1}{2} \|A w_1 + e_1 b_1\|^2 + \tfrac{c_1}{2} y^T y + \tfrac{c_2}{2} z^T z \\ s.t \quad & (B w_1 + e_2 b_1) + y = e_2 \\ & (C w_1 + e_3 b_1) + z = e_3 \end{aligned} \qquad (8)$$

$$\underset{w_2,b_2}{Min} \quad \frac{1}{2}\|Bw_2 + e_2b_2\|^2 + \frac{c_3}{2}y^Ty + \frac{c_4}{2}z^Tz$$
$$s.t \quad (Aw_2 + e_1b_2) + y = e_1 \tag{9}$$
$$(Cw_2 + e_3b_2) + z = e_3$$

$$\underset{w_3,b_3}{Min} \quad \frac{1}{2}\|Cw_3 + e_3b_3\|^2 + \frac{c_5}{2}y^Ty + \frac{c_6}{2}z^Tz$$
$$s.t \quad (Aw_3 + e_1b_3) + y = e_1 \tag{10}$$
$$(Bw_3 + e_2b_3) + z = e_2$$

The objective function of QPP (8) makes class A close to $xw_1 + b_1 = 0$, constraints make class B and class C close to $xw_1 + b_1 = 1$. This modification allows to remove ε threshold and create a new decision function. By substituting the equality constraints into the QPP, the QPP can be represented by lagrange vectors:

$$L = \frac{1}{2}\|Aw_1 + e_1b_1\|^2 + \frac{c_1}{2}\|e_2 - (Bw_1 + e_2b_1)\|^2 + \frac{c_2}{2}\|e_3 - (Cw_1 + e_3b_1)\|^2 \tag{11}$$

Setting the gradient of (11) with respect to w1 and b1 to zero:

$$A^T(Aw_1 + e_1b_1) + c_1B^T(Bw_1 + e_2b_1 - e_2) + c_2C^T(Cw_1 + e_3b_1 - e_3) = 0 \tag{12}$$

$$e_1^T(Aw_1 + e_1b_1) + c_1e_2^T(Bw_1 + e_2b_1 - e_2) + c_2e_3^T(Cw_1 + e_3b_1 - e_3) = 0 \tag{13}$$

Arranging (12) and (13) in matrix form and let $E = [A \quad e_1]$, $F = [B \quad e_2]$, $G = [C \quad e_3]$:

$$\begin{bmatrix} w_1 \\ b_1 \end{bmatrix} = (c_1F^TF + E^TE + c_2G^TG)^{-1}(c_1F^Te_2 + c_2G^Te_3) \tag{14}$$

Similarly, the other solutions of QPP can be shown as follows:

$$\begin{bmatrix} w_2 \\ b_2 \end{bmatrix} = (c_3E^TE + F^TF + c_4G^TG)^{-1}(c_3E^Te_1 + c_4G^Te_3) \tag{15}$$

$$\begin{bmatrix} w_3 \\ b_3 \end{bmatrix} = (c_5E^TE + G^TG + c_6F^TF)^{-1}(c_5E^Te_1 + c_6F^Te_2) \tag{16}$$

3.2 Kernel ALST-KSVC

We also considering resolve nonlinear ALST-KSVC by introducing the kernel function [17–19]. And the QPPs of the Kernel ALST-KSVC can be modified in the same way are given as following:

$$\underset{w_1,b_1}{Min} \quad \tfrac{1}{2}\|K(A,D^T)w_1 + e_1 b_1\|^2 + \tfrac{c_1}{2}y^T y + \tfrac{c_2}{2}z^T z$$
$$s.t \quad (K(B,D^T)w_1 + e_2 b_1) + y = e_2$$
$$(K(C,D^T)w_1 + e_3 b_1) + z = e_3 \tag{17}$$

$$\underset{w_2,b_2}{Min} \quad \tfrac{1}{2}\|K(B,D^T)w_2 + e_2 b_2\|^2 + \tfrac{c_3}{2}y^T y + \tfrac{c_4}{2}z^T z$$
$$s.t \quad (K(A,D^T)w_2 + e_1 b_2) + y = e_1$$
$$(K(C,D^T)w_2 + e_3 b_2) + z = e_3 \tag{18}$$

$$\underset{w_3,b_3}{Min} \quad \tfrac{1}{2}\|K(C,D^T)w_3 + e_3 b_3\|^2 + \tfrac{c_5}{2}y^T y + \tfrac{c_6}{2}z^T z$$
$$s.t \quad (K(A,D^T)w_3 + e_1 b_3) + y = e_1$$
$$(K(B,D^T)w_3 + e_2 b_3) + z = e_2 \tag{19}$$

Where $D = [A; B; C]$, and the kernel function K is an arbitrary kernel. The solution of QPPs (17), (18) and (19) can be formulated as:

$$\begin{bmatrix} w_1 \\ b_1 \end{bmatrix} = \left(c_1 N^T N + M^T M + c_2 O^T O\right)^{-1}\left(c_1 N^T e_2 + c_2 O^T e_3\right) \tag{20}$$

$$\begin{bmatrix} w_2 \\ b_2 \end{bmatrix} = \left(c_3 M^T M + N^T N + c_4 O^T O\right)^{-1}\left(c_3 M^T e_1 + c_4 O^T e_3\right) \tag{21}$$

$$\begin{bmatrix} w_3 \\ b_3 \end{bmatrix} = \left(c_5 M^T M + O^T O + c_6 N^T M\right)^{-1}\left(c_5 M^T e_1 + c_6 N^T e_2\right) \tag{22}$$

Where $M = [K(A,D) \quad e_1]$, $N = [K(B,D) \quad e_2]$ and $O = [K(C,D) \quad e_3]$. So the solution of size of Kernel ALST-KSVC is $(l_1 + l_2 + l_3 + 1) \times (l_1 + l_2 + l_3 + 1)$.

3.3 Decision Function

As shown in Subsects. 3.1 and 3.2, ALST-KSVC evaluates all training data into the "1-versus-1-versus-1" structure. So for a new testing point xi, the decision function of Linear ALST-KSVC and Kernel ALST-KSVC by the following:

$$f(x_i) = \begin{cases} 1 & if \ x_i w_1 + e b_1 < x_i w_2 + e b_2 \ and \ x_i w_1 + e b_1 < x_i w_3 + e b_3 \\ 2 & if \ x_i w_2 + e b_2 < x_i w_1 + e b_1 \ and \ x_i w_2 + e b_2 < x_i w_3 + e b_3 \\ 3 & if \ x_i w_3 + e b_3 < x_i w_1 + e b_1 \ and \ x_i w_3 + e b_3 < x_i w_2 + e b_2 \end{cases} \tag{23}$$

$$f(x_i) = \begin{cases} 1 & if \ K(x_i,D^T)w_1 + e b_1 < K(x_i,D^T)w_2 + e b_2 \ and \ K(x_i,D^T)w_1 + e b_1 < K(x_i,D^T)w_3 + e b_3 \\ 2 & if \ K(x_i,D^T)w_2 + e b_2 < K(x_i,D^T)w_1 + e b_1 \ and \ K(x_i,D^T)w_2 + e b_2 < K(x_i,D^T)w_3 + e b_3 \\ 3 & if \ K(x_i,D^T)w_3 + e b_3 < K(x_i,D^T)w_1 + e b_1 \ and \ K(x_i,D^T)w_3 + e b_3 < K(x_i,D^T)w_2 + e b_2 \end{cases} \tag{24}$$

4 Experiments

In this section, we conduct experiments for image classification problems to evaluate the ALST-KSVC approach. In addition, in order to further compare with LST-KSVC method, we still use UCI datasets to verify our method.

4.1 Datasets

MNIST and USPS are pixel datasets. Amazon and Webcam extract 800-dimensional SURF features from the original image, and then use PCA [18] to reduce dimensions. we also conduct some experiments on five benchmark datasets: Iris, Teaching evaluation, Balance, Wine, and Transfusion from the UCI machine learning Repository. The detailed statistics with category and pattern numbers in Table 1.

Table 1. Statistics of the nine benchmark image/UCI datasets

Dataset	Type	Patterns	Attributes	Classes
MNIST	Image	2000	256	10
USPS	Image	1800	256	10
Amazon	Image	268	128	3
Caltech	Image	361	128	3
Iris	UCI	150	4	3
Teaching	UCI	151	5	3
Balance	UCI	625	4	3
Wine	UCI	178	13	3
Transfusion	UCI	748	4	3

- **MNIST**: MNIST handwritten digit dataset. The datasets consists of 2000 images randomly selected samples from MNIST.
- **USPS**: The US Postal handwritten digit dataset. The datasets consists of 1800 images randomly selected samples from USPS.
- **Amazon**: Images downloaded from online merchants. The datasets consists of 268 images randomly selected samples from Amazon [19].
- **Caltech**: A standard database for object recognition. The datasets consists of 361 images randomly selected samples from Caltech.

4.2 Experimental Results

Parameter Selection: The performance of SVMs family methods depends heavily on the choices of parameters. So in our comparative experiments, we first set the parameters range, and then obtain the optimal parameters by grid search. In addition, we set $c_1 = c_3 = c_5$, $c_2 = c_4 = c_6$ to reduce the complexity of searching c parameter. The c parameters were selected from the range of $\{2^i | i = -4, -2, -, 1, 2, 4, 6, 8\}$, ε parameter was selected from $\{0.1, 0.2, 0.3, 0.4\}$, kernel function type was selected from {"No kernel", "Linear", "Sigmoid", "Poly", "Gauss"} and the gauss kernel parameter γ was selected from $\{2^i | i = -4, -2, -, 1, 2, 4, 6, 8\}$, the other kernel function of parameter is $1/m$.

Implementation Step: First, enter the target domain of the training data set and parameters. Second, the matrix of solutions w and b are updated according to Eqs. 14, 15, 16 or Eqs. 20, 21, 22, then the K-classes separation model is established. Final, using the decision function Eqs. 23 or 24 to obtain the category of current sample.

Image Datasets

To illustrate the effect of classifier on the image classification results, image datasets has been evaluated. From the result of Table 2, ALST-KSVC achieves a significant improvement to recognition accuracy, with **20.0%** relative gain. Different classifier employ different structure and decision function, LST-KSVC uses the "1-versus-versus-1" and threshold, but ALST-KSVC uses the "1-versus-versus-1" and minimum model output to decision. So this improvement mainly derives from the new structure. As shown in Fig. 1, we take MNIST three-class image data as an example. LST-KSVC needs to set threshold of ε in Fig. 1a. If the output of model is greater than the ε, then the data belongs to category A. If the output of the model is less than the ε, it belongs to category B. The remaining patterns belong to category C. Because of its structural characteristics, ALST-KSVC does not need threshold in Fig. 1b. It only needs to compare the outputs of the three models and whose output value is small then the pattern belongs to the corresponding category.

Table 2. Performance comparison of 3-classes in image datasets

Dataset	MNIST	USPS	Amazon	Caltech
LST-KSVC	$0.3\ 2^8\ 2^2$	$0.1\ 2^8\ 2^6$	$0.4\ 2^{-4}\ 2^{-1}$	$0.4\ 2^{-2}\ 2^2$
ε, c1, c2	No kernel	No kernel	Gauss(2^{-4})	Poly
kernel(γ)	72.66%	82.35%	63.64%	64.38%
acc				
ALST-KSVC	$2^{-4}\ 2^{-4}$	$2^{-4}\ 2^{-4}$	$2^{-4}\ 2^{-4}$	$2^{-4}\ 2^{-4}$
c1, c2	No kernel	No kernel	Gauss(2^{-4})	Gauss(2^{-4})
kernel(γ)	**92.19%**	**94.77%**	**90.91%**	**84.93%**
acc				

Table 3. Performance comparison of 10-classes in image datasets

Dataset	MNIST	USPS
ALST-KSVC	$2^{-4}\ 2^{-4}\ 2^{-4}\ 2^{-4}\ 2^{-4}$	$2^{-4}\ 2^{-4}\ 2^{-4}\ 2^{-2}\ 2^{-2}$
c	$2^{-4}\ 2^{-4}\ 2^{-4}\ 2^{-4}\ 2^{-4}$	$2^{-4}\ 2^{-4}\ 2^{-2}\ 1\ 2^{-4}$
kernel(γ)	No kernel	No kernel
acc	**85.71%**	**86.30%**

ALST-KSVC implements K-classes requires stack QPP and add constraints of corresponding class, so only K classifiers are constructed compared to LST-KSVC. We conduct 10-classes on MNIST and USPS datasets in Table 3. Since Amazon and Caltech only have 3-classes, so it does not appear in the Table 3.

UCI Datasets

The classification accuracies of ALST-KSVC and the five UCI datasets are illustrated in Table 1. We observe that ALST-KSVC achieves much better performance than the LST-KSVC with statistical significance. The average classification accuracy of ALST-KSVC on the five datasets is **78.44%** and the performance improvement is **9.32%** compared to LST-KSVC. This verifies that ALST-KSVC can construct more effective and robust model representation for classification tasks. Figure 2 takes Iris dataset as an example. From Fig. 2a, it can be seen that LST-KSVC can not classify the third category clearly by using the "1-versus-1versus-rest" structure and threshold decision function. But the use of ALST-KSVC can be clearly classified, thus achieving batter accuracy as show in Fig. 2b.

Table 4. Performance comparison of 3-classes in UCI datasets

Dataset	Iris	Wine	Teaching	Balance	Transf.
LST-KSVC	$0.1\ 2^{-4}\ 2^2$	$0.4\ 1\ 2^4$	$0.1\ 2^{-4}\ 1$	$0.2\ 2^4\ 2^{-2}$	$0.1\ 2^2\ 2^{-4}$
ε, c1, c2	Gauss (2^{-4})	Poly	Gauss (2^{-2})	Linear	Sigmoid
kernel(γ)	73.33%	91.89%	61.29%	89.68%	76.00%
acc					
ALST-KSVC	$2^{-4}\ 2^{-4}$	$2^{-4}\ 1$	$2^{-4}\ 2^{-4}$	$2^6\ 2^{-4}$	$2^2\ 2^2$
c1, c2	Gauss(2^{-4})	Poly	Gauss (2^{-4})	Poly	Poly
kernel(γ)	**100.00%**	**100.00%**	67.74%	**94.44%**	**76.67%**
acc					

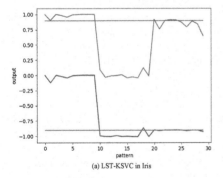
(a) LST-KSVC in Iris

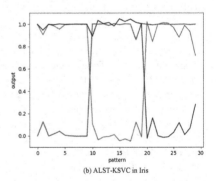
(b) ALST-KSVC in Iris

Fig. 2. Illustration of multi-class LST-KSVC (a) and ALST-KSVC (b) in Iris datasets. The horizontal axis represents the input pattern and the vertical axis represents the output of the model.

5 Conclusion

In this paper, we formulate a advanced version of LST-KSVC for few-shot classification. The proposed method evaluates all the training data into "1-versus-1-versus-1" structure by modify the quadratic programming problems of LST-KSVC and propose a new decision function to suit the structure. This allows ALST-KSVC only constructs K classifiers based on the new structure. Comprehensive experiments on MNIST, USPS, Amazon, Caltech image datasets by randomly selected and Iris, Teaching evaluation, Balance, Wine, Transfusion UCI datasets demonstrate that ALST-KSVC achieve superior performance compare with LST-KSVC. In the few-shot classification business scene, we will use the prior knowledge to achieve better classification accuracy by transfer learning in the future.

References

1. Simonyan, K., Zisserman, A.: Very deep convolutional networks for large-scale image recognition. In: International Conference on Learning Representations, May 2015
2. Ioffe, S., Szegedy, C.: Batch normalization: accelerating deep network training by reducing internal covariate shift. In: International Conference on International Conference on Machine Learning. JMLR.org (2015)
3. He, K., Zhang, X., Ren, S., et al.: Deep residual learning for image recognition (2015)
4. Huang, G., Liu, Z., Laurens, V.D.M., et al.: Densely connected convolutional networks (2016)
5. Kumar, M.A., Gopal, M.: Least squares twin support vector machines for pattern classification. Expert Syst. Appl. 36(4), 7535–7543 (2009)
6. Shao, Y.H., Deng, N.Y., Yang, Z.M.: Least squares recursive projection twin support vector machine for classification. Int. J. Mach. Learn. Cybern. 45(6), 2299–2307 (2012)
7. Lecun, Y., Bottou, L., Bengio, Y., et al.: Gradient-based learning applied to document recognition. Proc. IEEE 86(11), 2278–2324 (1998)
8. Wu, Y.C., Lee, Y.S., Yang, J.C.: Robust and efficient multiclass SVM models for phrase pattern recognition. Pattern Recogn. 41(9), 2874–2889 (2008)
9. Liu, R., Wang, Y., Baba, T., et al.: SVM-based active feedback in image retrieval using clustering and unlabeled data. Pattern Recogn. 41(8), 2645–2655 (2008)
10. Jayadeva, Khemchandani, R., Chandra, S.: Twin support vector machines for pattern classification. IEEE Trans. Pattern Anal. Mach. Intell. 29(5), 905–910 (2007)
11. Shao, Y.H., Zhang, C.H., Wang, X.B., et al.: Improvements on twin support vector machines. IEEE Trans. Neural Netw. 22(6), 962–968 (2011)
12. Bottou, L., Cortes, C., Denker, J.S., et al.: Comparison of classifier methods: a case study in handwritten digit recognition. In: International Conference on Pattern Recognition (1994)
13. Krebel, U.: Pairwise classification and support vector machines. In: Advances in Kernel Methods. MIT Press (1999)
14. Angulo, C., Parra, X., Català, A.: K-SVCR. A support vector machine for multi-class classification. Neurocomputing 55(1–2), 57–77 (2003)
15. Xu, Y., Guo, R., Wang, L.: A twin multi-class classification support vector machine. Cogn. Comput. 5(4), 580–588 (2013)
16. Nasiri, J.A., Charkari, N.M., Jalili, S.: Least squares twin multi-class classification support vector machine. Pattern Recogn. 48(3), 984–992 (2015)
17. Cristianini, N.: An introduction to support vector machines and other kernel-based learning methods. Kybernetes 32(1), 1–28 (2001)
18. Moore, B.C.: Principal component analysis in linear systems: controllability, observability, and model reduction. IEEE Trans. Autom. Control 26(1), 17–32 (2003)
19. Saenko, K., Kulis, B., Fritz, M., Darrell, T.: Adapting visual category models to new domains. In: Daniilidis, K., Maragos, P., Paragios, N. (eds.) ECCV 2010. LNCS, vol. 6314, pp. 213–226. Springer, Heidelberg (2010). https://doi.org/10.1007/978-3-642-15561-1_16

LLN-SLAM: A Lightweight Learning Network Semantic SLAM

Xichao Qu and Weiqing Li[✉]

Nanjing University of Science and Technology, Nanjing, China
li_weiqing@njust.edu.cn

Abstract. Semantic SLAM is a hot research subject in the field of computer vision in recent years. The mainstream semantic SLAM method can perform real-time semantic extraction. However, under resource-constrained platforms, the algorithm does not work properly. This paper proposes a lightweight semantic LLN-SLAM method for portable devices. The method extracts the semantic information through the matching of the Object detection and the point cloud segmentation projection. In order to ensure the running speed of the program, lightweight network MobileNet is used in the Object detection and Euclidean distance clustering is applied in the point cloud segmentation. In a typical augmented reality application scenario, there is no rule to avoid the movement of others outside the user in the scene. This brings a big error to the visual positioning. So, semantic information is used to assist the positioning. The algorithm does not extract features on dynamic semantic objects. The experimental results show that the method can run stably on portable devices. And the positioning error caused by the movement of the dynamic object can be effectively corrected while establishing the environmental semantic map.

Keywords: Semantic SLAM · Object detection · Point cloud segmentation · Augmented reality

1 Introduction

The problem solved by SLAM system is how to establish the environment model in the process of movement through sensor information, while estimating the problem of its own motion [10]. If the sensor is a camera, it is called "visual SLAM". For visual SLAM, the algorithm inputs the image data acquired by the camera and outputs the location of the camera with the map point of the environment. Although the image data contains a lot of scene information, the mainstream SLAM algorithm can only calculate and output a set of sparse or dense point data in a static environment. These data can provide accurate scene location information in static scenes, but cannot provide higher-level semantic information to the environment for understanding and sensing of the application or to remove the effects of dynamic objects.

This work was realized by a student. This work is supported by National Key R&D Program of China (2018YFB1004904) and Nation Key Technology Research and Development of china during the "13th Five Year Plan": 41401010203, 315050502, 31511040202.

© Springer Nature Switzerland AG 2019
Z. Cui et al. (Eds.): IScIDE 2019, LNCS 11936, pp. 253–265, 2019.
https://doi.org/10.1007/978-3-030-36204-1_21

At present, the development of deep learning methods in the fields of Object recognition and Object detection provides a method for extracting semantic information of the environment. For example, the Object detection method based on convolution neural network (CNN) can automatically extract the features from the input image, realize the classification of objects in the image and refine the bounding box. The sensor data image frame processed by the visual SLAM contains various kinds of object information. Therefore, the semantic information of the deep learning combined with the position information provided by the SLAM algorithm can extract the semantic information of the scene and establish a scene semantic map.

However, considering the portability requirements of wearable or mobile devices, high performance GPU is difficult to carry. A semantic segmentation method based on MobileNet-ssd model and point cloud segmentation is proposed in this paper. The semantic error is reduced by means of 3D data projection matching. Combining the accuracy of visual slam positioning with the advantages of deep neural network in semantic extraction, the accuracy of slam algorithm is improved, and fast semantic map construction based on lightweight model is realized. The main algorithm flowchart of this paper is as follows (Fig. 1):

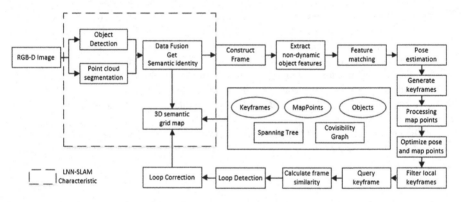

Fig. 1. LLN-SLAM main algorithm flowchart

2   Semantic SLAM

2.1   Related Work

Semantic SLAM means that the SLAM system not only obtains the geometric structure information in the environment, but also identifies the objects dates like position, posture, functional attributes and other semantic information, so as to cope with complex scenes and complete more intelligent service tasks [7]. The advantages of semantic SLAM are [1]: (1) The traditional SLAM method assumes the assumption of static environment, while the semantic SLAM can predict the movable properties of objects (people, cars, etc.). (2) The similar object knowledge representation in the semantic SLAM can be shared, and the scalability and storage efficiency of the SLAM system is improved by maintaining the shared knowledge base. (3) Semantic SLAM can realize intelligent interaction, such as augmented reality interaction conforms to real world physical laws.

Li and Belaroussi [8] perform 3D semantic mapping by means of monocular key frame image semantic segmentation. The author uses LSD-SLAM as a framework, combines CNN for organic fusion, and selects key frames for deep learning to achieve semantic segmentation.

Bowman, Atanasov [9] rely on low-level geometric features for the traditional SLAM method: point line surface and so on. These methods do not add semantic tags to the landmarks observed in the environment.

This paper focuses on the generation of semantic maps in resource constrained environment. The semantic information of the scene is extracted to generate the semantic map, the ORBSLAM2 is used to estimate the attitude of the RGB-D camera. And the semantic information is used to assist SLAM to eliminate the influence of dynamic objects (Fig. 2).

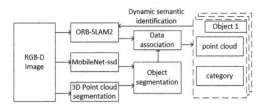

Fig. 2. LLN-SLAM system overview

2.2 Semantic Extraction

Although the image Object detection method runs faster, but because its recognition target is identified by a rectangular frame instead of the contour of the object, the loss of the depth scale and the gap between the rectangular frame and the contour of the object bring a large error to the semantic information. In this paper, the object is segmented by means of point cloud segmentation. The segmentation result is combined with the Object detection result to reduce the semantic error.

Object Detection Based on Lightweight Network

In the construction of semantic maps, it is necessary to obtain environmental semantic information. Semantics refers to the annotation of images or point clouds. The current main semantic acquisition methods are semantic segmentation and Object detection methods. The algorithm based on point cloud data runs slower and the overall recognition is not high. The image semantic segmentation method can classify every pixel in the image with accurate semantic information, but it is difficult to guarantee the real-time performance of the program without relying on GPU acceleration due to the prediction and classification of every pixel and the large amount of calculation.

In the construction of semantic map, we need the algorithm to provide original semantic information for the system. Because of the limitation of real-time performance of SLAM algorithm, the method for obtaining semantic information is required to run at a high speed. Therefore, this paper adopts End to End Object detection method: SSD [11] for semantic information extraction on key frames of SLAM system. In addition, the Mobilenet model [12] is a lightweight deep network model for mobile applications. Considering the resource limitations of wearable or mobile devices, this paper finally designs the network model with reference to the MobileNet-ssd framework. The VOC data is used as the training data sample, the classification number is 21, and the model size after training is only about 20M, which satisfies the hardware limitations of the device.

Point Cloud Segmentation Based on Euclidean Clustering

At present, the method based on object point cloud segmentation mainly consists of two parts: the segmentation of the object and the environment, and the segmentation between the objects. Because the planar color features in the indoor environment are more obvious, the rgb-based sampling random consistency algorithm is adopted in this paper, which divides the plane in the scene by the normal direction and color features of each point.

Based on the real laboratory scene and the TUM-RGBD data set, the structure point cloud is generated for each frame of data, and the plane segmentation effect is tested. The result of the scene plane segmentation is as follows (Fig. 3):

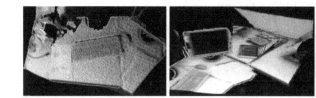

Fig. 3. Plane segmentation result (identified by outline)

Mainstream segmentation methods between objects include: regional growth method, Euclidean distance clustering and connected component analysis method [13]. Euclidean distance clustering has great advantages in algorithm robustness and running speed. Therefore, Euclidean distance clustering method is used to segment object point cloud (Fig. 4).

Fig. 4. Object point cloud segmentation result (color identification)

Semantic Extraction Based on Projection Matching

When the object point cloud is segmented, since the sensor input image size (width * height) of the visual SLAM is known. The object point cloud can be projected into the 2d plane by calculation. The pixel coordinates of each point are calculated as follows:

$$\text{cloud.point}[i].\text{pixel\_x} = \text{cloud.indices}[i] \% \text{width} \qquad (1)$$

$$\text{cloud.point}[i].\text{pixel\_y} = \text{cloud.indices}[i] / \text{width} \qquad (2)$$

After projecting the point cloud to the pixel coordinates, by comparing the maximum and minimum values of the point cloud in the x and y directions, the 2d bounding box of the object point cloud in pixel coordinates can be obtained. Then, the back-projection 2D bounding box of the point cloud is matched with the circumscribed rectangular bounding box output by the Object detection algorithm. And the object point cloud which is the most similar to the object detected by the Object detection is recorded, and the 3D target object with the label is generated (Fig. 5).

Fig. 5. The left picture shows the projection matching result (red: point cloud projection bounding box, green: Object detection result bounding box). The right picture shows the point cloud semantic fusion result. (Color figure online)

In this paper, IOU [14], the center distance of the frame, and the size difference are used to calculate the matching similarity.

Intersection over Union (IOU) is proposed in the evaluation system of Object detection, which calculates the overlap rate between the target area generated by the model and the original marked real area. The higher the correlation between the predicted range and the true range, the higher the value. This comparison is migrated to the comparison between the Object detection result rectangle and the point cloud segmentation projection rectangle:

$$IOU = \frac{Object\ detection\ box \cap segmentation\ box}{Object\ detection\ box \cup segmentation\ box} \tag{3}$$

In addition, the distance similarity is described by the distance between the centers of the two frames p1 and p2, and the size similarity is described by the area gap.

$$Center\ distance = \sqrt{\left((p1.x - p2.x)^2 + p1.y - p2.y^2\right)} \tag{4}$$

$$Area\ gap = \frac{max(Lengths\ in\ the\ row)}{min(Lengths\ in\ the\ row)} * \frac{max(Lengths\ in\ the\ col)}{min(Lengths\ in\ the\ col)} \tag{5}$$

$$Matching\ similarity = \frac{IOU}{Center\ distance * Area\ gap} \tag{6}$$

3   LLN-SLAM and Global Semantic Database Update

3.1   LLN-SLAM

Nowadays, Typical SLAM systems still have some common problems:

a. It is easily affected by dynamic objects, and it is hard to eliminate the noise influence of dynamic objects in the scene on camera displacement calculation.
b. The semantic information cannot be extracted from the scene, and only the map containing the geometric information can be generated. Taking ORBSLAM [2] as an example, it generates maps with Sparse map points and key frames.
c. The inefficiency of establishing a three-dimensional map of SLAM, especially in a wide range of scenarios, real-time performance cannot be guaranteed.

Aiming at the above problems, a real-time semantic LNN SLAM system based on ORBSLAM2 is designed in this paper, which combines semantic extraction and 3D mapping modules, and eliminates the noise caused by dynamic objects in the scene through semantic information (Fig. 6).

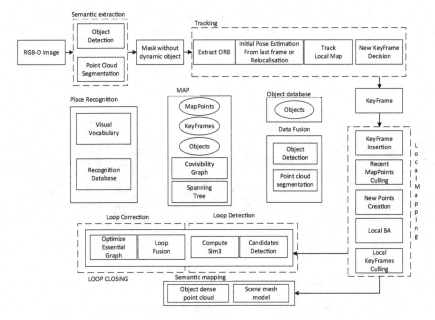

Fig. 6. LLN-SLAM framework

The association between key frames, map points and semantic information in this paper is imitated to the construction method of the relationship between key frames and graph points in ORBSLAM (Fig. 7).

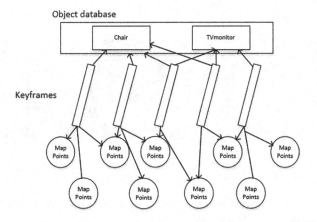

Fig. 7. The relationship between key frames, map points, and objects

After completing the construction relationship of objects, key frames and map feature points, the influence of dynamic objects on key frames eliminated by semantic information. The dynamic objects in the image are distinguished and labeled by the semantic extraction method introduced in Sect. 2.3 of the article. For example, points in the back projection contour box which detected as people are marked as unavailable. And feature points are not extracted from the contours of these dynamic objects, thereby improving the accuracy of SLAM positioning (Fig. 8).

Fig. 8. Original feature extraction (left), Semantic graph (middle), After feature screening (right)

3.2 Semantic Map Storage Structure Design

When the semantic database update is completed, the system records the obtained local semantic point cloud. Due to the limited resources of the wearable device, it is costly for the scene to save all the dense point clouds of the scene. The application only cares about the semantic point cloud information. For other dense point cloud information of the scene, the application only needs to obtain the surface condition and facilitate the interaction between the application and the environment. Therefore, the article only retains the dense point cloud of local semantics, and adopts a greedy triangle meshing method for the scene background to generate the global massive dense point cloud data into a single grid (Fig. 9).

Fig. 9. Semantic grid map

The method of mesh generation removes the useless information in the dense point cloud and generates a continuous surface model (Fig. 10).

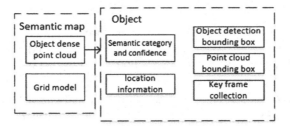

Fig. 10. Semantic map storage structure

3.3 Global Semantic Database Update

Since the input of the algorithm is the rgb and depth data collected by the camera each time, it may be that only a part of the actual object is collected at this moment. How to segment the semantic target according to the instance in the global model is a very important problem. This paper solves this problem by designing a global semantic database based on prior knowledge.

Since the target detection method model is based on dataset training and the object categories in the scene are known. This paper designs the object size information as a priori knowledge.

Whenever the algorithm updates the new 3D fusion data, it passes the data to the global semantic map database. If there is no object of the same type as the new data in the current database, insert it directly; otherwise, iterate through each object of the same type, find the nearest one to the new data center store, and calculate the distance between the two centers. If the distance is smaller than the size of the a priori object, the two parts are considered to be the same object in space, and the object information in the data is updated; if the distance exceeds the size of the object, it is considered to be the same object at different positions, and the new data is directly put into the database.

4   Experiment and Result Analysis

4.1   LNN SLAM Semantic Accuracy Test

In order to measure the efficiency of the algorithm, this paper tests the running efficiency of different algorithms on the same data set, taking the time and accuracy of each frame as the standard. We compare the algorithm of this paper with the high-precision Object detection method Faster RCNN, YOLO and semantic segmentation method FCN, Mask RCNN method on VOC dataset (Table 1).

LNNSLAM is based on the MobileNet, which enables the ideal speed requirements to be achieved on the CPU. Lightweight model requires little hardware resources.

The experimental platform is a portable computer, equipped with windows10 system, CPU model is i5-8400, no GPU. The experimental results are as follows:

Table 1. Semantic precision experiment results

Semantic segmentation	MIOU	Frame time (MS)	Model size
FCN [4]	62.2%	2851	64M
Mask-RCNN [3]	63.1%	18125	513M
LNN SLAM	58.7%	92	22M
Object detection	Accuracy	Frame time (MS)	Model size
YOLO v3 [5]	65.4%	212	236M
Faster-RCNN [6]	70.17%	2154	114M
LNN SLAM	68.2%	92	22M

4.2   LNN SLAM Positioning Accuracy Test

In this paper, the tracking and positioning accuracy of the semantic SLAM designed in this paper is tested. By comparing the generation trajectories of different SLAM systems under the same data set sequence. In order to compare static and dynamic scenarios, this paper chooses: TUM's fr1/room static scene and fr3/walking_xyz, fr3/walking_rpy dynamic scene RGB-D data (Table 2; Figs. 11, 12 and 13):

Table 2. Camera track RMSE

TUM data set [25]	Camera track RMSE (m*10^{-2})		
	RGBD-SLAM	ORB-SLAM	LNN SLAM
fr1/room	9.8412	6.3240	6.3303
fr3/walking_xyz	88.0750	58.8164	3.1949
fr3/walking_rpy	94.1421	62.2741	12.6499

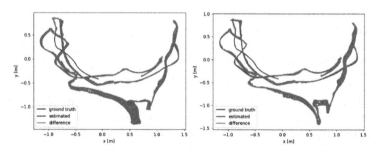

Fig. 11. Trajectory of ORBSLAM (left) and LNNSLAM (right) on fr1/room dataset

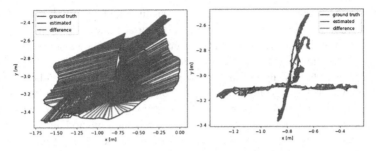

Fig. 12. Tracking of ORBSLAM (left) and LNNSLAM (right) on the fr3/walking_xyz dataset

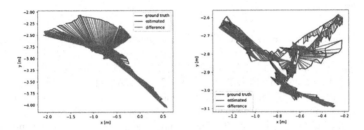

Fig. 13. Tracking of ORBSLAM (left) and LNNSLAM (right) on the fr3/walking_rpy dataset

In the static environment, the test results show that the trajectories of ORBSLAM and LNN SLAM are very similar, and their RMSE indicators are very close. However, since the data set still contains a person standing in the scene, LNN SLAM does not extract the feature point information of the person. Therefore, because of the difference in the number of feature points, ORB SLAM is better than LNN SLAM in the static environment. But the effect is not obvious.

The accuracy of RGBDSLAM [15] is lower because it calculates the camera pose through RANSAC, its calculation accuracy is lower than that of using the minimum reprojection error algorithm.

In the dynamic scene, only the trajectory of the LNNSLAM is close to the real trajectory. The other two SLAM systems have larger errors. Because LNNSLAM can ensure that the system feature points and map points are based on static objects, the positioning is accurate. However, once most of the feature points of ORBSLAM are marked on the dynamic object, it will cause a large deviation error in the camera positioning position.

5   Conclusion

In this paper, a lightweight and high-precision real-time semantic SLAM method is designed for wearable portable devices in a resource-constrained environment. By acquiring the camera displacement through SLAM and extracting the semantics through Object detection and point cloud segmentation algorithm, the semantic map of the user's environment can be constructed in real time and run stably on the mobile device. The semantics can also assist the positioning accuracy of the SLAM calculation, and greatly improve the positioning accuracy of the dynamic scene SLAM.

The paper verifies the effectiveness of the algorithm in laboratory and TUM and other indoor data sets. The algorithm can run at 14 frames per second, and the algorithm is robust. However, the algorithm operation rate and robustness of the outdoor scene need further research. In addition, there are still many improvements in the system of this paper, such as optimizing the point cloud segmentation method and perfecting the matching similarity calculation method of the fusion data to improve the accuracy of semantic information. These will be refined and improved in the next experiments.

References

1. Salas-Moreno, R.F.: Dense semantic SLAM. Doctoral dissertation, Imperial College London (2014)
2. Mur-Artal, R., Montiel, J.M.M., Tardos, J.D.: ORB-SLAM: a versatile and accurate monocular SLAM system. IEEE Trans. Robot. **31**(5), 1147–1163 (2015)
3. He, K., Gkioxari, G., Dollár, P., Girshick, R.: Mask R-CNN. In: Proceedings of the IEEE International Conference on Computer Vision, pp. 2961–2969 (2017)
4. Long, J., Shelhamer, E., Darrell, T.: Fully convolutional networks for semantic segmentation. In: Proceedings of the IEEE Conference on Computer Vision and Pattern Recognition, pp. 3431–3440 (2015)
5. Redmon, J., Farhadi, A.: Yolov3: An incremental improvement. arXiv preprint arXiv:1804.02767 (2018)
6. Girshick, R.: Fast R-CNN. In: Proceedings of the IEEE International Conference on Computer Vision, pp. 1440–1448 (2015)
7. Yu, J.S., Wu, H., Tian, G.H., et al.: Semantic database design and semantic map construction of robots based on the cloud. Robot **38**(4), 410–419 (2016)
8. Li, X., Ao, H., Belaroussi, R., Gruyer, D.: Fast semi-dense 3D semantic mapping with monocular visual SLAM. In: 2017 IEEE 20th International Conference on Intelligent Transportation Systems (ITSC), pp. 385–390. IEEE (2017)
9. McCormac, J., Handa, A., Davison, A., Leutenegger, S.: Semanticfusion: dense 3D semantic mapping with convolutional neural networks. In: 2017 IEEE International Conference on Robotics and Automation (ICRA), pp. 4628–4635. IEEE (2017)
10. Davison, A.J., Reid, I.D., Molton, N.D., Stasse, O.: MonoSLAM: real-time single camera SLAM. IEEE Trans. Pattern Anal. Mach. Intell. **6**, 1052–1067 (2007)
11. Liu, W., et al.: SSD: single shot multibox detector. In: Leibe, B., Matas, J., Sebe, N., Welling, M. (eds.) ECCV 2016. LNCS, vol. 9905, pp. 21–37. Springer, Cham (2016). https://doi.org/10.1007/978-3-319-46448-0_2
12. Howard, A.G., et al.: Mobilenets: Efficient convolutional neural networks for mobile vision applications. arXiv preprint arXiv:1704.04861 (2017)
13. Trevor, A.J., Gedikli, S., Rusu, R.B., Christensen, H.I.: Efficient organized point cloud segmentation with connected components. In: Semantic Perception Mapping and Exploration (SPME) (2013)
14. Nowozin, S.: Optimal decisions from probabilistic models: the intersection-over-union case. In: Proceedings of the IEEE Conference on Computer Vision and Pattern Recognition, pp. 548–555 (2014)

15. Endres, F., Hess, J., Sturm, J., Cremers, D., Burgard, W.: 3-D mapping with an RGB-D camera. IEEE Trans. Robot. **30**(1), 177–187 (2014)
16. Bowman, S.L., Atanasov, N., Daniilidis, K., Pappas, G.J.: Probabilistic data association for semantic SLAM. In: 2017 IEEE International Conference on Robotics and Automation (ICRA), pp. 1722–1729. IEEE (2017)
17. Ma, L., Stückler, J., Kerl, C., Cremers, D.: Multi-view deep learning for consistent semantic mapping with RGB-D cameras. In: 2017 IEEE/RSJ International Conference on Intelligent Robots and Systems (IROS), pp. 598–605. IEEE (2017)
18. DeTone, D., Malisiewicz, T., Rabinovich, A.: Toward geometric deep SLAM. arXiv preprint arXiv:1707.07410 (2017)
19. Zhou, T., Brown, M., Snavely, N., Lowe, D.G.: Unsupervised learning of depth and ego-motion from video. In: Proceedings of the IEEE Conference on Computer Vision and Pattern Recognition, pp. 1851–1858 (2017)
20. Kendall, A., Grimes, M., Cipolla, R.: Posenet: a convolutional network for real-time 6-DOF camera relocalization. In: Proceedings of the IEEE International Conference on Computer Vision, pp. 2938–2946 (2015)
21. Yang, N., Wang, R., Stuckler, J., Cremers, D.: Deep virtual stereo odometry: leveraging deep depth prediction for monocular direct sparse odometry. In: Ferrari, V., Hebert, M., Sminchisescu, C., Weiss, Y. (eds.) ECCV 2018. LNCS, vol. 11212, pp. 835–852. Springer, Heidelberg (2018). https://doi.org/10.1007/978-3-030-01237-3_50
22. Carvalho, L.E., von Wangenheim, A.: 3D object recognition and classification: a systematic literature review. Pattern Anal. Appl. 1–50 (2019)
23. Tekin, B., Sinha, S.N., Fua, P.: Real-time seamless single shot 6D object pose prediction. In: Proceedings of the IEEE Conference on Computer Vision and Pattern Recognition, pp. 292–301 (2018)
24. Brachmann, E., Rother, C.: Learning less is more-6D camera localization via 3D surface regression. In: Proceedings of the IEEE Conference on Computer Vision and Pattern Recognition, pp. 4654–4662 (2018)
25. Sturm, J., Burgard, W., Cremers, D.: Evaluating egomotion and structure-from-motion approaches using the TUM RGB-D benchmark. In: Proceedings of the Workshop on Color-Depth Camera Fusion in Robotics at the IEEE/RJS International Conference on Intelligent Robot Systems (IROS) (2012)

Meta-cluster Based Consensus Clustering with Local Weighting and Random Walking

Nannan He and Dong Huang[✉]

College of Mathematics and Informatics,
South China Agricultural University, Guangzhou, China
henannan11@hotmail.com, huangdonghere@gmail.com

Abstract. Consensus clustering has in recent years become one of the most popular topics in the clustering research, due to its promising ability in combining multiple weak base clusterings into a strong consensus result. In this paper, we aim to deal with three challenging issues in consensus clustering, i.e., the high-order integration issue, the local reliability issue, and the efficiency issue. Specifically, we present a new consensus clustering approach termed **m**eta-**c**luster based **c**onsensus **c**lustering with **l**ocal weighting and **r**andom walking (MC^{3}LR). To ensure the computational efficiency, we use the base clusters as the graph nodes to construct a cluster-wise similarity graph. Then, we perform random walks on the cluster-wise similarity graph to explore its high-order structural information, based on which a new cluster-wise similarity measure is derived. To tackle the local reliability issue, all of the base clusters are assessed and weighted according to the ensemble-driven cluster index (ECI). Finally, a locally weighted meta-clustering process is performed on the newly obtained cluster-wise similarity measure to build the consensus clustering result. Experimental results on multiple datasets have shown the effectiveness and efficiency of the proposed approach.

Keywords: Consensus clustering · Ensemble clustering · High-order integration · Local weighting · Random walk

1 Introduction

Consensus clustering, also known as ensemble clustering or cluster ensembles, is the process of combining multiple base clusterings into a probably better clustering result [2–11]. It has proved to be a very powerful clustering technique with its capability of finding clusters of arbitrary shapes, dealing with data noise (or outliers), producing robust clusterings, and coping with data from multiple sources (or views).

In recent years, the consensus clustering technique has drawn increasing attention, and many consensus clustering methods have been developed [2–11]. These existing consensus clustering methods can be classified into three main

© Springer Nature Switzerland AG 2019
Z. Cui et al. (Eds.): IScIDE 2019, LNCS 11936, pp. 266–277, 2019.
https://doi.org/10.1007/978-3-030-36204-1_22

categories. The first category is the pair-wise co-occurrence based methods [3, 22], which typically construct a co-association matrix by considering the frequency that two objects appear in the same cluster among the multiple base clusterings [3]. By using the co-association matrix as the similarity matrix, the agglomerative clustering algorithm can then be performed to obtain the consensus clustering result [3]. Yi et al. [22] further considered the uncertain entries in the co-association matrix, and refined the co-association matrix by the matrix completion technique. The second category is the median partition methods [11, 20], which formulate the consensus clustering problem into an optimization problem, and aim to find a median clustering (or partition) such that the similarity between this clustering and the multiple base clusterings is maximized. To tackle the median partition problem, Topchy et al. [20] exploited the EM algorithm to obtain an approximate solution. Huang et al. [11] cast the median partition problem into a binary linear programming problem, and solve it by the factor graph model. The third category of consensus clustering is the graph partitioning based methods [2, 19]. Strehl and Ghosh [19] treated each cluster in the base clusterings as a hyper-edge, and proposed three hyper-graph partitioning based consensus clustering methods. Fern and Brodley [2] built a hybrid bipartite graph with both clusters and objects treated as the nodes, and partition this graph by the METIS algorithm [2] to obtain the final consensus clustering result.

These methods [2, 3, 11, 19, 20, 22] deal with the consensus clustering problem from different technical perspectives. However, most of them are still faced with three challenging issues, namely, (i) how to integrate the high-order structural information in ensembles, (ii) how to estimate the local reliability of different ensemble members and weight them accordingly, and (iii) how to maintain high efficiency in the meantime.

Recently, some efforts have been made to (partially) address the above-mentioned three issues. For example, Huang et al. [8] presented the normalized crowd agreement index (NCAI) to evaluate and weight the base clusterings, where each base clustering is viewed as an individual and is assigned a (global) weight accordingly. To go from global weighting to local weighting, Huang et al. [6] estimated the local uncertain of the clusters in each base clustering by an entropic criterion, where each cluster is viewed as a local region in the base clustering and then weighted by the newly designed ensemble-driven cluster index (ECI) [6]. The ECI index is further applied to the meta-clustering algorithm, which leads to the locally weighted meta-clustering (LWMC) algorithm [12] for consensus clustering. Despite the advance in the local weighting mechanism, these methods [6, 12] still cannot tackle the high-order integration issue. In a graph, the direct links between nodes correspond to the first-order relationship. To go beyond the first-order relationship, Iam-On et al. [13] exploited the second-order structural information (i.e., the common neighborhood information) to refine the co-association matrix, and proposed the weighted connected triple (WCT) method. Breaking through the second order to explore the higher-order relationship and maintaining the high efficiency at the same time, Huang et al. [9] preprocessed the original data objects into a smaller number of microclusters and performed random walks on the microcluster similarity graph. Based on the probability trajectories of the random walkers, a new similarity measure termed probability trajectories based similarity (PTS) is derived and then two novel consensus clustering algorithms are designed [9]. To investigate the high-order information at an even higher level of granularity than the microclusters, Huang et al. [4] further treated each base cluster as a graph node and performed random walks on the cluster-wise similarity graph, so as to obtain a new cluster-wise similarity matrix by integrating higher-order structural information in ensembles [4]. By random walking at the microcluster-level [9] or at the cluster-level [4], the higher-order information can be investigated in a very efficient manner; but a limitation to these methods [4,9] lies in that they neglect the local reliability issue in the base clusterings.

Despite the significant progress, each of these methods [4,6,8,9,12,13] is only able to address one or two of the aforementioned three issues (i.e., the high-order integration issue, the local reliability issue, and the efficiency issue). Few, if not none, of the existing methods are capable of tackling all these three issues in consensus clustering at the same time.

In light of this, this paper aims to address these three challenging issues in a new meta-cluster based consensus clustering approach based on local weighting and random walking (MC^3LR). Note that the MC^3LR approach incorporates the local weighting strategy [6] and the cluster-wise similarity propagation (by random walk) scheme [4] in a unified consensus clustering framework. Specifically, we first construct a similarity graph between base clusters, and conduct random walks on the clusters to derive a new cluster-wise similarity measure. Based on the new cluster-wise similarity measure, all the base clusters can be partitioned by normalized cut (NCut) [18] to obtain a set of meta-clusters. Each meta-cluster consists of a certain number of base clusters. With the weight of each base cluster computed according to the entropic criterion, the final consensus clustering can be achieved by the locally weighted majority voting. Experiments conducted on several benchmark datasets demonstrate the effectiveness and efficiency of the proposed MC^3LR approach.

The rest of the paper is organized as follows. Section 2 describes the proposed consensus clustering approach. Section 3 provides the experimental results. Section 4 concludes this paper.

2   Proposed Approach

In this section, we first formulate the consensus clustering problem in Sect. 2.1. Then, we build the cluster-wise similarity graph in Sect. 2.2, and conduct random walks on this graph in Sect. 2.3. Finally, we partition the base clusters into multiple meta-clusters, and obtain the consensus clustering by locally weighted majority voting in Sect. 2.4.

2.1 Formulation

The purpose of consensus clustering is to combine multiple base clusterings into a better clustering result. Given a dataset with N data objects, denoted as $\mathcal{X} = \{x_1, x_2, \cdots, x_N\}$, where x_i is the i-th object, each base clustering is a partition of the data objects in $\mathcal{X}$. Let

$$\Pi = \{\pi^1, \pi^2, \cdots, \pi^M\} \tag{1}$$

be the set of M base clusterings, where

$$\pi^m = \{C_1^m, C_2^m, \cdots, C_{n^m}^m\} \tag{2}$$

is the m-th base clustering in Π, C_i^m is the i-th cluster in π^m, and n^m is the number of clusters in π^m. By treating the ensemble Π as input, the objective of consensus clustering is to fuse the information of the M base clusterings in Π to achieve a final clustering result π^*.

2.2 Cluster-Wise Similarity Graph

Each base clustering consists of a certain number of clusters. For clarity, we can represent the set of all clusters in Π as follows:

$$\mathcal{C} = \{C_1, C_2, \cdots, C_{n^c}\} \tag{3}$$

where C_i is the i-th cluster and n^c is the total number of clusters in the ensemble Π. It is obvious that $n^c = n^1 + n^2 + \cdots + n^m$.

By treating each cluster as a node, we first construct a cluster-wise similarity graph, denoted as

$$G = \{\mathcal{V}, \mathcal{L}\}, \tag{4}$$

where $\mathcal{V} = \mathcal{C}$ is the node set and $\mathcal{L}$ is the link set. The link weight (i.e., similarity) between two clusters is defined by the Jaccard coefficient [15]. Given two clusters C_i and C_j, their Jaccard coefficient is computed as

$$Jaccard(C_i, C_j) = \frac{|C_i \bigcap C_j|}{|C_i \bigcup C_j|}, \tag{5}$$

where $|\cdot|$ denotes the number of elements (objects) in a set. Then, the similarity matrix of the graph G can be defined as

$$E = \{e_{ij}\}_{n^c \times n^c}, \tag{6}$$
$$e_{ij} = Jaccard(C_i, C_j). \tag{7}$$

With the cluster-wise similarity graph constructed, in the next section, we will proceed to explore its high-order structural information by random walk.

2.3   Random Walk Propagation on Clusters

The random walk is a dynamic process that at each step transits from one node to each of its neighbors with a certain probability [9]. Before conducting the random walk on the cluster-wise similarity graph, we first need to define the transition probability matrix for it. That is

$$P = \{p_{ij}\}_{n^c \times n^c},\tag{8}$$

$$p_{ij} = \frac{e_{ij}}{\sum_{k \neq i} e_{ik}},\tag{9}$$

where p_{ij} denotes the probability of a random walker transiting from node i to node j in one step. Starting from the one-step transition probability matrix, we can compute the T-step transition probability matrix as follows:

$$P^{(T)} = \begin{cases} P, & \text{if } T = 1, \\ P \times P^{(T-1)}, & \text{otherwise.} \end{cases}\tag{10}$$

We denote the i-th row in $P^{(T)}$ as $p_{i:}^{(T)}$, which corresponds to the probability of node i transiting each of the other nodes after T steps. It is recognized that the different steps of random walks can reflect the different scales of graph structure information [9]. Here, we represent the *probability trajectory* [9] of node i from step 1 to step T as a $T \cdot n^c$-tuple, that is

$$PT_i = [p_{i:}^{(1)}, p_{i:}^{(2)}, \cdots, p_{i:}^{(T)}],\tag{11}$$

Thereafter, the similarity between two nodes (i.e., two clusters) can be recomputed by considering the similarity of their probability trajectories. Note that any similarity measure can be used to compute the similarity of probability trajectories. In this paper, we adopted the cosine similarity, that is

$$Cosine(PT_i, PT_j) = \frac{<PT_i, PT_j>}{\sqrt{<PT_i, PT_i> \cdot <PT_j, PT_j>}},\tag{12}$$

where $< \cdot >$ denotes the inner product of two vectors. Then, the new cluster-wise similarity matrix can be represented as

$$\tilde{E} = \{\tilde{e}_{ij}\}_{n^c \times n^c},\tag{13}$$

$$\tilde{e}_{ij} = Cosine(PT_i, PT_j).\tag{14}$$

Thus, by exploiting the probability trajectories, the higher-order structural information can be integrated into the cluster-wise similarity by means of the random walks of different steps (or scales).

2.4   Consensus by Locally Weighted Majority Voting

With the new cluster-wise similarity matrix $\tilde{E}$, we can perform the NCut algorithm [18] on the graph derived from this new similarity matrix, and partition the set of base clusters into a certain number of *meta-clusters*. The obtained set of meta-clusters is represented as

$$\mathcal{MC} = \{MC_1, MC_2, \cdots, MC_k\}, \tag{15}$$

where MC_i is the i-th meta-cluster and k is the number of meta-clusters. Note that each meta-cluster is a set of clusters, and our final aim is to obtain a partition of the original data points. To achieve this aim, a common strategy is to map the meta-clustering result from the cluster-level to the object-level by (cluster-wise) majority voting. However, in the majority voting process, the different reliability of clusters is often neglected.

To assess the reliability of different clusters, we adopt the local weighting strategy in [6]. First, we use entropy to evaluate the uncertainty of each cluster by considering how this cluster *agrees* with other clusters in the ensemble. Specifically, the entropy (or uncertainty) of a cluster, say, C_i, w.r.t. those clusters in the base clustering π^m is computed as follows [6]:

$$H^m(C_i) = - \sum_{C_j^m \in \pi^m} \frac{|C_i \bigcap C_j^m|}{|C_i|} \log_2 \frac{|C_i \bigcap C_j^m|}{|C_i|}. \tag{16}$$

With the assumption that different base clusterings are independent of each other, we can compute the entropy of C_i w.r.t. the entire ensemble as follows [6]:

$$H^{\Pi}(C_i) = \sum_{m=1}^{M} H^m(C_i). \tag{17}$$

Note that $H^{\Pi}(C_i) \in [0, +\infty)$ indicates the uncertainty of cluster C_i by considering how it agrees with the clusters in different base clusterings. When the objects in cluster C_i are in the same cluster in all of the M base clusterings, its uncertainty $H^{\Pi}(C_i)$ reaches its minimum value.

With the uncertainty estimation, we proceed to compute the ensemble-driven cluster index (ECI) [6] to quantify the local reliability of each cluster. That is

$$ECI(C_i) = e^{-\frac{H^{\Pi}(C_i)}{M}} \tag{18}$$

Obviously, it holds that $ECI(C_i) \in (0, 1]$ for any cluster C_i in the ensemble Π.

After that, we can use the ECI measure to design the locally weighted majority voting mechanism for achieving the final clustering result. Given an object $x_i \in \mathcal{X}$ and a meta-cluster $MC_j \in \mathcal{MC}$, the locally weighted voting score of x_i w.r.t. MC_j is defined as

$$Score(x_i, MC_j) = \frac{1}{|MC_j|} \sum_{C_k \in MC_j} w(C_k) \cdot \delta(x_i, C_k), \tag{19}$$

$$w(C_k) = ECI(C_k), \tag{20}$$

$$\delta(x_i, C_k) = \begin{cases} 1, & \text{if } x_i \in C_k, \\ 0, & \text{otherwise,} \end{cases} \tag{21}$$

where $|MC_j|$ denotes the number of clusters in MC_j. Then, each object will be assigned to the meta-cluster with the highest voting score. That is

$$MetaCls(x_i) = \arg \max_{MC_j \in \mathcal{MC}} Score(x_i, MC_j). \tag{22}$$

Each object is assigned to a meta-cluster through locally weighted majority voting. By treating the objects in the same meta-cluster as a final cluster, the final consensus clustering result can thereby be obtained.

3   Experiments

In this section, we conduct experiments on multiple datasets to evaluate the proposed MC³LR approach against several other consensus clustering approaches.

3.1   Datasets and Experimental Setting

In our experiments, six datasets are used, i.e., *Wine, Ecoli, Cardiotocography (CTG), MNIST, Gisette*, and *Letter Recognition (LR)*. The *MNIST* dataset is from [14], while the others from the UCI repository [1]. The details of these datasets are given in Table 1.

Table 1. Details of the benchmark datasets.

Dataset	#Object	#Class	Dimension
Wine	178	3	13
Ecoli	336	8	7
CTG	2,126	10	21
MNIST	5,000	10	784
Gisette	7,000	2	5,000
LR	20,000	26	16

To compare different consensus clustering methods, an ensemble of $M = 20$ base clusterings will be generated by k-means with the number of clusters randomly selected in $[K, min(\sqrt{N}, 100)]$, where K is the true number of classes and N is the number of objects in each dataset. With an ensemble of base clusterings generated at each time, we will run each test method twenty times, and compare their average performances. In our MC³LR method, the parameter T will be set to 20 on all datasets. Additionally, the normalized mutual information (NMI) [19] is used as the evaluation measure to compare the clustering results of different methods. Note that a greater NMI indicates a better clustering.

Table 2. Average NMI scores (over 20 runs) by different consensus clustering methods. The best score in each row is highlighted in bold.

Dataset	MCLA	EAC	KCC	SEC	PTGP	ECC	LWGP	ECPCS	MC^{3}LR
Wine	0.822	0.863	0.860	0.861	0.868	0.833	**0.883**	0.879	**0.883**
Ecoli	0.492	0.583	0.496	0.518	0.503	0.506	0.591	0.595	**0.597**
CTG	0.247	0.262	0.233	0.244	0.252	0.236	0.262	**0.269**	**0.269**
MNIST	0.583	0.619	0.509	0.457	0.636	0.500	0.622	0.638	**0.644**
Gisette	0.417	0.270	0.173	0.121	0.471	0.292	0.397	0.470	**0.480**
LR	0.386	0.383	0.349	0.331	0.392	0.357	0.392	**0.393**	0.393
Avg. Rank	6.50	5.00	7.83	7.50	4.00	7.17	3.17	2.00	**1.00**

3.2   Comparison with Other Consensus Clustering Methods

In this section, we compare our MC^{3}LR method with eight consensus clustering methods, namely, evidence accumulation clustering (EAC) [3], meta-clustering algorithm (MCLA) [19], k-means based consensus clustering (KCC) [21], spectral ensemble clustering (SEC) [16], probability trajectory based graph partitioning (PTGP) [9], entropy based consensus clustering (ECC) [17], locally weighted graph partitioning (LWGP) [6], and ensemble clustering by propagating cluster-wise similarities (ECPCS) [4] associated with meta-cluster (MC) based consensus function. Note that the number of clusters for each consensus clustering method is set to the true number of classes in each dataset.

As shown in Table 2, our MC^{3}LR method achieves the best NMI scores on all of the six datasets, and obtains an average rank of 1.00, while the second and third best methods, i.e., ECPCS and LWGP, obtains average ranks of 2.00 and 3.17, respectively. The results in Table 2 have shown the advantage of the proposed method over the other consensus clustering methods.

3.3   Robustness to Ensemble Size M

In this section, we compare the NMI performances of different consensus clustering methods as the ensemble size varies from 10 to 50. As shown in Fig. 1, our MC^{3}LR method yields quite consistent performance under different ensemble sizes, and achieves overall the best performance on the six datasets.

3.4   Sensitivity of Parameter T

In our MC^{3}LR method, the parameter T denotes the length of the random walk trajectory (as described in Sect. 2.3). We test the sensitivity of parameter T in our method on the six datasets. As shown in Fig. 2, our MC^{3}LR method exhibits stably good performance as the value of parameter T goes from $2^0 = 1$ to $2^5 = 32$. Empirically, moderate values of parameter T, e.g., in the range of [10, 30], are suggested. In this paper, we use $T = 20$ on all of the six datasets.

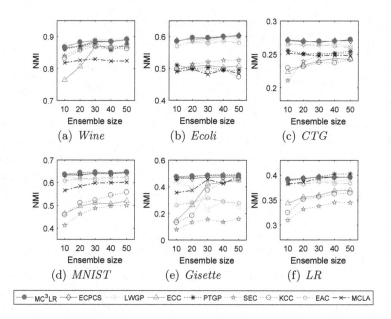

Fig. 1. The NMI performances of different consensus clustering methods with varying ensemble size M.

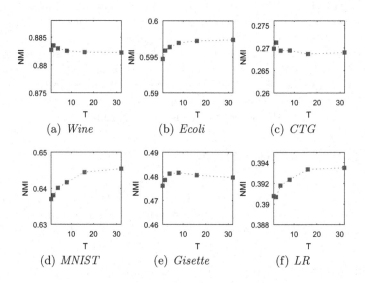

Fig. 2. The NMI performance of our MC³LR method with varying parameter T.

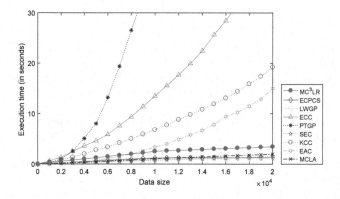

Fig. 3. The time costs of different consensus clustering methods on the *LR* dataset as the data size goes from 0 to 20,000.

3.5 Time Cost

In this paper, we test the time efficiency of different consensus clustering methods on the *LR* dataset. All experiments are conducted on a computer with an Intel i7-6700K CPU and 64 GB memory.

To compare the time costs under different data sizes, we perform the test methods on differently-sized subsets of the *LR* dataset, and report their time costs as the data size goes from 0 to 20,000. As shown in Fig. 3, our MC^{3}LR method is slightly slower than ECPCS, LWGP, SEC, and MCLA, but much faster than EAC, KCC, ECC, and PTGP, to process the entire *LR* dataset.

To summarize, the proposed MC^{3}LR method outperforms the other consensus clustering methods on most of the datasets (as shown in Table 2 and Fig. 1), while maintaining very competitive time efficiency (as shown in Fig. 3).

4 Conclusion

In this paper, we present a novel consensus clustering algorithm termed MC^{3}LR to simultaneously address three key issues in the consensus clustering research, namely, the high-order integration issue, the local reliability issue, and the efficiency issue. Specifically, we build a cluster-wise similarity graph by using base clusters as nodes, and perform random walks on this graph so as to derive a new cluster-wise similarity measure with higher-order information incorporated. To address the local reliability issue, we adopt the ECI index to estimate the reliability of each base cluster, and then perform the locally weighted meta-clustering on the newly obtained cluster-wise similarity measure to achieve the final consensus clustering result. Experiments have been conduced on six real-world datasets, which demonstrate the superiority of our MC^{3}LR method over the other consensus clustering methods.

Acknowledgments. This work was supported by NSFC (61976097 & 61602189).

References

1. Bache, K., Lichman, M.: UCI machine learning repository (2017). http://archive.ics.uci.edu/ml

2. Fern, X.Z., Brodley, C.E.: Solving cluster ensemble problems by bipartite graph partitioning. In: Proceedings of International Conference on Machine Learning (ICML) (2004)

3. Fred, A.L.N., Jain, A.K.: Combining multiple clusterings using evidence accumulation. IEEE Trans. Pattern Anal. Mach. Intell. **27**(6), 835–850 (2005)

4. Huang, D., Wang, C., Peng, H., Lai, J., Kwoh, C.: Enhanced ensemble clustering via fast propagation of cluster-wise similarities. IEEE Trans. Syst. Man Cybern.: Syst. (2018, in press). https://doi.org/10.1109/TSMC.2018.2876202

5. Huang, D., Wang, C., Wu, J., Lai, J., Kwoh, C.K.: Ultra-scalable spectral clustering and ensemble clustering. IEEE Trans. Knowl. Data Eng. (2019, in press). https://doi.org/10.1109/TKDE.2019.2903410

6. Huang, D., Wang, C.D., Lai, J.H.: Locally weighted ensemble clustering. IEEE Trans. Cybern. **48**(5), 1460–1473 (2018)

7. Huang, D., Lai, J.H., Wang, C.D.: Exploiting the wisdom of crowd: a multi-granularity approach to clustering ensemble. In: Proceedings of International Conference on Intelligence Science and Big Data Engineering (IScIDE), pp. 112–119 (2013)

8. Huang, D., Lai, J.H., Wang, C.D.: Combining multiple clusterings via crowd agreement estimation and multi-granularity link analysis. Neurocomputing **170**, 240–250 (2015)

9. Huang, D., Lai, J.H., Wang, C.D.: Robust ensemble clustering using probability trajectories. IEEE Trans. Knowl. Data Eng. **28**(5), 1312–1326 (2016)

10. Huang, D., Lai, J.H., Wang, C.D., Yuen, P.C.: Ensembling over-segmentations: from weak evidence to strong segmentation. Neurocomputing **207**, 416–427 (2016)

11. Huang, D., Lai, J., Wang, C.D.: Ensemble clustering using factor graph. Pattern Recogn. **50**, 131–142 (2016)

12. Huang, D., Wang, C.D., Lai, J.H.: LWMC: a locally weighted meta-clustering algorithm for ensemble clustering. In: Proceedings of International Conference on Neural Information Processing (ICONIP), pp. 167–176 (2017)

13. Iam-On, N., Boongoen, T., Garrett, S., Price, C.: A link-based approach to the cluster ensemble problem. IEEE Transactions on Pattern Analysis and Machine Intelligence **33**(12), 2396–2409 (2011)

14. LeCun, Y., Bottou, L., Bengio, Y., Haffner, P.: Gradient-based learning applied to document recognition. Proc. IEEE **86**(11), 2278–2324 (1998)

15. Levandowsky, M., Winter, D.: Distance between sets. Nature **234**, 34–35 (1971)

16. Liu, H., Liu, T., Wu, J., Tao, D., Fu, Y.: Spectral ensemble clustering. In: Proceedings of ACM SIGKDD International Conference on Knowledge Discovery and Data Mining, pp. 715–724 (2015)

17. Liu, H., Zhao, R., Fang, H., Cheng, F., Fu, Y., Liu, Y.Y.: Entropy-based consensus clustering for patient stratification. Bioinformatics **33**(17), 2691–2698 (2017)

18. Shi, J., Malik, J.: Normalized cuts and image segmentation. IEEE Trans. Pattern Anal. Mach. Intell. **22**(8), 888–905 (2000)

19. Strehl, A., Ghosh, J.: Cluster ensembles: a knowledge reuse framework for combining multiple partitions. J. Mach. Learn. Res. **3**, 583–617 (2003)

20. Topchy, A., Jain, A.K., Punch, W.: Clustering ensembles: models of consensus and weak partitions. IEEE Trans. Pattern Anal. Mach. Intell. **27**(12), 1866–1881 (2005)

21. Wu, J., Liu, H., Xiong, H., Cao, J., Chen, J.: K-means-based consensus clustering: a unified view. IEEE Trans. Knowl. Data Eng. **27**(1), 155–169 (2015)
22. Yi, J., Yang, T., Jin, R., Jain, A.K.: Robust ensemble clustering by matrix completion. In: Proceedings of IEEE International Conference on Data Mining (ICDM) (2012)

Robust Nonnegative Matrix Factorization Based on Cosine Similarity Induced Metric

Wen-Sheng Chen[1,2], Haitao Chen[1], Binbin Pan[1,2(✉)], and Bo Chen[1,2]

[1] College of Mathematics and Statistics, Shenzhen University, Shenzhen, China
{chenws,pbb,chenbo}@szu.edu.cn
[2] Guangdong Key Laboratory of Media Security, Shenzhen University,
Shenzhen 518060, People's Republic of China

Abstract. Nonnegative matrix factorization (NMF) is a low-rank decomposition based image representation method under the nonnegativity constraint. However, a lot of NMF based approaches utilize Frobenius-norm or KL-divergence as the metrics to model the loss functions. These metrics are not dilation-invariant and thus sensitive to the scale-change illuminations. To solve this problem, this paper proposes a novel robust NMF method (CSNMF) using cosine similarity induced metric, which is both rotation-invariant and dilation-invariant. The invariant properties are beneficial to improving the performance of our method. Based on cosine similarity induced metric and auxiliary function technique, the update rules of CSNMF are derived and theoretically shown to be convergent. Finally, we empirically evaluate the performance and convergence of the proposed CSNMF algorithm. Compared with the state-of-the-art NMF-based algorithms on face recognition, experimental results demonstrate that the proposed CSNMF method has superior performance and is more robust to the variation of illumination.

Keywords: Nonnegative matrix factorization · Face recognition · Cosine similarity induced metric

1 Introduction

NMF is a promising nonnegative feature representation method and has been successfully applied to a variety of tasks such as face recognition [1], clustering [2] and hyper-spectral data unmixing [3,4] etc. NMF aims to approximately decompose a nonnegative data matrix into two low-rank matrices (called a basis image matrix and a feature matrix) under nonnegativity constraints. Two kinds multiplicative update rules (MUR) of NMF [5,6], namely FNMF and KLNMF proposed by Lee *et al.* are based on the measures of Frobenius-norm and KL-divergence respectively. NMF has a meaningful visual interpretation since the basis images of NMF are shown to be some local features of facial image [6]. This means that NMF is capable of learning parts-based image representation.

© Springer Nature Switzerland AG 2019
Z. Cui et al. (Eds.): IScIDE 2019, LNCS 11936, pp. 278–288, 2019.
https://doi.org/10.1007/978-3-030-36204-1_23

Subsequently, a large number of variants of NMF have been proposed [7–10]. Cai *et al.* developed a graph regularized NMF (GNMF) [7] to preserve the manifold structure of data. Li *et al.* proposed an orthogonal NMF (ONMF) [8] by imposing orthogonality constraints on factor matrices. But GNMF and ONMF are sensitive to outliers because of using Frobenius norm or KL-Divergence. To handle this problem, Du *et al.* [9] presented a NMF method with correntropy induced metric (CIMNMF). In addition, Li *et al.* [10] developed a robust structured NMF (RSNMF) algorithm using a semi-supervised matrix decomposition strategy. RSNMF learns a robust discriminative representation by leveraging the block-diagonal structure and the $l_{2,p}$-norm loss function. In the case of noise and outliers, empirical results indicate that these methods with correntropy or $l_{2,p}$-norm loss function surpass NMF approaches with Frobenius-norm or KL-divergence. Nevertheless, the measurements of the above-mentioned methods cannot preserve dilation-invariant. It may leads to undesirable performance under the variations of illumination.

In this paper, we propose a novel robust nonnegative matrix factorization by means of the cosine similarity induced metric (CSNMF). This metric satisfies both rotation-invariant and dilation-invariant and thus the proposed CSNMF method is robust to scale-change illumination and variation. We first construct an auxiliary function according to the objective function with cosine similarity induced metric. And then, the update rules of CSNMF are derived by solving the stable point of the auxiliary function. The property of the auxiliary function theoretically guarantees the convergence of our CSNMF algorithm. Finally, we evaluate the proposed CSNMF method by experiments on both convergence and face recognition. The results indicate the effectiveness and superior performance of CSNMF method.

The rest of this paper is organized as follows. In Sect. 2, we briefly introduce the related work. The proposed CSNMF approach is given in Sect. 3. The experimental results on face recognition are reported in Sect. 4. The final conclusions are drawn in Sect. 5.

2   Related Work

This section will briefly introduce some related algorithms, such as NMF [5], CIMNMF [9] and RSNMF [10] etc.

2.1   NMF

Let $X \in R_+^{m \times n}$ be a non-negative matrix containing n column training samples with dimension m. NMF aims to decomposes X into two non-negative factors $W \in R_+^{m \times r}$ and $H \in R_+^{r \times n}$ such that their product is approximately equal to the original data matrix X, namely

$$X \approx WH,$$

where parameter r is the number of features, matrices W and H are called basis image matrix and feature matrix respectively. Based on the following two kinds of measurements,

$$D_F(X\|WH) = \frac{1}{2}\|X - WH\|_F^2 = \frac{1}{2}\sum_{ij}(X_{ij} - (WH)_{ij})^2,$$

and

$$D_{KL}(X\|WH) = \sum_{ij}(X_{ij}log\frac{X_{ij}}{(WH)_{ij}} - X_{ij} + (WH)_{ij}),$$

two MURs of NMF are given in [5] respectively as follows:

$$H_{a\mu} \leftarrow H_{a\mu}\frac{(W^\top X)_{a\mu}}{(W^\top WH)_{a\mu}}, \quad W_{ia} \leftarrow W_{ia}\frac{(XH^\top)_{ia}}{(WHH^\top)_{ia}}$$

and

$$H_{a\mu} \leftarrow H_{a\mu}\frac{\sum_i W_{ia}X_{i\mu}/(WH)_{i\mu}}{\sum_k W_{ka}}, \quad W_{ia} \leftarrow W_{ia}\frac{\sum_\mu H_{a\mu}X_{i\mu}/(WH)_{i\mu}}{\sum_v H_{av}}.$$

2.2 CIMNMF

CIMNMF algorithm [9] is to minimize the following objective function:

$$D_{CIM}(X\|WH) = 1 - \frac{1}{NM}\sum_i\sum_j g(X_{ij} - \sum_k W_{ik}H_{kj}, \sigma),$$

where $g(e, \sigma) = \frac{1}{2\pi\sigma}exp(-\frac{e^2}{2\sigma^2})$. Its update formulae are as below:

$$Q_{i\mu} \leftarrow exp(-\frac{(X - WH)_{i\mu}^2}{2\sigma^2}),$$

$$H_{a\mu} \leftarrow H_{a\mu}\frac{(W^\top(X \odot Q))_{a\mu}}{(W^\top(WH \odot Q))_{a\mu}},$$

$$W_{ia} \leftarrow W_{ia}\frac{((Q \odot X)H^\top)_{ia}}{((Q \odot WH)H^\top)_{ia}},$$

$$\sigma^2 \leftarrow \frac{1}{2NM}\sum_i\sum_j(X_{ij} - (WH)_{ij})^2.$$

2.3 RSNMF

Robust structured NMF (RSNMF) utilizes both label and unlabel information to learn NMF representation, which is based on a block-diagonal structure and the $l_{2,p}$-norm loss function. The objective function of RSNMF is defined by

$$\|X - WH\|_{2,p}^p + \lambda\|I \odot H\|_F^2,$$

where $\|X\|_{2,p} = (\sum_i \|X._i\|_2^p)^{1/p}$ and I is a block-diagonal matrix. The iterative formulae of RSNMF are as follows:

$$H_{a\mu} \leftarrow H_{a\mu} \frac{(W^T X D)_{a\mu}}{(W^T W H D)_{a\mu} + \lambda(I \odot H)_{a\mu}},$$

$$W_{ia} \leftarrow W_{ia} \frac{(X D H^T)_{ia}}{(W H D H^T)_{ia}},$$

where D is a diagonal matrix with $D_{kk} = \frac{p}{2\|z_k\|_2^{2-p}}$, and $Z = X - WH$.

3   The Proposed CSNMF Approach

Besides above mentioned methods, a lot of NMF-based algorithms are based on the metrics which are not dilation-invariant. So, they are not robust to the scale-change illuminations. This section will propose a novel CSNMF method to avoid this problem.

3.1   Cosine Similarity Induced Metric

Given two matrices $A \in R^{m \times n}$ and $B \in R^{m \times n}$, the cosine similarity induced metric is defined as:

$$D_{CS}(A\|B) = 1 - \frac{\sum_{i,j} A_{ij} B_{ij}}{\|A\|_F \|B\|_F}.$$

We can easily verify that the metric $D_{CS}(A\|B)$ possesses the property that $D_{CS}(A\|B) = D_{CS}(kA\|lB)$, for any positive scalars $k, l \in R_+$. This means that $D_{CS}(A\|B)$ is a dilation-invariant metric.

3.2   The CSNMF Model

The objective function of CSNMF is established as follows:

$$F_{CS}(W, H) = D_{CS}(X\|WH),$$

which can be rewritten as:

$$F_{CS}(W, H) = 1 - \frac{tr(X^\top W H)}{tr^{\frac{1}{2}}(X^\top X) tr^{\frac{1}{2}}(H^\top W^\top W H)}.$$

Hence, the CSNMF model is as below:

$$\min_{W \geq 0, H \geq 0} F_{CS}(W, H) = 1 - \frac{tr(X^\top W H)}{tr^{\frac{1}{2}}(X^\top X) tr^{\frac{1}{2}}(H^\top W^\top W H)}. \tag{1}$$

By solving optimization problem (1), we acquire the following update rules of CSNMF:

$$H_{a\mu} \leftarrow H_{a\mu} \frac{(W^\top X)_{a\mu}(tr(H^\top W^\top W H) - (W^\top W H)_{a\mu} H_{a\mu})}{(W^\top W H)_{a\mu}(tr(X^\top W H) - (W^\top X)_{a\mu} H_{a\mu})}, \tag{2}$$

$$W_{ia} \leftarrow W_{ia} \frac{(XH^\top)_{ia}(tr(H^\top W^\top WH) - (WHH^\top)_{ia}W_{ia})}{(WHH^\top)_{ia}(tr(X^\top WH) - (XH^\top)_{ia}W_{ia})}. \tag{3}$$

Although there are subtraction operations in the update rules (2) and (3), the final update results are not less than zeros. For $H_{a\mu} \geq 0$, it is because that $tr(H^\top W^\top WH) = \sum_{ij}(W^\top WH)_{ij}H_{ij} \geq (W^\top WH)_{a\mu}H_{a\mu}$ and $tr(X^\top WH) = \sum_{ij}(W^\top X)_{ij}H_{ij} \geq (W^\top X)_{a\mu}H_{a\mu}$. While for (3), we can similarly show that $W_{ia} \geq 0$ after updating.

3.3   Theoretical Analysis on CSNMF

This subsection discusses how to get the update rules of CSNMF and analyzes its convergence as well. To this end, we need auxiliary function technique.

Definition 1. $G(H, H^t)$ *is called an auxiliary function of* $F(H)$, *if it holds that*

$$G(H, H^t) \geq F(H) \quad and \quad G(H^t, H^t) = F(H^t).$$

One can obtain the following Lemma by a simple derivation.

Lemma 1. *If* $G(H, H^t)$ *is an auxiliary function of* $F(H)$, *then* $F(H)$ *is non-increasing under the update rule*

$$H^{t+1} = arg\min_H \ G(H, H^t).$$

We denote $F_1(H)$ and $F_2(W)$ as $F_1(H) = F_{CS}(W, H)$ with fixed W and $F_2(W) = F_{CS}(W, H)$ with fixed H, respectively. The solution (W, H) of the optimal problem (1) is obtained by solving two convex optimization subproblems: $\min_{H\geq0} F_1(H)$ and $\min_{W\geq0} F_2(W)$. We focuses on dealing with the first subproblem on H. The second subproblem is omitted here.

It is known that for fixed W,

$$F_1(H) = 1 - \frac{tr(X^\top WH)}{tr^{\frac{1}{2}}(X^\top X)tr^{\frac{1}{2}}(H^\top W^\top WH)}. \tag{4}$$

The crucial step is to construct an auxiliary function of $F_1(H)$. For this purpose, we have the following theorem.

Theorem 1. *Let*

$$G(H, H^t) = 1 - \frac{tr(X^\top WH)}{tr^{\frac{1}{2}}(X^\top X)(\sum_{ik} \frac{(W^\top WH^t)_{ik}}{H^t_{ik}} H^2_{ik})^{\frac{1}{2}}},$$

then $G(H, H^t)$ *is an auxiliary function of* $F_1(H)$.

Proof. By direct computation, it holds that $G(H^t, H^t) = F(H^t)$. Next we need to prove that $G(H, H^t) \geq F(H)$. To do this, we have

$$\sum_{ij}(W^\top W)_{ij}H_{ik}H_{jk} \leq \sum_i \frac{(W^\top WH^t)_{ik}}{H^t_{ik}} H^2_{ik}.$$

Therefore,

$$tr(H^\top W^\top WH) = \sum_{ij}(W^\top W)_{ij}(\sum_{k} H_{ik}H_{jk}) \leq \sum_{ik} \frac{(W^\top WH^t)_{ik}}{H^t_{ik}} H^2_{ik}.$$

Above inequality implies that

$$F_1(H) \leq G(H, H^t).$$

It concludes that $G(H, H^t)$ is an auxiliary function of $F_1(H)$.

To obtain the update rule (2) of entry $H_{a\mu}$, we need an auxiliary matrix Q. The auxiliary matrix $Q = (Q_{ij})_{r\times n}$ is defined by $Q_{a\mu} = H_{a\mu}$ and $Q_{ij} = H^t_{ij}$ for all (i,j) except $(i,j) = (a, \mu)$. If let $H = Q$, then the auxiliary function $G(H, H^t)$ can be rewritten as

$$G(Q, H^t) = 1 - \frac{tr(X^\top WQ)}{tr^{\frac{1}{2}}(X^\top X)(\sum_{ik} \frac{(W^\top WH^t)_{ik}}{H^t_{ik}}Q^2_{ik})^{\frac{1}{2}}}.$$

This means that $\frac{\partial G(H,H^t)}{\partial H_{a\mu}} = \frac{\partial G(Q,H^t)}{\partial Q_{a\mu}}$. We denote α by

$$\alpha = tr^{\frac{1}{2}}(X^\top X)(\sum_{ik} \frac{(W^\top WH^t)_{ik}}{H^t_{ik}}Q^2_{ik})^{\frac{3}{2}},$$

and let $\frac{\partial G(Q,H^t)}{\partial Q_{a\mu}} = 0$, it yields that

$$\frac{1}{\alpha}(\frac{(W^\top WH^t)_{a\mu}}{H^t_{a\mu}}Q_{a\mu}tr(X^\top WQ) - (W^\top X)_{a\mu}\sum_{ik} \frac{(W^\top WH^t)_{ik}}{H^t_{ik}}Q_{ik}{}^2) = 0.$$

Form the definition of Q, we have

$$(\sum_{(i,k)\neq(a,\mu)}(W^\top X)_{ik}H^t_{ik})\frac{(W^\top WH^t)_{a\mu}}{H^t_{a\mu}}H_{a\mu} - (W^\top X)_{a\mu}\sum_{(i,k)\neq(a,\mu)}(W^\top WH^t)_{ik}H^t_{ik} = 0.$$

Thereby, the following formula is obtained from above equation directly,

$$H_{a\mu} = H^t_{a\mu}\frac{(W^\top X)_{a\mu}(tr(H^{t\top}W^\top WH^t) - (W^\top WH^t)_{a\mu}H^t_{a\mu})}{(W^\top WH^t)_{a\mu}(tr(X^\top WH^t) - (W^\top X)_{a\mu}H^t_{a\mu})}.$$

Lemma 1 guarantees that the objective function $F_1(H)$ is monotonically non-increasing under the update rule (2). The update rule of W can be similarly obtained using the auxiliary function strategy. Also, $F_2(H)$ is monotonically decreasing under the update rule (3). So, the proposed CSNMF algorithm is convergence.

4   Experimental Results

The proposed algorithm is evaluated on two face databases, namely FERET and Yale B face databases. Four state of the art NMF-based algorithms, such as KLNMF [5], FNMF [5], CIMNMF [9], and RSNMF [10], are implemented for comparisons.

4.1   Face Image Databases

The FERET database contains of 14,126 images that includes 1199 individuals. The variations in FERET database involves pose, illumination, facial expression and aging [12]. We select a subset including 720 images of 120 individuals (6 images for each) to evaluate the algorithms.

The Yale B database contains 5850 face images of ten persons each captured under 585 viewing conditions (9 poses 65 illumination conditions). Our experiment settings are similar with [13]. In our experiments, we use images under 45 illumination conditions. These 4050 images (9 poses $\times$ 45 illumination conditions $\times$ 10 individuals) are divided into four subsets according to the angle the light source direction makes with the camera axis, namely subset 1 (0 to 12, seven images per pose), subset 2(up to 25, 12 images per pose), subset 3 (up to 50, 12 images per pose), and subset 4 (up to 77, 14 images per pose).

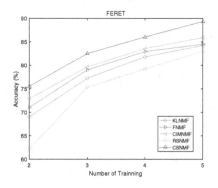

Fig. 1. Comparisons on FERET database

Table 1. Mean accuracy (%) versus Training Number (TN) on FERET database

TN	2	3	4	5
KLNMF	69.04	77.25	81.83	84.25
FNMF	71.06	79.02	82.96	84.50
CIMNMF	72.92	79.67	83.58	85.92
RSNMF	62.25	75.33	79.21	83.00
CSNMF	**75.54**	**82.53**	**86.04**	**89.33**

4.2   Results on FERET Database

We randomly select TN ($TN = 2, 3, \cdots, 5$) images from each person as training sample while the rest $(6 - TN) \times 120$ images are used for testing. The experiments are repeated ten times and the average accuracies are recorded for comparison. The results of mean accuracies are tabulated in Table 1 and plotted in Fig. 1. It can be seen that the accuracies of all approaches will be improved as the number of training samples increases. In detail, the recognition rate of our CSNMF method increases from 75.54% with $TN = 2$ to 89.33% with $TN = 5$, while the accuracies of KLNMF, FNMF, CIMNMF and RSNMF raise from 69.04%, 71.06%, 72.92% and 62.25% with $TN = 2$ to 84.25%, 84.50%, 85.92% and 83.00% with $TN = 5$, respectively. Our CSNMF approach surpasses KLNMF, FNMF, CIMNMF and RSNMF in all cases.

4.3   Results on Yale B Database

In this subsection, we will evaluate the performance of the proposed algorithm when the face images are undergone greater variations in illumination and pose.

4.3.1   Fixed Pose with Illumination Variations

This part evaluates the performance on the variations of illumination for each fixed pose. Two images/person from each subset are randomly selected for training (2×4 images for training per individual), and rest of the images from the four subsets are used for testing (37 images for testing per individual). The experiments are repeated ten times and the average accuracies of rank 1 are recorded and shown in the Fig. 2 (left). It can be seen from Fig. 2 (left) that the proposed CSNMF method outperforms other four methods under all illumination variations. If we consider the overall average accuracy for all poses, Fig. 2 (right) shows the recognition accuracy of rank 1 to rank 3. It indicates that our CSNMF method increases from 96.21% (rank1) to 99.15% (rank3), while KLNMF, FNMF, CIMNMF and RSNMF increase from 93.39%, 94.40%, 92.48%, and 92.87% (rank1) to 97.69%, 98.15%, 97.65%, and 96.81% (rank3), respectively. The results show that the proposed method is more robust to the change of illumination and outperforms the other four methods under all illumination variations.

4.3.2   Variations on Pose and Illumination

This section reports the results when both pose and illumination are changed. We randomly select 720 images (10 persons × 9 poses × 4 subsets × 2 images) for training. The rest of the 3330 images (10 persons × 9 poses × 37 images) are used for testing. The experiments are run ten times and the average accuracies (rank 1 to rank 3) are recorded and shown in Fig. 3. It can be seen that the recognition accuracy of our method increases from 88.63% with rank 1 to 96.64% with rank 3. The recognition accuracies of KLNMF, FNMF, CIMNMF and RSNMF ascend from 86.15%, 88.01%, 84.52%, and 84.67% with rank 1

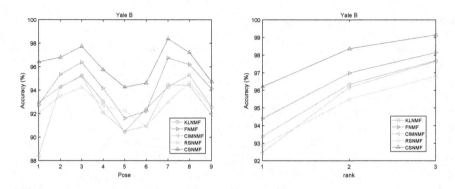

Fig. 2. Performance comparison with fixed pose and different illumination (left) and the entire average accuracy (rank 1–rank 3) of all nine poses (right).

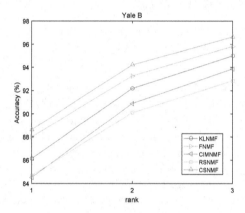

Fig. 3. Performance evaluation on pose and illumination.

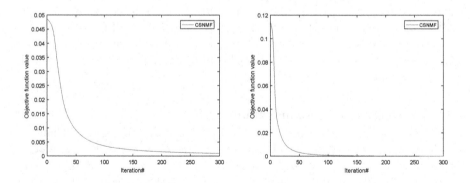

Fig. 4. Convergence curve of CSNMF. FERET (Left) and Yale B (Right)

to 94.99%, 95.82%, 93.87%, and 92.85% with rank 3, respectively. The results demonstrate that our method also achieves the best performance.

4.4 Convergence Verification

In this section, we experimentally validate the convergence of the proposed CSNMF algorithm. The convergence curves are shown in Fig. 4, where the x-axis denotes the iteration number and the y-axis is the value of objective function. We see that our algorithm usually converges within 300 iterations.

5 Conclusion

This paper proposes a new cosine similarity induced metric based non-negative matrix factorization (CSNMF) approach to overcome the problem caused by illumination. We derive the iterative formulas of CSNMF by means of auxiliary function strategy. The proposed algorithm is shown to be convergence from both theoretical and empirical aspects. Experiments on face recognition demonstrate that our approach achieves encouraging results under illumination variations.

Acknowledgements. This paper was partially supported by the Interdisciplinary Innovation Team of Shenzhen University and NNSF of China (Grants 61272252) and NSF of Guangdong Province (2018A030313364). We would like to thank the US Army Research Laboratory and Yale University for providing the facial image databases.

References

1. Li, S.-Z., Hou, X., Zhang, H., Cheng, Q.: Learning spatially localized, parts-based representation. In: Proceedings of International Conference on Computer Vision and Pattern Recognition, vol. 1, pp. 606–610 (2001)
2. Xu, W., Liu, X., Gong, Y.: Document clustering based on non-negative matrix factorization. In: Proceedings of ACM SIGIR, pp. 267–273 (2003)
3. Jia, S., Qian, Y.: Constrained nonnegative matrix factorization for hyperspectral unmixing. IEEE Trans. Geosci. Remote Sens. **47**(1), 161–173 (2009)
4. Feng, X.-R., Li, H.-C., Li, J.: Hyperspectral unmixing using sparsity-constrained deep nonnegative matrix factorization with total variation. IEEE Trans. Geosci. Remote Sens. **56**(10), 6245–6257 (2018)
5. Lee, D.-D., Seung, H.-S.: Algorithm for non-negative matrix factorization. In: Proceedings of NIPS (2001)
6. Lee, D.-D., Seung, H.-S.: Learning the parts of the objects by nonnegative matrix factorization. Nature **401**(6755), 788–791 (1999)
7. Cai, D., He, X., Han, J.: Graph regularized non-negative matrix factorization for data representation. IEEE Trans. Pattern Anal. Mach. Intell. **33**(8), 1548–1560 (2011)
8. Li, Z., Wu, X., Peng, H.: Nonnegative matrix factorization on orthogonal subspace. Pattern Recogn. Lett. **31**(9), 905–911 (2010)

9. Li, X., Shen, Y.: Robust nonnegative matrix factorization via half-quadratic minimization. In: 2012 IEEE 12th International Conference on Data Mining. IEEE Computer Society (2012)

10. Li, Z., Tang, J., He, X.: Robust structured nonnegative matrix factorization for image representation. IEEE Trans. Neural Netw. Learn. Syst. **29**(5), 1947–1960 (2018)

11. He, Z., Xie, S., Zdunek, R.: Symmetric nonnegative matrix factorization: algorithms and applications to probabilistic clustering. IEEE Trans. Neural Netw. **22**(12), 2117–2131 (2011)

12. Phillips, P.-J., Moon, H., Rizvi, S.-A., Rauss, P.-J.: The FERET evaluation methodology for face-recognition algorithms. IEEE Trans. Pattern Anal. Mach. Intell. **22**(10), 1090–1104 (2000)

13. Georghiades, A.-S., Belhumeur, P.-N., Kriegman, D.-J.: From few to many: illumination cone models for face recognition under variable lighting and pose. IEEE Trans. Pattern Anal. Mach. Intell. **23**(6), 643–660 (2001)

Intellectual Property in Colombian Museums: An Application of Machine Learning

Jenny Paola Lis-Gutiérrez[1](✉) ⓘ, Álvaro Zerda Sarmiento[1] ⓘ,
and Amelec Viloria[2] ⓘ

[1] Universidad Nacional de Colombia, Bogotá, Colombia
jplisg@unal.edu.co
[2] Universidad de la Costa, Barranquilla, Colombia
aviloria7@cuc.edu.co

Abstract. The purpose of this research is to answer the following guiding question: how can the behavior of museum networks in Colombia be predicted with respect to the protection of intellectual property (copyright, confidential information and use of patents, domain names, industrial designs, use of trademarks) and the interaction of different types of proximity (geographical, organizational, relational, cognitive, cultural and institutional), based on the use of supervised learning algorithms?

Among the main findings are that the best learning algorithms to predict the behavior of networks, considering different target variables are the AdaBoost, the naive Bayes and CN2 rule inducer.

Keywords: Proximity · Intellectual property · Intellectual property management · Museum · Museum networks · Machine learning

1 Introduction

Cultural industries have gained recognition in the last 20 years, thanks to the fact that creativity began to be considered as an economic resource and it was possible to associate them with positive effects on economic development, in scenarios of: (i) globalization, (ii) rapid progress of knowledge and information, and (iii) knowledge-based economies. According to [1] creative and cultural industries are responsible for a very significant share of employment and growth in developed countries. Within the cultural industries are museums.

In Colombia, the organization of museum entities is structured in two types of networks: thematic and territorial, but it must be considered that geographical proximity is not the only form of interaction of museums. It is therefore necessary to identify, study and interpret new ways of interrelation between museum entities. Considering this context, it was established that the foundations of the French School of Proximity (FSP) could be used [2, 3]. FSP considers that there are other types of proximity, in addition to geographical (static) proximity [4, 5]. These additional classes correspond to non-geographical (or dynamic) proximity [6–9], which includes: (i) cultural, (ii) institutional, (iii) relational or social, (iv) organizational and (v) cognitive or technological.

© Springer Nature Switzerland AG 2019
Z. Cui et al. (Eds.): IScIDE 2019, LNCS 11936, pp. 289–301, 2019.
https://doi.org/10.1007/978-3-030-36204-1_24

However, there is another element that must be considered: Management of Intellectual Property (MIP) in museums. The MIP corresponds to the administration and management of intangible and material goods. It includes the construction and constant updating of an inventory of goods subject to intellectual property protection, the establishment and implementation of intellectual property policy, the design and use of licensing practices, the management of digital rights, the design and use of communication and marketing plans, and the understanding of the environment [10, 11].

Therefore, the purpose of this research is to answer the following guiding question: how to predict the behavior of museum networks in Colombia with respect to the protection of intellectual property (copyright, confidential information and use of patents, domain names, industrial designs, use of trademarks) and the interaction of different types of proximity (geographical, organizational, relational, cognitive, cultural and institutional), based on the use of supervised learning algorithms?

2 Context

2.1 Museum Networks in Colombia

In Colombia there are 27 museum networks, divided into thematic and territorial, which bring together more than 500 of the 768 museums in the country [12]:

- Thematic networks[1]: Acopazoa - Colombian Association of Parks, Zoos and Aquariums, Network of Community Museums, Astronomy Museum Network of Colombia, Network of Science and Natural History Museums, Liliput Museum Network, Network of Gold Museums of the Bank of the Republic, University Museums Network, National Network of Science and Technology Museums, National Network of Medical Museums, National Network of Botanical Gardens.
- Territorial networks[2]: Museums of Bogota, Network of Museums of Cartagena de Indias, networks in the departments of Antioquia, Atlántico, Bolívar, Boyacá, Caldas, Cauca, Córdoba, Cundinamarca, Guainía, Huila, Magdalena, Nariño, Norte de Santander, Quindío, Risaralda, Santander, Tolima, Valle del Cauca.

According to the [12], in Colombia only 18.7% of municipalities (281) had at least one museum.

2.2 Recent Studies on Intellectual Property

Based on the theory of the evolutionary games [13], they analyze the behavior of intellectual property cooperation between government, industry, university, and

---

[1] According to Resolution 1975 of 2013 of the Ministry of Culture, a thematic network is an "organizational form that links the different agents of the museum sector according to the themes of interest, common administrative forms and typologies of collection" (article 1).

[2] In accordance with Resolution 1975 of 2013 of the Ministry of Culture, a territorial network is an "organizational form that links the different agents of the museum sector located in the different territories of the country" (article 1).

research units, as well as factors influencing the market mechanism and administrative oversight. The work is based on evolutionary replication dynamics equations that study stable multi-stakeholder evolutionary strategies. With respect to the education sector, [14] identified what is required for a successful management of the GPI: (i) be interested in exploiting university patents, i.e., have a technical and market assessment (ii) that the university's knowledge is of interest to the productive sector; (iii) have financial mechanisms for investments; (iv) be able to cover the costs of patent management and maintenance.

[15] and [16] emphasized the importance of IPG training for companies and universities, including technological intelligence and prospecting exercises, licensing challenges and forms of negotiation, in a gamification approach. In this same line is the work of [17] who identified that one of the main concerns of organizations in developing countries is the gap between the creation of IP protected products and their commercialization [18], so they propose an IPG model with 5 phases and 15 processes. Another work that proposes a methodology for IPM in science, technology and innovation projects is that of [19].

Another focus in recent literature on IPG corresponds to its relationship with the creative or cultural industries. [20], for example, identified that open IP-based business models can also be profitable, as in the case of video games or 3D printing [21]. Regarding crowdsourcing, the work of [22] illustrates four approaches to managing intellectual property (passive, possessive, persuasive, and prudent) when firms make use of micro-patronage [23]. Approaches depend on two dimensions of IP acquisition and limitation of responsibilities. Liabilities correspond to low interest in acquiring and limiting; possessive to high interest in acquiring and low to limit; prudent to low interest in acquiring and high to limit; and persuasive to high interest in acquiring and limiting.   '

2.3   Application of Machine Learning in Museums

[24] applied an automatic learning model for evaluating the interaction of visitors to a museum with the museum's exhibition modules. This work was also related to that of [25], which sought to emphasize the need for an enhanced digital cultural experience based on the attractiveness and personalization of each visitor, through the incorporation of cutting-edge technologies in various areas of information technology, such as digital documentation, augmented reality, statistics and automatic learning with recommendation systems.

Also applied to the subject of museums is the work of [26], which applies techniques of automatic learning and deep learning, such as Neuronal Convolutional Networks (CNN), for the classification of images and data, cultural heritage. In the same sense of heritage protection and classification is the document by [27], in which they propose a framework that incorporates computational methods (digital image processing, text analysis in several languages and 3D modeling), to facilitate improved data archiving, search and analysis.

This same horizon was worked by [28] who shows the need for articulation between research centers or researchers in information science, computer science and computer engineering, to make the digital cultural record more useful and usable. [29]

for their part, succeeded in showing the feasibility of identifying artistic paintings through the use of automatic learning, which would make it easier to detect counterfeits. The same happened in the work of [30], who through three-dimensional models and automatic learning, achieved the classification of Mexican ceramic figures of the ancient west.

For their part, [31] analyzed the impact of automatic learning on improving the user experience in museums, finding that traditional applications were associated with artistic authentication, guidance and virtual reality, but that it could be extended to other aspects related to interaction with visitors. [32] mentioned the advantages of automatic learning in museums, such as: prediction of visitor numbers; improvements in planning and resource allocation; identification of members most likely to renew, upgrade or cancel or their memberships; increased resources through e-commerce; classification of collections and testimonies. [33] analyzed the case of El Nuevo Rembrandt, a portrait that was made by a computer from the analysis of the artist's works, and in turn studies the implications for copyright of these new works.

Finally, we will mention the work of [34], who analyze the advantages of organizational microblogging for event marketing. They applied an automatic learning algorithm and performed Kendall's Tau test, looking for a correlation between the degree of online activity of an account and the frequency of on-site visits. The findings showed how organizations can use social networking tools to engage the public more fully in social and cultural events.

3 Method

In this section, the data, variables and method used for the analysis are presented.

3.1 Data

The initial purpose of the study was to perform an analysis for all museum networks in Colombia, however, only 9 responded to the instrument that was applied. The instrument consisted of 32 questions and 59 items (Table 1), of which 20 sought to measure different types of proximity (4 items for each type).

The collection of information took place between 11 November 2016 and 6 August 2017. It was only possible to obtain information from 9 geographical entities, none of the thematic networks responded to the questionnaire. The networks that responded were: Red de Museos de Santander, Mesa de Museos de Bogotá, Red de Museos de Córdoba, Red de Museos de Bolívar, Red de Museos de Antioquia, Red de Museos de Atlántico, Red de museos de Cauca, Red de Museos del Valle del Cauca, Red de Museos de Nariño. These nine networks include 237 Colombian museums (Table 2).

Table 1. Items by instrument section.

Section	Number of items
Characterization information	9
Interaction and proximity	41
Intellectual property	9
Total	59

Table 2. Basic information about museum networks.

Network name	Number of museums	Number of active museums	Percentage of active museums	Number of leading museums	Percentage of leading museums
Cauca museum network	7	7	100%	1	4,29%
Museums of Bogota	10	10	100%	10	100%
MUVAC	37	16	43,2%	1	2,70%
Nariño network of museums	30	20	66,7%	5	16,67%
Antioquia museum network	88	15	17%	6	6,82%
Bolívar network of museums	6	6	100%	6	100%
Cordoba network of museums	9	4	44,4%	1	11,11%
Santander museum network	32	12	37,5%	6	18,75%
Atlantic museum network	18	9	50%	3	16,67%
Total	237	99	41,8%	39	16,46%

3.2 Variables Used

The variables that were analyzed in the document were the following: (i) Copyright protection, (ii) protection of confidential information and use of patents, (iii) domain name protection, (iv) protection of industrial designs, (v) use of trademarks; (vi) geographical proximity, (vii) organizational proximity, (viii) relational proximity, (ix) cognitive proximity, (x) cultural proximity, (xi) institutional proximity.

3.3 Algorithms Used

For information processing, several methods (algorithms) of supervised learning were applied, which are summarized in Table 3. The selection of automatic learning for data analysis is due to: (i) allows the application of algorithms that do not require normal data, i.e. they are non-parametric; (ii) algorithms are calibrated with few data and even work with missing data or complex interactions; (iii) cross validation and specifically leave one out were used for prediction. This technique is suggested for the case of few data since each observation is part of both the training set and the prediction set, and therefore, it is possible to reduce the variability of the data.

Table 3. Supervised learning algorithms.

Method	Synthesis
AdaBoost	It is an adaptive learning algorithm that combines weak classifiers and adapts to the hardness of each training sample, achieving robust classifiers. Allows binary and multi-class classification
Random forest	Uses a set of decision trees for the classification and projection. It is a non-parametric procedure which can be used when: (i) there are correlated variables; (ii) few data; (iii) complex interactions; (iv) missing data
Support Vector Machines (SVM)	Allows the binary classification and multi-class. Makes use of regression analysis and classification
Neural network	Makes use of the algorithm for multi-layer perceptron (MLP) with retropropagation. Its advantage is that nonlinear models can be learned
KNN	This mechanism is for the recognition of non-parametric patterns; it is based on the nearest training instances, being an algorithm of k nearest neighbors. It is also known as lazy learning
Naive Bayes	A quick and simple probabilistic classifier based on the Bayes theorem, with the assumption of independence of features. It is assumed that the absence or presence of a characteristic is not related to the presence or absence of another property. Its advantage is that it requires little data for training
Learning algorithm of the decision tree	It is used for discrete and continuous data. It is based on the prediction of the value of a target variable based on various input variables

4   Results

This section shows the application of automatic learning to the information obtained in the surveys carried out in the 9 museum networks. Information was collected on 183 different items. In this case, two models were used: using "target" variables:

- Question 25: Do the museums in the network have an explicit and public policy on intellectual property? The possible answers were: less than 25% of the museums in the network (1), between 25 and 49% of the museums in the network (2); half of the museums in the network (3); between 51 and 75% of the museums in the network More than 75% of the museums in the network (4); more than 75% of the museums in the network (5) and does not apply (0).
- Question 24, associated with the existence of strategies or possibilities for intellectual property protection in museums. The answer was dichotomous: yes (1), no (0).

Different methods (algorithms) of supervised learning were applied, which would make it possible to replicate what has been learned in the past to new data.

Figure 1 shows the different learning and prediction algorithms used and Fig. 2 shows the precision calculations of the learning and prediction algorithms using question 25 as the target variable (percentage of museums with an explicit and public IP protection policy). In this case, none of them has a performance higher than 60%. It should be clarified that two mechanisms were used to calculate the sample for the evaluation is called Leave-one-out, which is a cross validation, leaving in each case one of the subjects outside, ie for each iteration there is only one sample for testing and the rest of the data are training. In addition to the Leave-one-out, the option of random cross validation was used. This method consisted of randomly dividing the data set into training and testing, using 100 iterations. The adjustment is obtained from the arithmetic mean of the values obtained for each of the iterations.

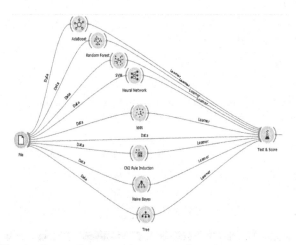

Fig. 1. Image of the model representation using Orange [35].

Test & Score					
Settings					
Sampling type: Leave one out					
Target class: Average over classes					
Scores					
Method	**AUC**	**CA**	**F1**	**Precision**	**Recall**
kNN	0.321	0.667	0.533	0.444	0.667
Tree	0.452	0.556	0.476	0.417	0.556
SVM	0.000	0.667	0.533	0.444	0.667
Random Forest	0.000	0.667	0.533	0.444	0.667
Neural Network	0.024	0.556	0.476	0.417	0.556
Naïve Bayes	0.976	0.333	0.185	0.133	0.333
CN2 rule inducer	0.321	0.111	0.148	0.222	0.111
AdaBoost	0.179	0.556	0.476	0.417	0.556

Test & Score					
Settings					
Sampling type: Shuffle split, 100 random samples with 75% data					
Target class: Average over classes					
Scores					
Method	**AUC**	**CA**	**F1**	**Precision**	**Recall**
kNN		0.690	0.563	0.476	0.690
Tree		0.387	0.394	0.402	0.387
SVM		0.690	0.563	0.476	0.690
Random Forest		0.690	0.563	0.476	0.690
Neural Network		0.463	0.453	0.444	0.463
Naïve Bayes		0.250	0.119	0.078	0.250
CN2 rule inducer		0.350	0.400	0.467	0.350
AdaBoost		0.640	0.586	0.541	0.640

Fig. 2. Calculation of learning and prediction algorithms using question 25 as target variable.

In the case of Fig. 3, the variable of interest was binary and consisted of the knowledge about the possibilities of intellectual property protection. Using random cross validation for 100 iterations, the methods with the best predictive value (considering both prediction and recalculation) were: AdaBoost (71.1% and 71.7%, respectively) and CN2 rule inducer (89.7% and 85%).

Test & Score

Settings

Sampling type: Stratified Shuffle split, 100 random samples with 75% data
Target class: Average over classes

Scores

Method	AUC	CA	F1	Precision	Recall
kNN	0.603	0.590	0.567	0.556	0.590
Tree	0.757	0.703	0.710	0.802	0.703
SVM	0.575	0.667	0.533	0.444	0.667
Random Forest	0.777	0.677	0.590	0.651	0.677
Neural Network	0.355	0.487	0.468	0.454	0.487
Naïve Bayes	0.625	0.433	0.354	0.790	0.433
CN2 rule inducer	0.887	0.850	0.854	0.897	0.850
AdaBoost	0.605	0.717	0.673	0.711	0.717

Test & Score

Settings

Sampling type: Leave one out
Target class: Average over classes

Scores

Method	AUC	CA	F1	Precision	Recall
kNN	0.500	0.444	0.410	0.381	0.444
Tree	0.750	0.667	0.667	0.833	0.667
SVM	0.222	0.667	0.533	0.444	0.667
Random Forest	0.000	0.667	0.533	0.444	0.667
Neural Network	0.333	0.556	0.476	0.417	0.556
Naïve Bayes	0.833	0.444	0.372	0.792	0.444
CN2 rule inducer	0.833	0.889	0.892	0.917	0.889
AdaBoost	0.500	0.667	0.533	0.444	0.667

Fig. 3. Calculation of learning and prediction algorithms using question 24 as target variable.

	IF conditions		THEN class	Distribution	Probabilities [%]	Quality	Length
0	Edad de la red≥6.0	→	P24PI=0.0	[5, 0]	86 : 14	-0.00	1
1	Número de museos≥30.0	→	P24PI=0.0	[1, 0]	67 : 33	-0.00	1
2	TRUE	→	P24PI=1.0	[0, 3]	20 : 80	-0.00	0

CN2 Rule Viewer   Sat Jun 09 18, 23:45:13

Data domain

Features: Edad de la red, Número de museos, P10aInteraccionFrecuenciaMail, P10bInteraccionFrecuenciaReunionPresencial, P10cInteraccionFrecuenciaReunionAsincronica, P10dInteraccionFrecuenciaReunionAsincronicaTIC, P10eInteraccionFrecuenciaCelular, P10fInteraccionFrecuenciaWhatsApp, P10gInteraccionFrecuenciaTelFijo, P10hInteraccionFrecuenciaVisitas, P10iInteraccionFrecuenciaRedesSociales, P10jInteraccionFrecuenciaOtro, P11aInteraccionPeriodicidadMail, P11bInteraccionPeriodicidadReunionPresencial, P11cInteraccionPeriodicidadReunionAsincronica, P11dInteraccionPeriodicidadReunionAsincronicaTIC, P11eInteraccionPeriodicidadCelular, P11fInteraccionPeriodicidadWhatsApp, P11gInteraccionPeriodicidadTelFijo, P11hInteraccionPeriodicidadVisitas, P11iInteraccionPeriodicidadRedesSociales, P11jInteraccionPeriodicidadOtro, P12aInteraccionUsoPagweb, P12bInteraccionUsoReunionRedSocial, P12cInteraccionUsoRedAcademica, P12dInteraccionUsoCodigoQr, P12eInteraccionUsoBlog, P12fInteraccionUsoPublicacionesconjuntas, ... (total: 178 features)
Meta attributes: P1CarNombreRed, Nom, Selected
Target: P24PI

Rule induction algorithm

Rule ordering: ordered
Covering algorithm: exclusive
Gamma: 0.7
Evaluation measure: entropy
Beam width: 5
Minimum rule coverage: 1
Maximum rule length: 5
Default alpha: 1.0
Parent alpha: 1.0

Induced rules

	IF conditions		THEN class	Distribution	Probabilities [%]	Quality	Length
	Edad de la red≥6.0	→	P24PI=0.0	[5, 0]	86 : 14	-0.00	1
1	Número de museos≥30.0	→	P24PI=0.0	[1, 0]	67 : 33	-0.00	1
2	TRUE	→	P24PI=1.0	[0, 3]	20 : 80	-0.00	

Fig. 4. CN2 rule viewer.

The CN2 rule inducer also qualifies as one of the robust classifiers, obtaining 91.7% accuracy and 88.9% recalculation. Figure 4 shows the output of the software, which shows that despite having 178 variables, the most significant for prediction using the algorithm were the age of the network and the number of museums.

An attempt was made to use the different algorithms to predict other variables. These results are shown in Table 4.

Table 4. Prediction results for different variables using machine learning

Variable	Cross validation leaving one outside	Simple sample cross validation with 100 iterations
Copyright	The only algorithm with precision capability was the AdaBoost with 62%	No algorithm had predictive capability greater than 60%
Protection of confidential information and use of patents	No algorithm had predictive capability greater than 60%	The only algorithm with predictive capability was naive Bayes, however, only achieved 68.2% accuracy
Domain name	No algorithm had predictive capability greater than 60%	No algorithm had predictive capability greater than 60%
Protection of industrial designs	No algorithm had predictive capability greater than 60%	No algorithm had predictive capability greater than 60%
Use of trademarks	The only algorithm with predictive capability were neural networks, however, only reached 55.6% accuracy	No algorithm had predictive capability greater than 60%
Geographical proximity	The only algorithm with predictive capability was the AdaBoost, however, only reached 66.7% accuracy	No algorithm had predictive capability greater than 60%
Organizational proximity	The only algorithm with predictive capability were neural networks, however, only reached 75.6% accuracy	No algorithm had predictive capability greater than 60%
Relational proximity	The only algorithm with predictive capability was using decision trees, however, only reached 75.6% accuracy	No algorithm had predictive capability greater than 60%
Cognitive proximity	The most accurate algorithms for this variable were: KNN (60.5%), decision tree (60.5%), random forest (60.5%) and CB2 rule (62.2%)	No algorithm had predictive capability greater than 60%

(continued)

Table 4. (*continued*)

Variable	Cross validation leaving one outside	Simple sample cross validation with 100 iterations
Cultural proximity	The only algorithm with predictive capability was naive Bayes, however, only reached 68.5% accuracy	The only algorithm with predictive capability was naive Bayes, however, only achieved 58% accuracy
Institutional proximity	No algorithm had predictive capability greater than 60%	No algorithm had predictive capability greater than 60%

5 Conclusions

This research sought to define the state of museum networks in Colombia and their strategies for managing intellectual property, according to the different types of proximity (static or dynamic). Applying automatic learning to the database obtained, it was found that for the variable of whether to know the IP protection mechanisms the best algorithms were: AdaBoost (71.1% and 71.7%, respectively) and CN2 rule inducer (89.7% and 85%). For copyright protection the best algorithm was AdaBoost (62%). For the Protection of confidential information and use of patents, the best algorithm was naive Bayes (68.2%). For geographical proximity the best algorithm was Ada-Boost (66.7%).

The only algorithm with predictive capability for organizational proximity was the neural network, however, only reached 75.6% accuracy. The algorithms with the best accuracy for cognitive proximity were: KNN (60.5%), decision tree (60.5%), random forest (60.5%) and CB2 rule (62.2%). The only algorithm with predictive capability for cultural proximity was naive Bayes, however, only reached 68.5% accuracy.

Within the recommendations or suggestions is emphasized in the following. It is necessary to promote training and registration campaigns aimed at the museum sector in Colombia, by the Superintendence of Industry and Commerce. This training would be aimed at people working directly in museums. A redefinition of the relationship between the museum sector and ICTs and intellectual property protection mechanisms is required. The Ministry of Culture of Colombia, the Ministry of Information and Communication Technologies of Colombia, and the Ministry of Commerce, Industry and Tourism of Colombia should participate jointly in this regard.

It is imperative to strengthen the system of information and indicators for the sector, so that it can be consulted publicly, and the information is timely and reliable. The above, to facilitate and promote the processes of research, knowledge creation and social appropriation in the museum sector. For this reason, it would be worthwhile to achieve greater articulation with higher education institutions, not only for the development of research, but also for the training of personnel. Some of the possible strategies could be: (i) search by the Ministry of Culture or the PFM for strategic allies in the regions for the development of diplomas or courses (virtual or face-to-face) in specific topics, a possible entity for this would be the Higher School of Public Administration or the SENA; (ii) search for support from international entities such as

ICOM and WIPO for the development or improvement of research and transfer capacities; (iii) design of a volunteer program in which teachers interested in the cultural sector can be involved, who can provide training or accompaniment to museum entities, could be articulated with a virtual mentoring plan; (iv) design of a sponsorship plan in which one or more national or local companies provide support to sector entities, within their corporate social responsibility programs.

Now, within the future works derived from the present investigation it is suggested, (i) advance work related to the access, conservation, dissemination and exploitation of works in the digital environment; (ii) study the challenges and opportunities of international practices in terms of IP protection, limitations and exceptions in the museum setting; (iii) investigate future paths on access (innovative accessibility solutions) to testimonies for people with disabilities in the digital environment; (iv) studies associated with the implications of works produced by artificial intelligence, on the copyright legislation of different countries.

References

1. Blanco-Valbuena, C.E., Bernal-Torres, C.A., Camacho, F., Díaz-Olaya, M.: Industrias Creativas y Culturales: Estudio desde el Enfoque de la Gestión del Conocimiento. Información tecnológica **29**(3), 15–28 (2018)
2. Balland, P.-A., Boschma, R., Frenken, K.: Proximity and innovation: from statics to dynamics. Reg. Stud. **49**(6), 907–920 (2015)
3. Lis-Gutiérrez, J.P.: La economía de la proximidad en la última década. Criterio Libre **14**(25), 247–269 (2016)
4. Torre, A.: Proximity relations at the heart of territorial development processes: from clusters, spatial conflicts and temporary geographical proximity to territorial governance. In: Torre, A., Wallet, F. (eds.) Regional Development and Proximity Relations, pp. 94–134. NRA – Agroparistech, París (2014)
5. Torre, A.: La figure du réseau: Dimensions spatiales et organisationnelles. Geographie Economie Societe **18**(4), 455–469 (2016)
6. Wang, J., Liu, X., Wei, Y., Wang, C.: Cultural proximity and local firms' catch up with multinational enterprises. World Dev. **60**, 1–13 (2014)
7. Talbot, D.: La dimension politique dans l'approche de la proximité. Géographie Economie Société **22**(2), 125–144 (2010)
8. Talbot, D.: Institutions, organisations et espace: les formes de la proximité. Habilitation à diriger des recherches. Université Montesquieu Bordeaux IV (2011)
9. Beaugency, A., Talbot, D.: Cognitive proximity and knowledge spillover in the avionics industry: an analysis from patents and scientific publications. Innovations **55**(1), 223–246 (2018)
10. Pantalony, R.E.: Managing Intellectual Property for Museums. WIPO, Ginebra (2013)
11. Lis-Gutiérrez, J.P., Viloria, A., Gaitán-Angulo, M., Balaguera, M.I., Rodríguez, P.A.: Museums and management of intellectual property. J. Control Theory Appl. **9**(44), 457–462 (2016)
12. Departamento Nacional de Planeación: Bases del plan nacional de desarrollo 2018–2022 pacto por Colombia. DNP, Bogotá (2019)

13. Yang, Z., Shi, Y., Li, Y.: Analysis of intellectual property cooperation behavior and its simulation under two types of scenarios using evolutionary game theory. Comput. Ind. Eng. **125**, 739–750 (2018)

14. García Galván, R.: Patentamiento universitario e innovación en México, país en desarrollo: teoría y política. Revista de la educación superior **46**(184), 77–96 (2017)

15. Holgersson, M., Tietze, F.: Sharing innovative and best-practice approaches for teaching intellectual property management: a workshop report. World Patent Inf. **49**, 75–76 (2017)

16. Yue, X.P.: Behavior of inter-enterprises patent portfolio for different market structure. Technol. Forecast. Soc. Change **120**, 24–31 (2017)

17. Gargate, G., Momaya, K.S.: Intellectual property management system: develop and self-assess using IPM model. World Patent Inf. **52**, 29–41 (2018)

18. Jeong, Y., Park, I., Yoon, B.: Identifying emerging Research and Business Development (R&BD) areas based on topic modeling and visualization with intellectual property right data. Technol. Forecast. Soc. Change **146**, 655–672 (2018)

19. Domínguez, M.M., Rodríguez, I.G., Pérez, M.G., Almeida, G.C., Velázquez, Y.T.: Metodología de gestión de la propiedad intelectual en los proyectos de ciencia, tecnología e innovación. Revista de Ciencias Médicas de Pinar del Río **22**(6), 1090–1102 (2018)

20. Erickson, K.: Can creative firms thrive without copyright? Value generation and capture from private-collective innovation. Bus. Horiz. **61**(5), 699–709 (2018)

21. Greco, M., Grimaldi, M., Cricelli, L.: Benefits and costs of open innovation: the BeCO framework. Technol. Anal. Strateg. Manag. **31**(1), 53–66 (2019)

22. De Beer, J., McCarthy, I.P., Soliman, A., Treen, E.: Click here to agree: managing intellectual property when crowdsourcing solutions. Bus. Horiz. **60**(2), 207–217 (2017)

23. Stobo, V., Erickson, K., Bertoni, A., Guerrieri, F.: Current best practices among cultural heritage institutions when dealing with copyright orphan works and analysis of crowdsourcing options. EnDOW (2018)

24. Rosales, R., et al.: Modelling the interaction levels in HCI using an intelligent hybrid system with interactive agents: a case study of an interactive museum exhibition module in Mexico. Appl. Sci. **8**(3), 446 (2018)

25. Charalampos, G., Dimitrios, C., Georgios, P., Tzoumanika, A.: Enhanced digital cultural experience. In: 2017 8th International Conference on Information, Intelligence, Systems & Applications (IISA), pp. 1–5. IEEE (2017)

26. Belhi, A., Bouras, A.: Towards a multimodal classification of cultural heritage. In: Qatar Foundation Annual Research Conference Proceedings, vol. 2018, no. 3, p. ICTPD1010. HBKU Press, Qatar (2018)

27. Yasser, A.M., Clawson, K., Bowerman, C.: Saving cultural heritage with digital make-believe: machine learning and digital techniques to the rescue. In: Proceedings of the 31st British Computer Society Human Computer Interaction Conference, p. 97. BCS Learning & Development Ltd. (2017)

28. Owens, T.: We have interesting problems: some applied grand challenges from digital libraries, archives and museums. In: Proceedings of the 18th ACM/IEEE on Joint Conference on Digital Libraries, p. 1. ACM (2018)

29. Nemade, R., Nitsure, A., Hirve, P., Mane, S.B.: Detection of forgery in art paintings using machine learning. Int. J. Innov. Res. Sci. Eng. Technol. **6**(5), 8681–8692 (2017)

30. Forrester-Sellers, J., Velasco, A., Pickering, R., Diaz, J.C.: Classifying ancient west Mexican ceramic figures using three-dimensional modelling and machine learning. In: Proceedings of the Conference on Electronic Visualisation and the Arts, pp. 19–24. BCS Learning & Development Ltd. (2017)

31. Majd, M., Safabakhsh, R.: Impact of machine learning on improvement of user experience in museums. In: Artificial Intelligence and Signal Processing Conference (AISP), pp. 195–200. IEEE (2017)
32. Clecko, B.: Examining the impact of artificial intelligence in museums. Museums and the Web. MW17 (2017)
33. Guadamuz, A.: La inteligencia artificial y el derecho de autor. Revista OMPI, 5 (2017)
34. Li, X., Duan, B.: Organizational microblogging for event marketing: a new approach to creative placemaking. Int. J. Urban Sci. 22(1), 59–79 (2018)
35. Demsar, J., et al.: Orange: data mining toolbox in Python. J. Mach. Learn. Res. 14(1), 2349–2353 (2013)

Hybrid Matrix Factorization
for Multi-view Clustering

Hongbin Yu[1] and Xin Shu[2(✉)]

[1] The School of Digital Media, Jiangnan University, Wuxi, China
alexander.yuu@hotmail.com
[2] College of Information Science and Technology,
Nanjing Agricultural University, Nanjing, China
xinshu@hotmail.com

Abstract. Multi-view clustering (MVC) has gained considerable attention recently. In this paper, we present a hybrid matrix factorization (HMF) framework which is a combination of the nonnegative factorization and the symmetric nonnegative matrix factorization for MVC. HMF can uncover linear and nonlinear manifold within multi-view dataset. In addition, HMF also learns weights for each view to characterize the contribution of each view to the final common clustering assignment. The proposed model can be solved by nonnegative least squares. Unlike previous approaches, our approach can obtain the clustering results straightforwardly due to the nonnegative constraints. We conduct experiments on multi-view benchmark datasets to verify the effectiveness of our proposed approach.

Keywords: Nonnegative matrix factorization · Symmetric matrix factorization · Multi-view clustering

1 Introduction

Data clustering is a fundamental research topic in computer vision and data mining. Various data clustering algorithms have been presented in the literature [11]. Among them nonnegative matrix factorization (NMF) based clustering and spectral clustering (SC) [20] are two typical examples. NMF was originally studied by [16]. It has been show that NMF with sum squared error cost function is equivalent to a relaxed k-means clustering. In addition, NMF with I-divergence cost function is equivalent to probabilistic latent semantic indexing [8]. Unlike NMF based clustering with the feature matrix as input, SC accepts similarity matrix as input, showing superior performance comparing to other algorithms [23]. Note that SC admits the continuous solution with mixed sign, which can not directly capture the clustering structure. Recently, nonnegative symmetric matrix factorization(SymNMF) has been presented in [13]. Unlike NMF, SymNMF is based on the similarity measure between data points and factorizes the symmetric matrix containing pairwise similarity vlaues to achieve the clustering assignment matrix.

Z. Cui et al. (Eds.): IScIDE 2019, LNCS 11936, pp. 302–311, 2019.
https://doi.org/10.1007/978-3-030-36204-1_25

Further, a hybrid clustering which combines the merits of NMF and SymNMF has been developed in [9].

Although NMF based approaches and SC have gained promising results, they are single-view approaches. In real applications, many datasets naturally appear in the form of multiple views or come from multiple sources [5]. Therefore, several multi-view clustering algorithms have been developed in the literature. Existing work can be roughly divided into two categories. Algorithms in the first category aims to find a unified low-dimensional embedding of multi-view data, then exploit classical clustering algorithms to derive clustering results. The canonical correlation analysis [3,7] approach and nonnegative matrix factorization [17] approach are two typical examples. The second category directly combines the information of different views in the clustering process. Typical examples include Multi-view k-means clustering [6] and co-EM clustering [2].

The above algorithms directly employ original data matrix as input. However, when data points are embedded in a nonlinear manifold, it is better to represent data points as graph. Based on this observation, multi-view clustering has also been investigated in the framework of spectral clustering [14,15,22,26]. A subspace co-training framework for multi-view data clustering has been proposed in [25]. In [24], a multiple graph based approach has been proposed for multi-view clustering. Based on a mixture of Markov chains defined on different views, a multiview spectral embedding and transductive inference approaches is presented in [26]. However, these approaches do not explicitly eliminate the view disagreement across different views. Recently, a co-training approach for multi-view spectral clustering has been presented in [15]. The main idea of [15] is to identify the clustering in one view and use to "label" the data in other views so as to modify the similarity matrix. [14] employs pairwise co-regularization to achieve a consistent clustering across different views. Similar to traditional spectral clustering, multi-view spectral clustering (MSC) depends crucially on the construction of graph Laplacian and the resulting eigenvectors that reflect the cluster structure in the data. In other words, these approaches require eigenvectors as input for k-means to generate the final cluster result.

In this paper, we propose a Hybrid Matrix Factorization (HMF) framework for multi-view clustering. HMF simultaneously factorizes the similarity matrix and the feature matrix of each view respectively. The final common cluster assignment can be obtained from the weighted sum of each view. Unlike traditional approaches, HMF employs nonnegative matrix factorization technique which naturally captures the cluster structure and produces the clustering assignment without additional clustering procedures. Our method can also learn the weights of each view adaptively which further refine the clustering results. In summary, our work has the following contributions

- We propose a hybrid matrix factorization (HMF) for multi-view clustering. HMF combines the merits of NMF and SymNMF which can yield the clustering assignment without additional clustering procedures.
- HMF can uncover the latent information by performing feature matrix and manifold structure matrix factorization simultaneously.

- An efficient iterative algorithm to has also presented to solve the optimization of HMF.

The rest of this paper is organized as follows. Section 2 reviews two important matrix factorization algorithms. We propose our approach in Sect. 3. Experimental results are presented in Sect. 4. Followed with conclusion in Sect. 5.

2   Overview of NMF and SymNMF

In this section, we review the two algorithms nonnegative matrix factorization (NMF) and symmetric nonnegative matrix factorization (SymNMF) [13]. In the next section, we show how to combine the merits of NMF and SymNMF and propose hybrid matrix factorization for multi-view data clustering.

2.1   NMF

Let $X = [x_1, x_2, \cdots, x_n] \in \mathbf{R}^{d \times n}$ denote the nonnegative data matrix. NMF aims to find two nonnegative matrix $W \in \mathbf{R}^{d \times k}$ and $H \in \mathbf{R}^{k \times n}$ such that the following expression is minimized

$$\|X - WH\|_F^2 \tag{1}$$

where $\| \cdot \|_F^2$ is the Frobenius norm. Since the objective function is not convex in W and H together, it is impractical to find the global minimum. [16] proposed the multiplicative update rules to iteratively minimize the objective function as follows

$$W_{ip} \leftarrow W_{ip} \frac{(XH^T)_{ip}}{(WHH^T)_{ip}}, \quad H_{pj} \leftarrow H_{pj} \frac{(X^T W)_{pj}}{(H^T W^T W)_{ip}}$$

2.2   SymNMF

Unlike traditional NMF works on the content, SymNMF works on the similarity matrix and generates the clustering results straightforwardly. Suppose A is the similarity matrix where A_{ij} measures the similarity between x_i and x_j. The following two types of similarity graph are suggested in [13]

- Full graph: Each pair (x_i, x_j) is connected by an edge, and the edge weight is computed by

$$A_{ij} = \exp\left(-\frac{\|x_i - x_j\|_2^2}{\sigma_i \sigma_j}\right) \tag{2}$$

- Sparse graph: Every node is connected to its k' nearest neighbors.

$$\hat{A}_{ij} = \begin{cases} A_{ij}, \text{if } x_i \in N_{k'}(x_j) \text{ or } x_j \in N_{k'}(x_i) \\ 0, \quad \text{otherwise} \end{cases} \tag{3}$$

where A_{ij} is defined in Eq. (2), $x_i \in N_{k'}(x_j)$ implies x_i is among the k' nearest neighbors of x_j.

We are now ready to present the formulation of SymNMF. Mathematically, SymNMF can be formulated as

$$\min_{H \geq 0} \|A - HH^T\|_F^2 \tag{4}$$

where H is a nonnegative matrix of size $n \times k$, and k is the number of clusters. Due to the nonnegativity constraints, H naturally captures the cluster structure. More precisely, let H_{ij} be the largest value in i-th row of H, then x_i can be assigned into j-th cluster. Further, a Newton-like algorithm that takes advantage of second-order information information was developed to get the solution (4).

Although SymNMF has gained promising results on clustering, it only works for single-view dataset. Further, SymNMF accepts graph as input which neglects the feature matrix. In the following section, we show how to combine SymNMF and NMF to deal with multi-view dataset.

3   HMF

3.1   Formulation

Let $X^{(\nu)} = [x_1^{(\nu)}, x_2^{(\nu)}, \cdots, x_n^{(\nu)}] \in \mathbf{R}^{d^{(\nu)} \times n}$ denote the examples in view ν and $S^{(\nu)}$ denote the similarity matrix for view ν. Similar to symNMF algorithm, we write $A^{(\nu)} \in \mathbf{R}^{n \times n}$ as the similarity matrix for each view. To perform the latent information discovery from both feature matrix and manifold structure matrix, we propose the following hybrid matrix factorization (HMF) formulation

$$\min_{H^{(\nu)}, U^{(\nu)}, H^*, \alpha^{(\nu)}} \sum_{\nu=1}^{M} \left\{ \|A^{(\nu)} - H^{(\nu)T}H^{(\nu)}\|_F^2 + \lambda\|X^{(\nu)} - U^{(\nu)}H^{(\nu)}\|_F^2 \right.$$
$$\left. + (\alpha^{(\nu)})^\gamma \|H^{(\nu)} - H^*\|_F^2 \right\} \tag{5}$$
$$s.t., U^{(\nu)} \geq 0, H^{(\nu)} \geq 0, H^* \geq 0, \sum_{\nu=1}^{M} \alpha^{(\nu)} = 1, \alpha^{(\nu)} \geq 0$$

where $H^{(\nu)}$ is clustering result for the ν-the view and H^* is the unified clustering result for all views, $\alpha^{(\nu)}$ is the weight factor for the ν-the view and γ is the parameter to control the weights distribution which indicates the contribution to the final clustering.

3.2   Optimization

In this section, we propose an iterative update procedure to solve the optimization problem (5). Specifically, we update one variable while keeping other variables fixed. The optimization is carried out in the following steps:

– Computing $U^{(\nu)}$. Removing the terms irrelevant to $U^{(\nu)}$, we have

$$\min_{U^{(\nu)}} \|H^{(\nu)T}U^{(\nu)T} - X^{(\nu)T}\|_F^2 \text{ s.t } U^{(\nu)} \geq 0 \tag{6}$$

which is a nonnegative least squares problem can be solved via the algorithm proposed in [12].

– Computing $H^{(\nu)}$. For ν-th view, $H^{(\nu)}$ is independent with other views, so we omit the superscript (ν) temporarily. By doing this, we have

$$\min_{H} \|A - H^T H\|_F^2 + \lambda \|X - UH\|_F^2 + \beta \|H - H^*\|_F^2 \qquad (7)$$
$$s.t. H \geq 0$$

where $\beta = (\alpha^{(\nu)})^\gamma$.

To make (7) feasible, we introduce an auxiliary variable W and rewrite the optimization problem (7) as follows

$$\min_{H,W} \|A - W^T H\|_F^2 + \lambda \|X - UH\|_F^2 + \beta \|H - H^*\|_F^2 + \eta \|H - W\|_F^2 \qquad (8)$$
$$s.t. H \geq 0, W \geq 0$$

With algebraic manipulation, the optimal solution for W and H are respectively characterized by the following problem

$$\min_{H \geq 0} \left\| \begin{bmatrix} W^T \\ \sqrt{\lambda}U \\ \sqrt{\eta}I_k \\ \sqrt{\tau}I_k \end{bmatrix} H - \begin{bmatrix} A \\ \sqrt{\lambda}X \\ \sqrt{\eta}H^* \\ \sqrt{\tau}W \end{bmatrix} \right\|_F^2 \qquad (9)$$

$$\min_{W \geq 0} \left\| \begin{bmatrix} H^T \\ \sqrt{\tau}I_k \end{bmatrix} W - \begin{bmatrix} A^T \\ \sqrt{\tau}H \end{bmatrix} \right\|_F^2 \qquad (10)$$

where I_k is an identity matrix of $k \times k$. The above two problems are nothing but nonnegative least squares which can be solved via the algorithm proposed in [12].

– Computing H^*. Taking the derivative of the objective function with respect to H^*, we have

$$\sum_{\nu=1}^{M} (\alpha^{(\nu)})^\gamma (H^{(\nu)} - H^*) = 0$$

Finally, we get

$$H^* = \frac{\sum_{\nu=1}^{M} (\alpha^{(\nu)})^\gamma H^{(\nu)}}{\sum_{\nu=1}^{M} (\alpha^{(\nu)})^\gamma} \qquad (11)$$

– Computing $\alpha^{(\nu)}$. Let $z^{(\nu)} = \|H^{(\nu)} - H^*\|$. The optimization problem for $\alpha^{(\nu)}$ is

$$\min \sum_{\nu=1}^{M} (\alpha^{(\nu)})^\gamma z^{(\nu)} \quad s.t. \sum_{\nu=1}^{M} \alpha^{(\nu)} = 1, \alpha^{(\nu)} \geq 0 \qquad (12)$$

The Lagrange function of (12) is

$$\mathcal{L}(\alpha^{(\nu)}, \lambda) = \sum_{\nu=1}^{M} (\alpha^{(\nu)})^\gamma z^{(\nu)} - \lambda(\sum_{\nu=1}^{M} \alpha^{(\nu)} - 1)$$

Setting the derivative of $\mathcal{L}(\alpha^{(\nu)}, \lambda)$ with respect to $\alpha^{(\nu)}$ to zero, we have

$$\alpha^{(\nu)} = \left(\frac{\lambda}{\gamma z^{(\nu)}}\right)^{\frac{1}{\gamma - 1}}$$

Substituting the resultant expression of $\alpha^{(\nu)}$ to the constraint $\sum_{\nu=1}^{M} \alpha^{(\nu)} = 1$, we get

$$\alpha^{(\nu)} = \frac{(\gamma z^{(\nu)})^{\frac{1}{1-\gamma}}}{\sum_{\nu=1}^{M} (\gamma z^{(\nu)})^{\frac{1}{1-\gamma}}} \tag{13}$$

By the above four steps, we alternatively update $H^{(\nu)}, H^*, U^{(\nu)}$ as well as $\alpha^{(\nu)}$ and repeat the process until the objective function becomes converged. We summarize our algorithm in Table 1. Note that once we get the consensus matrix H^*, the cluster label of data point i can be computed as $\arg\max_k H^*_{ij}$.

Table 1. HMF

Input:multi-view feature matrix $X^{(\nu)}$ and similarity matrix $\{A^{(\nu)}\}_{\nu=1}^{m}$, Cluster assignment matrix H^*

while not converged **do**
    For each $\nu \in \{1, 2, \cdots, m\}$**do**
        Obtain $U^{(\nu)}$ by solving (6)
        Obtain $H^{(\nu)}$ by solving (9) and (10)
        Update weight $\alpha^{(\nu)}$ by solving (13)
    end
    Update H^* by (11)
end

Output: Clustering assignment matrix H^*

4   Experiments

4.1   Experimental Setup

In this section, we conduct experiments on two multi-view datasets to show the effectiveness and efficiency of our proposed approach. In particular, details of the two multi-view datasets are as follows

- **Reuters**. The Reuters dataset [1] consists of documents that are written in five languages and their translations. We use a subset containing 200 documents of 6 clusters. Specifically, we use five views, i.e., English, French, German, Spanish and Italian.

– **Wiki.** The Wiki dataset [21] contains 2866 articles crawled from Wikipedia featured articles. It consists of a pair of an image and the associated text. Each image is described by a 128-dimensional bag-of-visual SIFT feature vector [18] and each text is represented by a 10-dimensional feature vector generated by latent Dirichlet allocation [4]. 90% of the data are selected for training and the rest are used as the query set.

Table 2. Important statistics of datasets

Data sets	# of data	# of views	# of cluster
Reuters	1200	5	6
Wiki	2866	2	10

The important statistics are listed in Table 2. We compare all the approaches on a number of evaluation measures. In particular, we report Precision, Recall, F-score, normalized mutual information (NMI), adjusted rand index (AR) and average entropy (Entropy). Detailed definition of there criteria can be found in [10,19]. Note that the larger the values of the first four criteria while the smaller the value of the last criterion, the better the performance.

In all the experiments, Gaussian kernel is employed to compute the graph similarity. The standard derivation of the kernel is estimated as the median of the pair-wise Euclidean distances between two data samples.

We compare our proposed approach with a number of baselines. Specifically, The compared methods are as follows

– **singleSP:** We employ SP [20] on each view and report the best results.
– **conSP:** Concatenating the features of each view and then running SP [20] on the concatenated feature matrix.
– **PairwiseMSC:** The pairwise co-regularized multi-view spectral clustering (MSC) proposed in [14].
– **CentroidMSC:** The centroid based co-regularized multi-view spectral clustering (MSC) proposed in [14].
– **HMF:** Multi-view clustering via weighted symmetric nonnegative matrix factorization which is proposed in this paper.

4.2   Results

Experimental results are presented from Tables 3 and 4. From Table 3, we see that HMF gains better results than other methods in terms of F-score, Recall, NMI and AR. The algorithm centroidMSC achieves better performance than other methods in terms of Precision and Entropy. Table 4 presents the clustering results on wiki dataset. We see that HMF achieves the best results on this dataset. Specifically, the results of HMF indicate an increase of %4 in terms of F-score, precision and AR. We even have an %10 improvement in term of Entropy.

Table 3. Clustering performance on Reuters

Method	F-score	Precision	Recall	Entropy	NMI	AR
singSP	0.3454	0.3221	0.3835	1.9026	0.2732	0.2033
conSP	0.3450	0.3132	0.3876	1.9075	0.2725	0.1978
pairwiseMSC	0.3487	0.3214	0.3910	1.8922	0.2755	0.2046
centroidMSC	0.3473	**0.3250**	0.3961	**1.8888**	0.2755	0.2048
RMKMC	0.3305	0.2613	0.4494	1.9331	0.2824	0.1526
HFM	**0.3759**	0.2944	**0.5200**	1.9328	**0.2834**	**0.2081**

Table 4. Clustering performance on wiki

Method	F-score	Precision	Recall	Entropy	NMI	AR
single view	0.4919	0.4786	0.5058	3.1282	0.5508	0.4286
SP	0.4758	0.4792	0.4730	1.5343	0.5291	0.4130
pairwiseMSC	0.4946	0.5014	0.4883	1.5533	0.5238	0.4344
centroidMSC	0.4866	0.4843	0.4893	1.5469	0.5272	0.4243
RMKMC	0.5058	0.4797	0.5349	1.5324	0.5434	0.4424
HMF	**0.5423**	**0.5493**	**0.5355**	**1.4253**	**0.5619**	**0.4878**

4.3 Convergence

We experimentally show the convergence property of our proposed algorithm on two datasets. Specifically, We compute the objective value of in Eq. (5) for each iteration. The objective value curve is plotted in Fig. 1. We observe that the objective value decreases sharply and then remains stable. Specifically, our algorithm can converge within 50 iterations on both two datasets. In our experiments, we set the maximum number of iterations to 50.

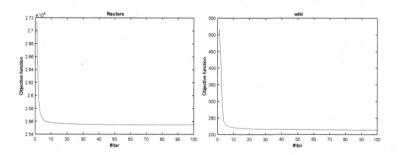

Fig. 1. Objective function value with respect to the number of iterations.

5   Conclusion

In this paper, we have proposed hybrid matrix factorization (HMF) for multi-view clustering. HMF can capture the cluster structure straightforwardly and learn weights of each view adaptively. We also introduce an optimization algorithm to iteratively solve the proposed optimization. Experimental results on multi-view datasets have delivered the effectiveness of our proposed approaches.

Acknowledgements. This work was supported by the National Natural Science Foundation of China (Grants No. 61602248) and the Natural Science Foundation of Jiangsu Province (Grants No. BK20160741).

References

1. Amini, M.R., Usunier, N., Goutte, C.: Learning from multiple partially observed views - an application to multilingual text categorization. Adv. Neural Inf. Process. Syst. **22**, 28–36 (2009)
2. Bickel, S., Scheffer, T.: Multi-view clustering. In: ICDM vol. 4, pp. 19–26 (2004)
3. Blaschko, M.B., Lampert, C.H.: Correlational spectral clustering. In: IEEE Conference on Computer Vision and Pattern Recognition, 2008. CVPR 2008, pp. 1–8. IEEE (2008)
4. Blei, D.M., Ng, A.Y., Jordan, M.I.: Latent dirichlet allocation. J. Mach. Learn. Res. **3**(Jan), 993–1022 (2003)
5. Blum, A., Mitchell, T.M.: Combining labeled and unlabeled data with co-training. In: Proceedings of the Eleventh Annual Conference on Computational Learning Theory, pp. 92–100 (1998)
6. Cai, X., Nie, F., Huang, H.: Multi-view k-means clustering on big data. In: Proceedings of the Twenty-Third international joint conference on Artificial Intelligence, pp. 2598–2604. AAAI Press (2013)
7. Chaudhuri, K., Kakade, S.M., Livescu, K., Sridharan, K.: Multi-view clustering via canonical correlation analysis. In: Proceedings of the 26th Annual International Conference on Machine Learning, pp. 129–136. ACM (2009)
8. Ding, C.H.Q., Li, T., Peng, W.: Nonnegative matrix factorization and probabilistic latent semantic indexing: equivalence, chi-square statistic, and a hybrid method. In: AAAI 2006 Proceedings of the 21st National Conference on Artificial Intelligence , vol. 1, pp. 342–347 (2006)
9. Du, R., Drake, B.L., Park, H.: Hybrid clustering based on content and connection structure using joint nonnegative matrix factorization. J. Global Optim. **74**, 1–17 (2017)
10. Hubert, L., Arabie, P.: Comparing partitions. J. Classif. **2**(1), 193–218 (1985). https://doi.org/10.1007/BF01908075
11. Jain, A., Murty, M., Flynn, P.: Data clustering: a review. ACM Comput. Surv. **31**(3), 264–323 (1999)
12. Kim, H., Park, H.: Nonnegative matrix factorization based on alternating nonnegativity constrained least squares and active set method. SIAM J. Matrix Anal. Appl. **30**(2), 713–730 (2008)
13. Kuang, D., Ding, C., Park, H.: Symmetric nonnegative matrix factorization for graph clustering. In: 12th SIAM International Conference on Data Mining, SDM 2012, pp. 106–117 (2012)

14. Kumar, A., Daumé, H.: A co-training approach for multi-view spectral clustering. In: Proceedings of the 28th International Conference on Machine Learning (ICML-11), pp. 393–400 (2011)

15. Kumar, A., Rai, P., Daume, H.: Co-regularized multi-view spectral clustering. In: Advances in Neural Information Processing Systems, pp. 1413–1421 (2011)

16. Lee, D.D., Seung, H.S.: Learning the parts of objects by non-negative matrix factorization. Nature **401**(6755), 788–791 (1999)

17. Liu, J., Wang, C., Gao, J., Han, J.: Multi-view clustering via joint nonnegative matrix factorization. In: Proceedings of SDM, vol. 13, pp. 252–260. SIAM (2013)

18. Lowe, D.G.: Distinctive image features from scale-invariant keypoints. Int. J. Comput. Vis. **60**(2), 91–110 (2004)

19. Manning, C.D., Raghavan, P., Schütze, H., et al.: Introduction to Information Retrieval, vol. 1. Cambridge University Press, Cambridge (2008)

20. Ng, A.Y., Jordan, M.I., Weiss, Y., et al.: On spectral clustering: Analysis and an algorithm. In: Advances in neural information processing systems, vol. 2, 849–856 (2002)

21. Rasiwasia, N., et al.: A new approach to cross-modal multimedia retrieval. In: Proceedings of the 18th ACM International Conference on Multimedia, pp. 251–260. ACM (2010)

22. de Sa, V.R.: Spectral clustering with two views. In: ICML Workshop on Learning with Multiple Views, pp. 20–27 (2005)

23. Shi, J., Malik, J.: Normalized cuts and image segmentation. IEEE Trans. Patt. Anal. Mach. Intell. **22**(8), 888–905 (2000)

24. Tang, W., Lu, Z., Dhillon, I.S.: Clustering with multiple graphs. In: Ninth IEEE International Conference on Data Mining, 2009. ICDM 2009, pp. 1016–1021. IEEE (2009)

25. Zhao, X., Evans, N., Dugelay, J.L.: A subspace co-training framework for multi-view clustering. Patt. Recogn. Lett. **41**, 73–82 (2014)

26. Zhou, D., Burges, C.J.C.: Spectral clustering and transductive learning with multiple views. In: International Conference on Machine Learning (2007)

Car Sales Prediction Using Gated Recurrent Units Neural Networks with Reinforcement Learning

Bowen Zhu, Huailong Dong, and Jing Zhang$^{(\boxtimes)}$

Department of Computer Science and Technology,
Nanjing University of Science and Technology,
No. 200 Xiaolingwei Street, Nanjing 210094, China
2391986553@qq.com, dh10010@163.com,
jzhang@njust.edu.cn

Abstract. In this paper, we propose a novel Gated Recurrent Units neural network with reinforcement learning (GRURL) for car sales forecasting. The car sales time series data usually have a small sample size and appear no periodicity. Many previous time series modeling methods, such as linear regression, cannot effectively obtain the best parameter adjustment strategy when fitting the final prediction values. To cope with this challenge and obtain a higher prediction accuracy, in this paper, we combine the GRU with the reinforcement learning, which can use the reward mechanism to obtain the best parameter adjustment strategy while making a prediction. We carefully investigated a real-world time-series car sales dataset in Yancheng City, Jiangsu Province, and built 140 GRURL models for different car models. Compared with the traditional BP, LSTM, and GRU neural networks, the experimental results show that the proposed GRURL model outperforms these traditional deep neural networks in terms of both prediction accuracy and training cost.

Keywords: Car sales prediction · BP · LSTM · Gated Recurrent Units · Reinforcement learning

1 Introduction

The prediction of time series has been widely studied in various fields such as industry, economy, agriculture, and biomedicine [1]. The modeling methods for time series are mainly divided into two categories: *linear modeling and nonlinear modeling* [2]. Linear models are suitable for predicting linear time series, including some time series prediction methods which are based on traditional statistics. However, the time series acquired in practical applications are generally nonlinear and non-stationary, so nonlinear models are proposed to solve these problems.

The nonlinear time series prediction [4] methods are represented by artificial neural networks [3], which is mainly divided into the backpropagation (BP) neural network and the recurrent neural network (RNN). Among them, the BP neural networks are a fully connected neural network based on the error backpropagation algorithm. However, it suffers the weaknesses of too many parameters and timing information cannot

© Springer Nature Switzerland AG 2019
Z. Cui et al. (Eds.): IScIDE 2019, LNCS 11936, pp. 312–324, 2019.
https://doi.org/10.1007/978-3-030-36204-1_26

being utilized. The recurrent neural networks are represented by the Long Short-Term Memory (LSTM) [10, 20], which exploits the information of the previous moment to influence the output of the current moment. Thus, it mines the timing information in the data. However, the gate mechanism of LSTM is so complicated that there is a delay when processing small data samples. As a variant of LSTM, the Gated Recurrent Units (GRU) neural networks [14] not only simplify the gate mechanism of LSTM but also preserve the core elements of LSTM. However, traditional GRU uses linear regression to fit the final predicted values. There is no guarantee that the most effective strategy will always be used for parameter adjustment.

This paper focuses on solving a real-world time series problem, predicting car sales based on historical data. This real-world problem is difficult because it has some specific characteristics. First, for an agent that aims to provide the car sales prediction, it usually collects car sales data of many different models. For example, in the dataset used in this paper, there are 140 car models in total. In addition, for each car model, it has a small sample size, and the data appear no periodicity due to the various influencing factors to the car sales. Considering the advantages and disadvantages of various neural networks and the characteristics of the real-world data, we choose GRU as the basic prediction model. At the same time, we notice that reinforcement learning [13, 19] can reward the positive parameter adjustment strategy and penalize the negative strategy by using the reward mechanism to obtain an optimized parameter adjustment strategy. Therefore, in this paper, we propose a gated recurrent units neural network with reinforcement learning (GRURL) model, which replaces the linear regression in the traditional GRU model with reinforcement learning, which can solve the problem that linear regression cannot always use the positive strategy to adjust parameters. We compare our GRURL mode with some popular deep learning models and the experimental results show some advantages of the proposed model.

The remainder of the paper is organized as follows: Sect. 2 reviews the related work of the time series prediction and its usage in car sales; Sect. 3 analyzes the characteristics of the car sales dataset and proposes a novel GRURL model; Sect. 4 presents the experimental results; and Sect. 5 concludes the paper.

2   Related Work

After the concept of deep learning was proposed, two types of networks are widely used in the prediction of time series. One is based on the feedforward neural networks. For example, Jo used a BP algorithm combined with the multi-layered sensing network to predict the multivariate time series [7]. Moody et al. proposed a radial basis function network for fast learning of the network [5]. This network is more straightforward than the BP neural network and has the advantages of fast and efficient learning. However, its performance depends on the center and the width of the basis function and the weight of the output layer. At the same time, one of the problems in the above two networks is that they cannot effectively learn the dependence between time. The RNN networks [6, 18] are thus proposed to solve the problem of dependence between time. Hochreiter et al. proposed the LSTM network which is widely used as a variant of the

RNN network in various time series predictions [8]. Duan et al. used the LSTM network to predict travel time [11] and obtained a good single-step prediction result.

Car sales prediction is of high values in today's commercial environment. Researchers have conducted a number of studies on car sales from the perspective of quantitative and qualitative. Qualitative methods include expert prediction, competent probability prediction, and other similar methods. For example, Cao et al. inferred the future development trend of the automobile industry from the rise of China's economy [15]. Quantitative methods can be divided into two categories. One is based on conventional prediction methods. For example, Katleho et al. proposed the autoregressive integrated moving average (AIMA) model to predict short term seasonal car sales [12]. Chen et al. established a grey theory prediction model, which can reflect the regularity of the sales data itself [16]. The other one is based on neural networks. For example, Li et al. proposed a prediction method combining the autoregressive-moving-average (ARMA) model and the radial basis function (RBF) neural network [17], which can reflect the linear relationship and nonlinear characteristics between sales data. Farahani et al. used artificial neural networks and analytical hierarchy process in car sales prediction, which first identified the factors that influence sales and then determine the interconnection between the data [9].

However, both qualitative and quantitative prediction methods rely on historical data with substantial delays. Also, the prediction granularity is coarse, as the annual data are generally used. However, the data used in this article is nonlinear, small in granularity, and low in latency. The model proposed in this paper targets the fine-grained prediction, which is more attractive in practice.

3   Data Analysis and the Proposed Method

This section first analyzes the characteristics of the car sales dataset and then selects features for prediction. Finally, a novel GRURL model is proposed for sales prediction.

3.1   Analysis of Car Sales Data Set

The dataset used in this article is from the real data of Yancheng car sales provided by Ali Tian-chi programming contest. It includes the monthly sales of more than 140 car models and 103,507 sales items from March 2015 to October 2017, which is equivalent to the three years of data volume. This subsection aims to discover the characteristics of these data to facilitate prediction. Through data analysis, we can find that car sales dataset has the following characteristics:

- There are 140 car models in total. Because of the different popularity of these models, the amount of the sales data of each mode varies greatly.
- The maximum number of car sales data for each model is no more than 100, and the minimum is less than 10.
- Sales data for each model appears no periodicity, and there is a dependency between time stamps.

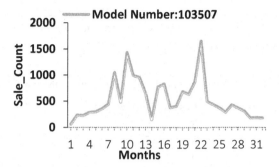

Fig. 1. Distribution characteristics of monthly sales of the regular samples

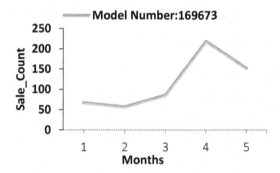

Fig. 2. Distribution characteristics of monthly sales of the irregular samples

In this paper, we refer to the number of sale data below 10 as irregular samples, and the others as regular samples. Figures 1 and 2 show the monthly sales distribution characteristics of the regular and the irregular samples, respectively. For the regular samples, we can observe the fluctuation of the sale volume against the months. For the irregular samples, because the amount of data is too small, any neural network does not work well. Thus, in this paper, we only focus on regular samples.

3.2 Feature Selection

Since our goal is to obtain the optimal predictive performance that can be applied to the real-world dataset, feature selection is essential. The question is which sequence of timestamps we need to use as input and output for prediction. Let us consider the following pattern. Assume that x_t represents sales for the t-th month. We use $x_t, x_{t+1}, \ldots \ldots x_{t+n}$ as input, x_{t+n+1} as output, and y_t as the actual sales of $t + n + 1$ months. Then, we use $\hat{x}_{t+n+1}$ to denote the predicted sales for the $t + n + 1$ month and it is recorded as $\hat{y}_t$. That is, the input and output styles are $\{x_t, x_{t+1}, \ldots \ldots x_{t+n}\} \rightarrow \hat{y}_t$. We need to determine a suitable value of n to make the difference between y_t and $\hat{y}_t$ reach a minimum. For this reason, the experiment to determine the value of n was carried out on 20 car models. Table 1 shows the effect of different n values for a model, where Length represents n value, and AVG_RMSE represents the average value of Root Mean Squared Error of two-month predicted sales. From Table 1, we found that $n = 3$ or $n = 2$ is the most suitable n value.

Table 1. AVG_RMSE with different n values for the same model

Length	AVG_RMSE
8	18.7809
7	17.2723
6	16.5579
5	12.6123
4	11.8259
3	11.0116
2	11.0324
1	82.3713

3.3   The Proposed GRURL Model

The structures of the proposed GRU with reinforcement learning (GRURL) model and the traditional GRU model are shown in Fig. 3.

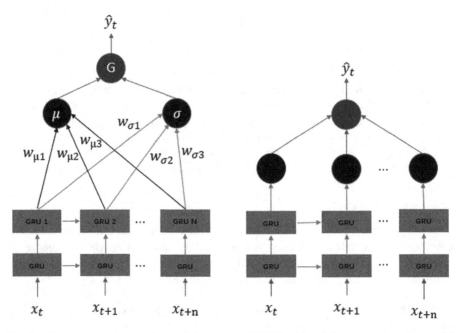

Fig. 3. The structures comparison of the proposed GRU with reinforcement learning model (left) and the traditional GRU model (right)

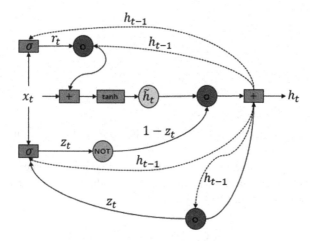

Fig. 4. The structure of a GRU cell

First, we briefly introduce the traditional GRU model, which has four layers. Two layers in the bottom are GRU cells. The black dots in the figure represent the full-link layer, which receives the output of the GRU cells and fits the output values of GRU cells into a one-dimensional predictive value by performing linear regression. At the top, the red dots represent the layer of the predicted value.

Similar to the traditional GRU model, our proposed GRURL model also consists of four layers. Two layers in the bottom of GRURL are the same as GRU. The back dots in GRURL represent the stochastic layer, and the red dot represents the output layer. Note that the definition of the output layer is different from the GRU model, where reinforcement learning is employed.

The basic element in the GRU layer in the GRU cell, whose structure is shown in Fig. 4. The output of a GRU cell is calculated as follows:

$$z_t = \sigma(x_t U^z + h_{t-1} W^z + b^z), \tag{1}$$

$$r_t = \sigma(x_t U^r + h_{t-1} W^r + b^r), \tag{2}$$

$$\tilde{h}_t = tanh(x_t U^h + (h_{t-1} \, o \, r_t) W^h + b^h), \tag{3}$$

$$h_t = (1 - z_t) \, o \, \tilde{h}_t + z_t \, o \, h_{t-1}, \tag{4}$$

where z, r, $\tilde{h}$, and h represent the update gate, the reset gate, the candidate value of the currently hidden node, and the output activation value of the currently hidden node, respectively.

In the stochastic layer, we have two neurons whose outputs are μ and σ, respectively. The weights on the links between these two neurons and the neurons on the second layer are denoted by $w_{\mu i}$ and $w_{\sigma i}$, respectively. The distributions of the parameters in the network obey the Gaussian distribution. Two outputs of the stochastic layer are calculated as follows:

$$\mu = \frac{e^{\beta_\mu \sum_{i=1}^n R_i w_{\mu i}} - e^{-\beta_\mu \sum_{i=1}^n R_i w_{\mu i}}}{e^{\beta_\mu \sum_{i=1}^n R_i w_{\mu i}} + e^{-\beta_\mu \sum_{i=1}^n R_i w_{\mu i}}}, \tag{5}$$

$$\sigma = \frac{e^{\beta_\sigma \sum_{i=1}^n R_i w_{\sigma i}} - e^{-\beta_\sigma \sum_{i=1}^n R_i w_{\sigma i}}}{e^{\beta_\sigma \sum_{i=1}^n R_i w_{\sigma i}} + e^{-\beta_\sigma \sum_{i=1}^n R_i w_{\sigma i}}}, \tag{6}$$

where β_μ and β_σ are both constants and used to adjust the output of μ and σ respectively.

In the output layer of the GRURL model, we introduce reinforcement learning to learn a stochastic policy. On this layer, we use a one-dimension Gaussian function to predict time series data, defined as follows:

$$\hat{y}_t = G(C, W, X) = \frac{1}{\sqrt{2\pi}\sigma} e^{-\frac{(C-\mu)^2}{2\sigma^2}}, \tag{7}$$

where C is a constant, which can be generated by a regular random number, W represents the weights $w_{\mu i}$ and $w_{\sigma i}$, and X represents inputs $\{x_t, x_{t+1}, \ldots, x_{t+n}\}$. The parameters $w_{\mu i}$ and $w_{\sigma i}$ are updated as follows:

$$D_i(t) = \frac{\partial ln \hat{y}_t}{\partial w_i} + \gamma D_i(t-1), \tag{8}$$

$$\begin{aligned}
\Delta w_{\mu i} &= (r_i - b) \cdot D_{\mu i}(t) \\
&= (r_i - b) \cdot \left(\frac{1}{\sigma^2}(C - \mu) \cdot (1 - \mu^2)\beta_\mu R_i + \gamma D_{\mu i}(t-1) \right)
\end{aligned} \tag{9}$$

$$\begin{aligned}
\Delta w_{\sigma i} &= (r_i - b) \cdot D_{\sigma i}(t) \\
&= (r_i - b) \cdot \left(\frac{(\sigma^2 - 1)\beta_\sigma R_i}{\sigma} \cdot \left(\frac{1}{2} - \frac{(C - \mu)^2}{\sigma^2} \right) + \gamma D_{\sigma i}(t-1) \right)
\end{aligned} \tag{10}$$

$$\Delta W = \left(\Delta w_{\mu i}, \Delta w_{\sigma i} \right), \tag{11}$$

$$W \leftarrow W + \alpha(1 - \gamma)\Delta W, \tag{12}$$

where α is a non-negative learning constant, r_i is a reward value, b is a constant which is used to make appropriate corrections to r_i and γ is a discount ($0 \le \gamma < 1$). In Eq. (8), we call $D_i(t)$ degree of absorption and use it to indicate the degree of the adjustment value at time step t that is affected by time step $t - 1$. In Eqs. (9) and (10), we use r_i and $D_{\mu i}(t)$, r_i and $D_{\sigma i}(t)$ to obtain the correction values of weights $w_{\mu i}$ and $w_{\sigma i}$, respectively. The main steps of reinforcement learning are listed in Algorithm 1.

4   Experiments

This section presents the experiments. First, we describe the experimental settings in detail. Then, we make performance comparisons among BP, LSTM, GRU, and the proposed GRURL model.

Algorithm 1. Reinforcement Learning (RL) in GRURL

Input: Observation X, b, γ, β_μ, β_σ, α
Output: Predicted value $\hat{y}$
1. Initialize the parameter value of each GRU cell, $w_{\mu i}$, and $w_{\sigma i}$.
2. **REPEAT** in each iteration
  FOR time step t from 1 to the last **DO**
    2.1 Calculate μ and σ by Eqs. (5) and (6)
    2.2 Predict $\hat{y}_t$ by Eq. (7)
    2.3 Calculate r_i based on $|\hat{y}_t - y_t|$
    2.4 Calculate $\Delta w_{\mu i}$, $\Delta w_{\sigma i}$ by Eqs. (8), (9) and (10)
    2.5 Update W by Eqs. (11) and (12)
  UNTIL $|\hat{y} - y|$ is smaller than a predefined value.
3. **RETURN** $\hat{y}$.

4.1   Experimental Settings

The data set includes monthly car sales data for 140 care models from January 2012 to October 2017. The starting dates car sales data of car models are different. We divided the data of each care model in the dataset into a training set and a test set. Since each car model usually has a small amount of data, to improve the performance of prediction models, we increased the amount of data in the training set. The training set includes the sales data of each car model from its beginning month to August 2017, and the test set is the sales data of each car model from September 2017 to October 2017. We trained the BP networks, the LSTM networks, the GRU networks, and the proposed GRURL networks on the same training set, and evaluated the performance of the four network models on the test set.

The hyperparameters of each network model are set as follows: For the BP networks, a four-layer network is set up. The input layer has three cells for input, the hidden layer is two layers with 20 cells of each layer, and the output layer has one cell to represent the predicted value. The activation function of the input layer and the output layer is a ReLU function, and the activation function of the hidden layer is a sigmoid function. The learning rate is set to 0.09. In the training phase, the BP algorithm is used for parameter adjustment. The batch size is set to 15 and the time step is set to three. For the traditional LSTM model, a two-layer LSTM network with one output layer is set up. The number of cells in each LSTM layers is 20, and the learning rate is 0.09. In the output layer, we used the full-link structure between the output layer and the second LSTM layer. The loss function in the output layer is the mean square error. For the traditional GRU model, a two-layer GRU network and one output layer are set up. The number of cells in each GRU layers is 20, and the learning rate is 0.09. In the output layer, we used the full-link structure between the output layer and the second GRU layer. The loss function of the output layer is also the mean square error. Both the LSTM and GRU models use the AdaGrad algorithm during the training phase with the batch size set to 15 and the time step set to 3.

For the proposed GRURL model, two GRU layers set are the same as the traditional GRU model. Parameter β_μ is set to 15 and parameter β_σ is set to 25. The initial values of $w_{\mu i}$ and $w_{\sigma i}$ are set to 0.2 and 0.002, respectively. Parameter α is set to 0.02. Parameter r_t is set according to the equation as follows:

$$r_t = \begin{cases} 0.03, & \text{if } |\hat{y}_t - y_t| \leq 10 \\ -0.03, & \text{if } |\hat{y}_t - y_t| > 10 \end{cases}. \tag{13}$$

In our study, we use two metrics Root Mean Square Error (RMSE) and Mean Absolute Percent Error (MAPE) to evaluate the performance of our network models in comparison, which are defined as follows:

$$\text{RMSE} = \sqrt{\frac{1}{num} \sum_{i=1}^{num} (y_i - \hat{y}_i)^2}, \tag{14}$$

$$\text{MAPE} = \left(\frac{100}{num}\right) \times \sum_{i=1}^{num} |(y_i - \hat{y}_i)/y_i.| \tag{15}$$

Besides, the running times of the model training are also compared. Because every network model will introduce randomness in the training process. For a fair comparison, each experiment was repeated ten times and the average metric values are reported. In the following subsection, we use AVG_RMSE to represent the average value of RMSE in the two months and use AVG_MAPE to represent the average value of the MAPE in the two months.

4.2   Experimental Results

For each regular sample, the maximum and minimum data sizes are 70 and 10, respectively. Since we have 140 car models. It is impossible to list all the comparison results on these 140 car models. To make a clearer presentation, we handled the experimental results a follows. We sorted all car model data according to their sample sizes. We divided each six car models into one segment and select one representative sample from each segment. That is, we only present the GRURL models for each representative car model of each segment. More precisely, we selected a car model with the amount of data close to the median as a representative car model and showed the experimental results on this car model data. Finally, we only show ten groups of experimental results. The comparison results of four network models in terms of AVG_RMSE, AVG_MAPE and the running time are shown in Figs. 5, 6 and Table 2, respectively.

From Figs. 5 and 6, we have the following observations: (1) The performance of the BP network model is obviously the worst one among the models in comparison.

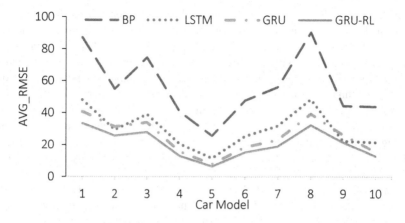

Fig. 5. RMSE predicted by BP, LSTM, GRU, and the proposed GRURL

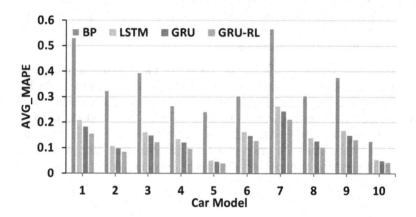

Fig. 6. MAPE predicted by LSTM, GRU, and the proposed GRURL

Table 2. The training time of four methods in comparison

Car model	BP	LSTM	GRU	GRURL
1	34.2269	11.8391	9.5983	12.6061
2	32.3448	11.5604	9.7403	12.9854
3	29.6712	11.2379	9.5967	12.1265
4	32.4937	11.0428	10.6605	12.7208
5	24.3223	11.6906	9.7211	11.9331
6	19.4172	10.9963	9.5887	12.8703
7	26.6551	10.9861	10.505	11.2169
8	27.1725	11.0436	9.7113	12.2937
9	24.0351	10.8618	9.5999	11.4266
10	36.7439	10.4036	9.8453	12.9604

This indicates that the BP network cannot deal with the dependence between times-tamps, which is critical in time series modeling. (2) The performance of the GRU model is better than that of the LSTM model. This indicates that GRU has a more straightforward strategy, which seems more effective on our real-world datasets. (3) The prediction accuracy of the proposed GRURL model is about 15% (absolute value) higher than that of the traditional GRU model. On our real-world dataset, compared with the linear regression which fits the final output in the traditional GRU, the reinforcement learning algorithm can adjust the parameters more effectively according to the dynamic changes of time series. Therefore, in both metrics of AVG_RMSE and AVG_MAPE, the proposed GRURL significantly outperforms the other methods.

From Table 2, we have the following observations: (1) The BP network has so many parameters to adjust that its training time is obviously the longest among all compared methods. (2) The training time of the GRU model is the shortest among all compared methods because it has the simplest structure. (3) Although compared with GRU the proposed GRURL needs more time to train models because it employs reinforcement learning, its running time is still only a little longer than that of the LSTM. Therefore, for our GRURL model, the performance improvement in the sac-rifice of some training time is worthwhile.

Discussion. As a variant of the LSTM model, the GRU model is also designed to capture the dependencies between different timestamps. Thus, the proposed GRURL is still based on GRU, which is suitable for the small data volume modeling, because it simplifies the "gate" mechanism of LSTM. On the other hand, to overcome the shortage of traditional GRU model that uses linear regression to fit the data and cannot makes a timely adjustment to the parameters when the input data are changing, we employ reinforcement learning. Reinforcement learning performs well in complex dynamics.

5   Conclusion

Achieving faster and more accurate car sales forecasts is critical to the automotive industry's participants. To improve the accuracy of car sales prediction in a real-world condition, this paper proposes a novel time series model GRURL, which combines the gated recurrent units neural network with reinforcement learning. The proposed model not only captures the dependencies between different timestamps like the traditional GRU model but also utilizes reinforcement learning to optimize the update of network parameters according to the dynamics of the input data. We applied the proposed model on the real-world problem of predicting car sales in Yancheng City based on the historical data, comparing with three traditional deep learning models. The experi-mental results demonstrate that the proposed GRURL network not only significantly outperforms the other deep learning networks in terms of RMSE and MAPE but also has only a little running time increment. Since there are some other factors that could influence the sales of cars, in the future, we would investigate how to include more features representing different factors into the model to obtain better prediction models.

Acknowledgment. This research has been supported by the National Natural Science Foundation of China (NSFC) under grants 91846104 and 61603186, the Natural Science Foundation of Jiangsu Province, China, under grants BK20160843, and the China Postdoctoral Science Foundation under grants 2017T100370.

References

1. Chatfield, C.: The Analysis of Time Series: An Introduction. Chapman and Hall/CRC, London (2003)
2. Anderson, T.W.: The Statistical Analysis of Time Series, vol. 19. Wiley, Hoboken (2011)
3. Benkachcha, S., Benhra, J., El Hassani, H.: Causal method and time series forecasting model based on artificial neural network. Int. J. Comput. Appl. **75**(7), 37–42 (2013)
4. Kantz, H., Schreiber, T.: Nonlinear Time Series Analysis, vol. 17. Cambridge University Press, Cambridge (2004)
5. Moody, J., Darken, C.J.: Fast learning in networks of locally-tuned processing units. Neural Comput. **1**(2), 281–294 (1989)
6. Zaremba, W., Sutskever, I., Vinyals, O.: Recurrent neural network regularization. arXiv preprint arXiv:1409.2329 (2014). https://arxiv.org/pdf/1409.2329.pdf
7. Jo, T.: VTG schemes for using back propagation for multivariate time series prediction. Appl. Soft Comput. **13**(5), 2692–2702 (2013)
8. Hochreiter, S., Schmidhuber, J.: Long short-term memory. Neural Comput. **9**(8), 1735–1780 (1997)
9. Farahani, D.S., Momeni, M., Amiri, N.S.: Car sales forecasting using artificial neural networks and analytical hierarchy process. In: The 5th International Conference on Data Analytics (DATA ANALYTICS), pp. 57–62. IARIA (2016)
10. Fu, R., Zhang, Z., Li, L.: Using LSTM and GRU neural network methods for traffic flow prediction. In: The 31st Youth Academic Annual Conference of Chinese Association of Automation (YAC), pp. 324–328. IEEE (2017)
11. Duan, Y., Lv, Y., Wang, F.Y.: Travel time prediction with LSTM neural network. In: The 19th IEEE International Conference on Intelligent Transportation Systems (ITSC), pp. 1053–1058. IEEE (2016)
12. Makatjane, K., Moroke, N.: Comparative study of holt-winters triple exponential smoothing and seasonal Arima: forecasting short term seasonal car sales in South Africa. Risk Gov. Control: Financ. Mark. Inst. **6**(1), 71–82 (2016)
13. Sutton, R.S., Barto, A.G.: Reinforcement Learning: An Introduction. MIT Press, Cambridge (2018)
14. Chen, X., Qiu, X., Zhu, C., Huang, X.: Gated recursive neural network for Chinese word segmentation. In: Proceedings of the 53rd Annual Meeting of the Association for Computational Linguistics and the 7th International Joint Conference on Natural Language Processing, pp. 1744–1753 (2015)
15. Cao, J.: Economic globalization and China's auto industry development. Manag. World **4**, 68–76 (2003)
16. Chen, H.: The application of grey theory in sales forecasting and investment decisions. Master thesis, Hefei University of Technology, Anhui, China (2008)
17. Li, X., Zhong, Q., Tong, L.: Hybrid forecasting method for automobile sale. J. Tianjin Univ. (Soc. Sci. Ed.) **8**(3), 175–178 (2006)
18. Giles, C.L., Kuhn, G.M., Williams, R.J.: Dynamic recurrent neural networks: theory and applications. IEEE Trans. Neural Netw. **5**(2), 153–156 (1994)

19. Dayan, P., Balleine, B.W.: Reward, motivation, and reinforcement learning. Neuron **36**(2), 285–298 (2002)
20. Gers, F.A., Eck, D., Schmidhuber, J.: Applying LSTM to time series predictable through time-window approaches. In: Dorffner, G., Bischof, H., Hornik, K. (eds.) ICANN 2001. LNCS, vol. 2130, pp. 669–676. Springer, Heidelberg (2001). https://doi.org/10.1007/3-540-44668-0_93

A Multilayer Sparse Representation of Dynamic Brain Functional Network Based on Hypergraph Theory for ADHD Classification

Yuduo Zhang, Zhichao Lian$^{(\boxtimes)}$, and Chanying Huang

Nanjing University of Science and Technology, Nanjing, China
zyd@njust.edu.cn, lzcs@163.com

Abstract. Nowadays, studies on the brain show that the resting brain is still dynamic, and the dynamics of brain functional connectivity remains to be proven, which is very important for the research and diagnosis of mental disorders. In this paper, we apply the Bayesian Connection Change Point Model (BCCPM) to perform dynamic testing on the brain. A sparse model is used to construct a hypergraph to represent the brain function connectivity network, and then the dictionary obtained by sparse learning is used to further extract the features of brain function network. The experimental results on ADHD data show that the accuracy of the proposed method has been improved. Meanwhile, we find that there are obvious differences in the sparse features values of the brain functional networks between patients and normal controls. In addition, the comparison between the proposed method with/without the BCCPM demonstrated the importance of dynamic detection further.

Keywords: ADHD · BCCPM · Hyper network · Sparse representation

1 Introduction

The relationship between research function and structural brain connections is critical for understanding and diagnosing neurological and psychiatric disorders. Functional magnetic resonance imaging (fMRI) is widely used in the study of a variety of mental disorders [1–5].

The brain function connection pattern changes overtime, even at rest [6, 7]. Therefore, brain function connection is also constantly changing for the resting state, the dynamic modeling of the resting state brain function connection is very important. In [8], Allen et al. used a sliding time window correlation method combined with k-means clustering of window correlation matrices to assess the dynamic connectivity of whole brain function. In [9], Zhang et al., based on sliding windows, performed an effective multi-view spectral clustering method by treating each dynamic function connector pattern as a view, proving that the motor resting network has substantially more time variability. However, the size of the sliding window is difficult to set. This paper used BCCPM [10] based on probability statistics to detect the dynamic change

© Springer Nature Switzerland AG 2019
Z. Cui et al. (Eds.): IScIDE 2019, LNCS 11936, pp. 325–334, 2019.
https://doi.org/10.1007/978-3-030-36204-1_27

points of time series. In our paper, the sparse characteristics of the hyper network extracted by our proposed method further prove the dynamics of the resting state brain.

The brain function network reflects the interrelationship between the various regions of the brain. Each region of interest (ROI) represents a node, and the degree of association between different ROIs represents edges. The traditional modeling methods of brain function networks often only indicate the degree of association between the two nodes. In [11], Li et al. used the local relationship of two adjacent nodes to represent the brain function connection state. However, each edge of the network may be constructed by multiple node interactions, not just the relationship between the two nodes. Edges constructed with only the degree of association between two nodes, it may lose meaningful information. In [12], Jie et al. used the theory of hyper network to model the brain function network for brain disease diagnosis. Generally, there are many ways to build a hyper network. Considering the convenience of calculation in this paper, we used a sparse representation method to represent the interaction among multiple nodes for constructing a directed hyper network.

The feature extraction of the hyper network has a great influence on the classification effect. Previous articles directly extracted local features between two ROIs [13]. In [14], Bahramf et al. used two BOLD signal at nodes to construct functional connectivity graph and analyzed the network structure without extracting the network structure features. Hiroyasu, Kohri and Hiwa used the correlation matrix between the multiple regions as the adjacency matrix of the graph and selected features to optimize the classification results [15]. These feature extraction methods can lead to incomplete descriptions of brain function connectivity states that result in the absence of critical information. In [16], Guo et al. used the method of constructing a resting-state brain function hyper network to improve classification performance. However, the division of a large number of ROI in fMRI preprocessing and the construction of hyper network for the subject can lead to the drawbacks of data dimensions being too large to handle. Therefore, our paper used the dictionary learning method to learn the feature representation of the constructed functional connection hyper network for improving classification performance and processing efficiency.

In this paper, we propose a new architecture that improves the accuracy of correct diagnosis for children with attention-deficit/hyperactivity disorder (ADHD). We preprocess rs-fMRI to get the average time series. Firstly, we use the BCCPM to complete the dynamic segmentation and find the dynamic changes of the brain. Secondly, we applied a sparse representation method to represent the interaction among multiple nodes for constructing a directed hyper network for each subject. Then a second layer dictionary learning is used to extract the hyper network features. Finally, the SVM/KELM classifier is used to complete the classification of ADHD patients and normal children. The whole process of our proposed method is shown in Fig. 1.

In the Sect. 3, we did some comparative experiments and analyzed the results in detail. In the Sect. 4, we made a summary.

2   Materials and Methods

2.1   Material Preparation

In our paper, the data set [10] has 68 subjects containing 45 normal children and 23 children with ADHD. We select 358 DICCCOLs [17] for each subject as network nodes and extract the rs-fMRI signal for each node. After this processing, we can get an average time series matrix of 270×358 for each subject. Then, dynamic detection is performed, and the average time series of each subject is segmented using BCCPM to find a time change point. The BCCPM estimates the change points by analyzing the joint probabilities among the nodes of brain networks between different time periods and statistically determine the boundaries of temporal blocks. After the BCCPM, each subject is divided into 2 to 3 segments of varying lengths, each of which is represented as $A \in R^{T \times 358}$. So in the next experimental processing, we treat each segment as a sample object. Therefore, we can get a new data set containing 127 normal children and 66 children with ADHD. This step of dynamic detection is very important, and the experimental analysis will explain in detail in the third part of the paper.

2.2   Construction of Hyper Network Based on the Hyper Graph Theory

The traditional graph theory only focuses on the relationship between two nodes, which loses a lot of information among multiple nodes. Therefore, we can use the theory of hypergraph to solve this problem. The hypergraph focuses on relationships between multiple nodes, and a hyper-edge can be represented as a collection of multiple nodes.

In order to express the correlation between 358 ROIs, we applied the theory of hyper graph to construct hyper network [12]. There are many methods for constructing hyper networks. Our paper used the sparse linear regression model with lasso method to construct hyper networks. In this method, 358 ROIs represent 358 nodes, and each hyper-edge will be represented as a linear combination of nodes. For weak correlation connections, the relationship coefficient will be set to zero.

In this paper, we combine the sparse linear regression model with lasso to build a hyper network and solve the L1-norm regularized least squares problem:

$$\min_x \frac{1}{2}||A_k x_k - y_k||_2^2 + \lambda||x_k||_1 \qquad (1)$$

Where $A \in R^{T \times K}$, $y_k \in R^{T \times 1}$, and $x_k \in R^{K \times 1}$. y_k denotes the average time series of k-th ROI. $A_k = \{y_1, \ldots, y_{k-1}, 0, y_{k+1}, \ldots, y_K\}$ indicates that the value of the average time series of the k-th ROI is set to zero. x_k is the item we are seeking, and denotes the weight vector of the degree of association between the k-th ROI and the remaining k − 1 ROIs. λ is the L1-norm regularization parameter. As the value of λ increases, the model becomes sparse. We use the SLEP-package [18] to optimize the function (1). In our experiments, the input λ value is from 0.1 to 0.9 with a step size of 0.1. λ controls the level of correlation of brain functional network connections, and weakly correlated edges will be set to zero. The actual regularization value used in the optimization equation is $\lambda \times \lambda_{max}$, where λ_{max} is the value at which the function should return a zero solution.

For each subject, we obtain the hyper-edge weight vector x of each brain region node by using lasso, a sparse linear regression model. And we combine these vectors to obtain a hyper-edge weight matrix $X = \{x_1, x_2, \ldots, x_{358}\} \in R^{358 \times 358}$ representing the degree of association between 358 ROIs. According to the value of λ, we can get multi-level hyper network for each subject, and each level denotes the correlation of different intensities.

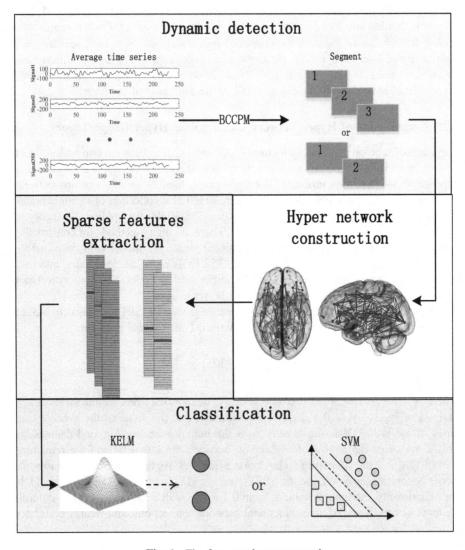

Fig. 1. The framework we proposed

2.3 Sparse Feature Extraction

In this paper, we assume that a brain functional network may consist of several subnetworks. Therefore, we use the sparse representation method to extract the sparse features of the hyper networks.

The sparse features of hypergraph network can be solved by:

$$\min_{D \in C} \lim_{n \to +\infty} \frac{1}{n} \sum_{i=1}^{n} \min_{\alpha_i} \left(\frac{1}{2} ||X_i - D\alpha_i||_2^2 + \lambda ||\alpha_i||_1 \right) \tag{2}$$

Where C is a constraint set for the dictionary, and D is the dictionary that we are aiming to learn. $X \in R^{m \times n}$ is our training set, m denotes one hyper network size and n denotes the number of training sample.

Since each subject of the original data set is divided into 2 to 3 segments, we can assume that the subjects in the new data set can be classified into five categories according to dynamic detection. In order to ensure that the training samples for dictionary learning are balanced and the dictionary obtained by learning has a certain generalization ability, we extract 80% of the samples from these five categories for training to obtain the dictionary. We used the spams-matlab-v2.6 tool-kit to solve Eq. (2). And learned a dictionary D which contains 50 elements. Then D can be used to sparsely represent each sample.

We use Eq. (3) to extract sparse features:

$$\min_{\alpha} \frac{1}{2} ||x - D\alpha||_2^2 + \lambda ||\alpha||_1 \tag{3}$$

Where $x \in R^{m \times n}$ is the full data set, m denotes one hyper network size and n denotes the number of samples to decompose. D is the dictionary that we learned from Eq. (2). α is the sparse feature that we are aiming to obtain. The spams-matlab-v2.6 tool-kit is used to solve the Eq. (2). After sparse representation, we can get a sparse feature matrix alpha $\in R^{50 \times 193}$. The 193 subjects are represented as matrix alpha, and each subject is composed of 50 elements.

Finally, we use different classifier to classify the normal children and children with ADHD.

3 Experimental Process and Results Analysis

In order to ensure the stability of the classification results, we designed a five-fold cross-validation experiment, and repeated the experiment 30 times to take the average performance of each fold. In this paper, we compared our proposed method with Jinli Ou [10] and Li [11], as shown in Table 1. In Table 1, our proposed hypergraph-based sparse representation method has achieved better results for both NC and ADHD comparing to existing methods. In addition, we applied different classifiers such as KELM and SVM and compared their performance as shown in Table 2. It is shown that both classifier obtained superior performances comparing to the existing methods.

Table 1. Yang Li, Jinli Ou and our paper proposed method classification results.

Method	Jinli Ou's method		Yang Li's method		Our proposed method	
CLASS	NC	ADHD	NC	ADHD	NC	ADHD
FOLD1	0.8889	1	0.9692	0.9769	0.9817	0.9723
FOLD2	0.8889	0.8	0.9808	0.9077	0.9853	0.9846
FOLD3	1	0.8	0.976	1	0.9831	0.9901
FOLD4	1	1	0.968	0.9615	0.9790	0.9549
FOLD5	1	1	0.988	0.9571	0.9894	0.9775
MEAN	0.95	0.92	0.9764	0.9606	0.9837	0.9759

Table 2. The classification results with different classifier.

Method	Proposed method with KELM		Proposed method with SVM	
CLASS	NC	ADHD	NC	ADHD
FOLD1	0.9817	0.9723	0.9960	0.9799
FOLD2	0.9853	0.9846	0.9841	0.9848
FOLD3	0.9831	0.9901	0.9907	0.9564
FOLD4	0.9790	0.9549	0.9818	0.9544
FOLD5	0.9894	0.9775	0.9775	0.9707
MEAN	0.9837	0.9759	0.9860	0.9692

To show the importance of dynamics and validate the segmentation results obtained by the BCCPM, we compared our proposed method with the proposed without dynamic detection of BCCPM. The classification results of the proposed method without BCCPM are shown in Table 3. Compared to the classification results using BCCPM in Table 2, we can find that the classification accuracy without using BCCPM significantly decreased no matter using the SVM or the KELM as the classification model. The results showed that the classification accuracy of the children with ADHD was greatly reduced without dynamic detection of the brain. This proves that the importance of dynamics and BCCPM can detect the dynamic change point accurately.

In addition, the hyper network sparse features extracted by our proposed method are shown in Figs. 2 and 3. After preprocessing and dynamic segmentation of the data set, each subject is divided into 2 or 3 segments. So we take the subjects with two segments as one class and the remaining part as the other class. Then we can extract four subjects from these two class respectively. As shown in Fig. 2, it displayed four people's hyper network sparse feature map, in which three columns represents a subject. The vertical axis represents 50 features. From the Fig. 2 we can see that the 50 features extracted by the same person are very similar, but the values of 50 features from the same person are different for different segments. As shown in row 34 of Fig. 2, the features of the first three columns are different in color. As shown in Fig. 3, it displayed four people's hyper network sparse feature map, in which two columns represents a subject. Figures 3 and 2 have the same characteristics. For instance, as shown in row 21 of Fig. 3, the features of the first two columns are different in color. It proves that the hyper network constructed by different segments of the same person has the same network form, but the weight of the hyper-edges are different, proving that the network is dynamic.

Table 3. The classification results without BCCPM.

Dynamic detection	Without BCCPM			
Classifier	KELM		SVM	
CLASS	NC	ADHD	NC	ADHD
FOLD1	0.8630	0.0883	0.9704	0.0217
FOLD2	0.8593	0.1700	0.9815	0.0083
FOLD3	0.8852	0.1250	0.9926	0.0233
FOLD4	0.8296	0.1400	0.9889	0.0233
FOLD5	0.8296	0.1250	0.9926	0.0283
MEAN	0.8533	0.1297	0.9852	0.0210

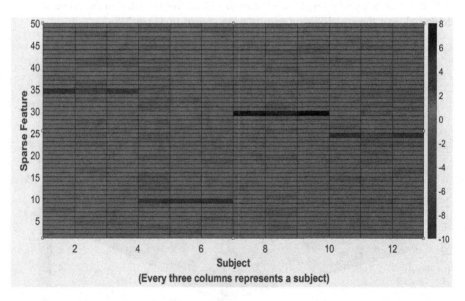

Fig. 2. Sparse feature map (every three columns represents a subject)

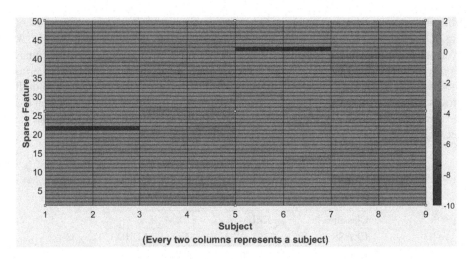

Fig. 3. Sparse feature map (every two columns represents a subject)

Classification results are affected by classifier parameters. As shown in Fig. 4, the SVM classification accuracy varies with the changes in parameters c and g. In order to find the best c and g, let c and g change within a certain range (in our paper $c = 2^{-2}, 2^{-1}, \ldots, 2^4; g = 2^{-4}, 2^{-3}, \ldots, 2^4$). After cross validation experiment, we get the best values of c and g. Finally, 30 times five-fold cross-validation experiments were performed, and the classification results were shown in Table 2.

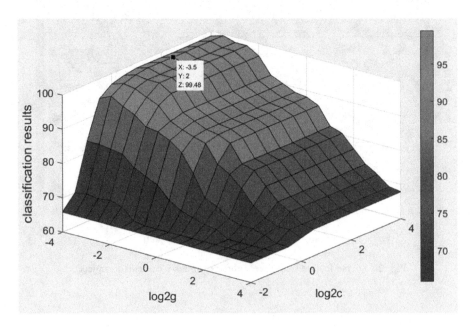

Fig. 4. SVM classification accuracy varies with parameters c and g

4 Conclusion

In this paper, we proposed a novel method to obtain a higher classification accuracy of ADHD patients than the existing methods. Combining the theory of hypergraphs, we constructed a brain-function-connected hyper network. Moreover, the sparse feature that we extracted from hyper network proved the dynamics of brain functional connectivity network. Finally, the sparse features help to obtain the stable classification results. We have designed some comparative experiments to confirm that our method is effective, which helps brain research and diagnosis of ADHD children. We hope that more scholars will study the dynamics of the resting brain in the future. This is conducive to human exploration of the brain and research on disease.

References

1. Song, M., Du, H., Wu, N., et al.: Impaired resting-state functional integrations within default mode network of generalized tonic-clonic seizures epilepsy. PLoS ONE **6**(2), 1729 (2011)
2. Rashid, B., Arbabshirani, M.R., Damaraju, E., et al.: Classification of schizophrenia and bipolar patients using static and time-varying resting-state FMRI brain connectivity. In: IEEE, International Symposium on Biomedical Imaging, pp. 645–657. IEEE (2015)
3. Wang, X., Ren, Y., Yang, Y., et al.: A weighted discriminative dictionary learning method for depression disorder classification using fMRI data. In: IEEE International Conferences on Big Data and Cloud Computing, pp. 618–623. IEEE (2016)
4. Golbabaei, S., Vahid, A., Hatami, J., et al.: Classification of Alzheimer's disease and mild cognitive impairment: machine learning applied to rs-fMRI brain graphs. In: 2016 23rd Iranian Conference on Biomedical Engineering and 2016 1st International Iranian Conference on Biomedical Engineering (ICBME). IEEE (2016)
5. Cicek, G., Akan, A., Orman, Z.: Classification of ADHD by using textural analysis of MR images. In: Medical Technologies National Congress, pp. 1–4 (2017)
6. Biswal, B., Yetkin, F.Z., Haughton, V.M., et al.: Functional connectivity in the motor cortex of resting human brain using echo-planar MRI. Magn. Reson. Med. **34**, 537–541 (1995)
7. Raichle, M.E., Mintun, M.A.: Brain work and brain imaging. Annu. Rev. Neurosci. **29**, 449–476 (2006)
8. Allen, E.A., Damaraju, E., Plis, S.M., et al.: Tracking whole-brain connectivity dynamics in the resting state. Cereb. Cortex **24**(3), 663–676 (2014)
9. Zhang, X., Li, X., Jin, C., et al.: Identifying and characterizing resting state networks in temporally dynamic functional connectomes. Brain Topogr. **27**(6), 747–765 (2014)
10. Lian, Z., Li, X., Xing, J., et al.: Exploring functional brain dynamics via a Bayesian connectivity change point model. In: 2014 IEEE 11th International Symposium on Biomedical Imaging, ISBI 2014. IEEE (2014)
11. Li, Y., Lian, Z., Li, M., et al.: ELM-based classification of ADHD patients using a novel local feature extraction method. In: IEEE International Conference on Bioinformatics & Biomedicine. IEEE (2017)
12. Jie, B., Wee, C.Y., Shen, D., et al.: Hyper-connectivity of functional networks for brain disease diagnosis. Med. Image Anal. **32**, 84–100 (2016). S1361841516300044
13. Ke, M., Duan, X., Zhang, F., Yang, X.: Complex brain network of patients with epilepsy using cascading failure analysis of fMRI data. In: 2015 IEEE/ACIS 16th International Conference on Software Engineering, Artificial Intelligence, Networking and Parallel/Distributed Computing (SNPD), Takamatsu, pp. 1–6 (2015)

14. Bahramf, M., Hossein-Zadeh, G.: Functional parcellations affect the network measures in graph analysis of resting-state fMRI. In: 2014 21th Iranian Conference on Biomedical Engineering (ICBME), Tehran, pp. 263–268 (2014)
15. Hiroyasu, T., Kohri, Y., Hiwa, S.: Sparse feature selection method by Pareto-front exploration—Extraction of functional brain network and ROI for fMRI data. In: 2017 IEEE Symposium Series on Computational Intelligence (SSCI), Honolulu, HI, pp. 1–8 (2017)
16. Guo, H., Li, Y., Xu, Y., et al.: Resting-state brain functional hyper-network construction based on elastic net and group lasso methods. Front. Neuroinform. 12, 25 (2018)
17. Zhang, X., Guo, L., Li, X., et al.: Characterization of task-free and task-performance brain states via functional connectome patterns. Med. Image Anal. 17(8), 1106–1122 (2013)
18. Liu, J., Ji, S., Ye, J.: SLEP: sparse learning with efficient projections. Comput. Sci. Eng. 61 (2011)

Stress Wave Tomography of Wood Internal Defects Based on Deep Learning and Contour Constraint Under Sparse Sampling

Xiaochen Du, Jiajie Li, Hailin Feng$^{(\boxtimes)}$, and Heng Hu

School of Information Engineering, Zhejiang A&F University, Hangzhou, China
hlfeng@zafu.edu.cn

Abstract. In order to detect the size and shape of defects inside wood using stress wave technology under sparse sampling, a novel tomography algorithm is proposed in this paper. The method uses instrument to obtain the stress wave velocity data by sensors hanging around the timber equally, visualizes those data, and reconstructs the image of internal defects with estimated velocity distribution. The basis of the algorithm is using deep learning to assist stress wave tomography to resist signal reduction. First, training CNN model with a large number of generated simulation samples and two-level defect location labeling, and detecting the defective region in wood. Second, using CNN detection results to assist tomography algorithm to precisely estimate the defective area with contour constraint including deepening and weakening operations. Both simulation and wood samples were used to evaluate the proposed method. Effect of CNN detection results on tomography and the shape of the imaging results were both analyzed. The comparison results show that the proposed method always can produce high quality reconstructions with clear edges, when the number of sensors is decreased from 12 to 6.

Keywords: Stress wave tomography · Wood internal defects · Deep learning · Contour constraint · Sparse sampling

1 Introduction

It is important to detect wood internal defects which can endanger the health of trees. The stress wave method has become the mainstream technology in wood non-destructive testing because of its low cost, portability, and harmlessness [1]. Using stress wave timing instruments, researchers can measure the transmission time of a manually produced sound impulse (stress wave). The concept of detecting defects using this method is based on the observation that stress wave propagation is sensitive to the presence of degradation in wood [2]. Stress wave velocity is directly related to the physical and mechanical properties of the wood, and defects result in a decrease in wood density or mass [3]. In general,

© Springer Nature Switzerland AG 2019
Z. Cui et al. (Eds.): IScIDE 2019, LNCS 11936, pp. 335–346, 2019.
https://doi.org/10.1007/978-3-030-36204-1_28

stress waves travel slower in decayed or deteriorated wood than in sound wood, and they also travel around hollows, increasing the transmission time between two testing sensors [4]. Based on this conclusion and signal acquisition of stress wave propagation velocity in wood cross sections, the horizontal distribution of the stress wave velocity in wood can be analyzed and images of wood internal defects can also be reconstructed.

The tomography algorithm is the key step of stress wave nondestructive testing technology, and many algorithms have been proposed in this field [5–7]. Among these algorithms, Ellipse-based spatial interpolation (EBSI for short) method for stress wave tomography has been developed in recent years [8–11]. Assuming that stress wave propagation in wood follows a straight path, a certain propagation ray can represent the velocity and path of a stress wave between a pair of sensors. As shown in Fig. 1a, after the acquisition of stress wave velocity signals, the velocity matrix which records the propagation velocity between any pair of sensors can be visualized as a graph of propagation rays. The red ray indicates that the wave velocity is slow, which means that the stress wave passes through the defective area. In contrast, the green ray indicates that the wave velocity is fast, which means that the stress wave passes through a healthy area. According to the principle of EBSI algorithm, as shown in Fig. 1b, each ray affects the surrounding area and the shape of the affected zone is elliptical. Then, the value of each grid cell in the affected zone is equal to the value of the corresponding ray. If a certain grid cell is in several affected zones simultaneously, the value of this grid cell will be determined by the value of those affected zones using different strategies.

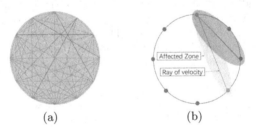

(a)   (b)

Fig. 1. An example of the rays graph and illustration of the EBSI (ellipse-based spatial interpolation) method.

The EBSI algorithm achieves relatively good experimental results in this research field, when the signals are adequate [10]. However, image reconstruction for stress wave tomography is a nonlinear problem, and the signals for tomography algorithms are usually inadequate. Moreover, in practical applications, it is often desirable to reduce the number of sensors in order to improve the efficiency of stress wave tomography. When the signal is further reduced, the existing stress wave tomography methods including EBSI tend to overestimate the defective area [12]. The objective of this study is to improve the quality of reconstructed internal defects in wood using stress wave technology under sparse sampling. We propose a novel stress wave tomography algorithm using deep learning and contour constraint (DLCC for short) to resist the reduction in the number of sensors. Then, the algorithm is applied to detect the size and shape of defects inside wood. Four specimens with defects of different size and position were tested, and the tomography results were analyzed in detail.

2   Proposed Method

2.1   Stress Wave Tomography Using Deep Learning and Contour Constraint

The rays graph can be regarded as the input of stress wave tomography algorithm using spatial interpolation such as EBSI. The color and length are two basic graphic features of a certain propagation ray. Because Convolution Neural Network (CNN for short) directly supports image as input data, we can try to use rays graphs as training samples, and use CNN to learn the spatial distribution of various forms of defects in rays graph. By considering this issue, a novel stress wave tomography method is presented in this paper. The basis of the proposed method is to make full use of the ability of CNN object detection to assist stress wave tomography with the detection results of defects in wood, and to restrain the contour of defective imaging area, so as to improve the problem that the existing stress wave tomography algorithms generally overestimate the defective area under sparse sampling. The framework of the proposed algorithm is shown in Fig. 2. The algorithm consists of two parts: sample training of propagation rays graph and stress wave tomography based on contour constraint.

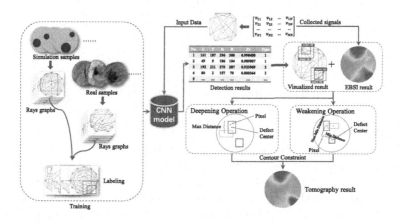

Fig. 2. Process flow of the proposed method.

SSD model is adopted in this paper, in order to take advantage of its ability in object detection. In addition, unlike common object detection applications, stress wave rays graphs are machine-generated images. Compared with natural images, the image pattern of rays graph is very uniform, without complex background, light and other interference factors. Therefore, MobileNet which is a lightweight CNN model is selected as the machine learning tool in this paper [13].

The first stage of the proposed algorithm is defects sample training. In order to ensure enough samples, we established a simulation sample library of wood internal defects rays graph. All samples are automatically generated in batches according to randomly defined defective location, number, shape and number of sensors by using the rays graph generation software (self-designed and developed). In order to ensure that the simulated rays graph is similar to the graph corresponding to the real defects in wood, single defective area pattern is set in 80% of the samples, and double defective areas pattern is set in the remaining 20% of the samples. In addition, the shape of internal defects in wood is generally similar to that of a circle, so the shape pattern of defects in simulated samples is set to a circle. The number of training samples in this paper is 10,000. Some of the auto-generated rays graphs in the sample library are shown in Fig. 3.

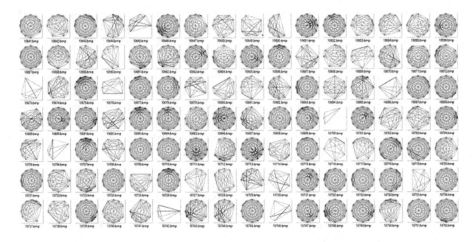

Fig. 3. Generated simulation rays graphs.

In order to facilitate the effective learning of CNN model, the exact positions of defective area corresponding to each rays graph are expressed as the upper-left coordinate and the lower-right coordinate of the rectangle corresponding to the circular virtual defects [14]. Through these two coordinate information, the defective region in each rays graph can be accurately located. In addition, two-level defect location labeling is designed in this paper. The second-level defect location information is the upper-left coordinate and lower-right coordinate of the rectangle corresponding to the defective shape described above, while the first-level defect location information is the upper-left coordinate and lower-right coordinate of the rectangle after reducing the rectangle to 60%. The first-level defect location information further reduces the defective region. The two-level labeling strategy can ensure that the CNN model can obtain the defective object detection results with a probability and locate the defective region as accurately as possible.

The second stage of the proposed algorithm is stress wave tomography using CNN object detection results after training. How to use the object detection results to help image reconstruction of defects in wood is a key problem. The basic idea of this algorithm is to use EBSI algorithm for stress wave tomography at first, then, use CNN detection results to constrain the contour of reconstructed defective area: to deepen the color of the grid cells within the boundaries of CNN detection results (high probability defective area), and to weaken the color of the grid cells outside the boundaries of CNN detection results.

The specific diagram of contour constraint strategy is shown in Fig. 4. In the process of color deepening, as shown in Fig. 4a, the blue box and the purple box are the visualization of CNN detection results, in which the blue box corresponds to the first-level defect location information detected, while the purple box corresponds to the second-level defect location information detected. All grid cells falling in these two rectangular boxes are high probability defective grid cells, which need color deepening. Definition of the contour constraint rate can be expressed as follows,

$$r = 1.0 - L_{\text{dis}}/L_{\text{max}} \tag{1}$$

where L_{dis} is the distance from the grid cell to the center point of the rectangular box (the center point of the defective area detected by CNN), and L_{max} is the edge length of the rectangular box where the grid cells to be deepened are located. The grid cells in all rectangular boxes are processed and their rates of contour constraint are recorded. On the other hand, for all grid cells outside the defective area detected by CNN, color weakening is needed. As shown in Fig. 4b, the corresponding contour constraint rate can be expressed as follows,

$$r = -L_{\text{min}}^2/L_{\text{image}} \tag{2}$$

where L_{min} is the distance between the imaging grid cell to be weakened and the corresponding center point of the nearest rectangular box, L_{image} is the diameter of the circular imaging area of the cross-section of wood sample. The grid cells outside any rectangular boxes are processed and their corresponding rates of contour constraint are recorded.

Then, contour constraint rates are applied to all grid cells in imaging region, which can be expressed as follows,

$$v' = \begin{cases} r \times (v_{\text{max}} - v) + v, r \leq 0 \\ v - (v - v_{\text{min}}) \times r, r > 0 \end{cases} \tag{3}$$

where v is the estimated velocity value of a certain grid cell obtained by EBSI method, v_{max} is the maximum value of all grid cells in imaging region, v_{min} is the minimum value of all grid cells, and v' is the velocity value of a certain grid cell after contour constrain. The contour constraint process can be repeated

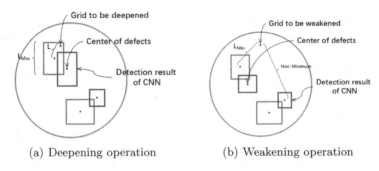

(a) Deepening operation   (b) Weakening operation

Fig. 4. Diagram of deepening operation and weakening operation. (Color figure online)

until the area ratio of the defective area in the two rectangular boxes reaches the determination ratio of CNN object detection corresponding to the rectangular box, respectively.

In summary, the steps of stress wave tomography of wood internal defects based on deep learning and contour constraint are as follows:

Step1: Establishing the training set. Generating enough stress wave rays graphs and labeling defective region using two-level defect location information.

Step2: Sample training based on SSD and MobileNet models using Tensorflow platform.

Step3: Normalizing all collected values of velocity rays with respect to the range of velocity values, visualizing them, and generating the rays graph.

Step4: Stress wave tomography using EBSI algorithm.

Step5: Defects detection by input rays graph. In the results returned by CNN, the results with two highest detection rates are selected as the first-level defect location information and the second-level defect location information respectively.

Step6: Updating the estimated velocity values of all grid cells using Formula 3.

Step7: Repeating Step 6 until the loop termination condition or maximum number of loops is reached.

Step8: Reconstructing the image of internal defects in wood using the estimated values of grid cells with a certain color scale.

2.2   Data Acquisition

In order to evaluate the effectiveness of the proposed method, both simulation data and experimental data were used for stress wave tomography. As shown in Fig. 5, two simulation test samples are designed to evaluate the algorithm, and the spatial distribution of defects includes overlapping double-circle distribution and double-edge semi-circle distribution.

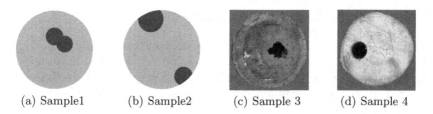

(a) Sample1   (b) Sample2   (c) Sample 3   (d) Sample 4

Fig. 5. Simulation data and real data.

Figure 5 also shows two real wood samples, including Pecan timber with cavity defects and Camphor timber with manual made cavity defects. The stress wave signal is collected by self-developed portable timber tomographic imaging instrument. As shown in Fig. 6, this instrument contains a stress wave signal processor box, several data wires, and 12 stress wave sensors. To detect a timber sample, sensors are equally fixed around the cross-section of a trunk section. When a sensor is knocked by the electronic hammer, all remaining sensors will receive the stress wave signal. Then, the signal processor analyzes the signals with DSP technology to obtain the stress wave transmission time.

Fig. 6. Experiment setup.

In order to achieve complete data, it is necessary to knock each sensor one by one until the propagation time of the stress wave between any two sensors is obtained. The distance between any two sensors is known, so the propagation velocity of the stress waves between any two sensors can also be calculated. Then, the velocity matrix of stress wave can be obtained, and the corresponding rays graph can also be visualized. For simulation samples, rays graph is generated based on corresponding graph of the defective distribution.

3   Experiment Results and Analysis

To unify the imaging conditions, the number of pixels in each grid cell is 1 pixel, and the resolution of reconstructed image is 200 * 200 pixels. In addition, to express the appropriate visual effects for the rays graph and the reconstructed image, red is used to represent low velocity values, and green is used to represents high velocity values.

3.1   Stress Wave Tomography Results Under Sparse Sampling

The problem discussed in this paper is specific, and EBSI is the state-of-the-art method in this research field. When the number of sensors is sufficient, such as 12, the imaging results of EBSI can be guaranteed [11]. Here, in order to show the superiority of our method with fewer sensors, we repeated EBSI to compared with ours. To test the imaging performance of the proposed method under sparse sampling, 12, 10, 8 and 6 sensors were used to perform tomography experiments for each sample, and the sensors were evenly distributed in the cross-section of wood sample clockwise. In addition, some training parameters, such as: width multiplier in training process is 1, epoch is 50, dropout is 0.001, and the number of training is 20235.

Table 1. Comparisons of tomography results of sample 1.

	12 sensors	10 sensors	8 sensors	6 sensors
Rays graph				
EBSI				
Ours				

The comparative experimental results of EBSI and our method are shown in Tables 1, 2, 3, and 4. It can be clearly found that the proposed algorithm achieves excellent tomographic effect. When the number of sensors decreases, the imaging effect of EBSI algorithm becomes worse. Especially for sample 1 and sample 4, when the number of sensors is reduced to 6, the reconstructed defects image of EBSI algorithm has been significantly distorted. In contrast, for all samples, when the number of sensors decreases, the proposed method maintains the shape of the reconstructed defects image, and no image distortion occurs. Furthermore, the negative effect of sparse sampling on the proposed method is almost non-existent. Regardless of the number of sensors, the tomography results of our

Table 2. Comparisons of tomography results of sample 2.

	12 sensors	10 sensors	8 sensors	6 sensors
Rays graph				
EBSI				
Ours				

algorithm differ little, and the location, shape and severity of defects can be accurately expressed.

Table 3. Comparisons of tomography results of sample 3.

	12 sensors	10 sensors	8 sensors	6 sensors
Rays graph				
EBSI				
Ours				

3.2 Effect of Object Detection Results on Tomography

In order to further analyze the positive influence of the object detection results of deep learning on contour constraints and their collaborative work in the proposed algorithm, we took sample 4 and sample 2 as examples to visualize the effect of the detection results on tomography.

From the visualization results shown in Fig. 7, it can be seen that the SSD and MobileNet models adopted in this paper can accurately locate the region of the single-circle defects, regardless of the number of sensors being 12, 10, 8 or 6. The blue box represents the first-order defect location returned by CNN, while the purple box represents the second-order defect location. Under four

Table 4. Comparisons of tomography results of sample 4.

	12 sensors	10 sensors	8 sensors	6 sensors
Rays graph				
EBSI				
Ours				

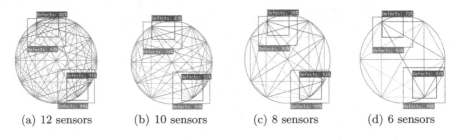

(a) 12 sensors   (b) 10 sensors   (c) 8 sensors   (d) 6 sensors

Fig. 7. Defects detection results of sample 4 (single-circle distribution).

(a) 12 sensors   (b) 10 sensors   (c) 8 sensors   (d) 6 sensors

Fig. 8. Defects detection results of sample 2 (double-circle distribution).

different sensor numbers, the probability of defects recognition corresponding to the purple box can reach 99%, while the probability of defects recognition corresponding to the smaller blue box can also be maintained in the range of 70%–90%. This result reflects that CNN has fully understood the rays pattern corresponding to the defective region through sample learning, and can locate the exact region of defects with great probability. Through the detection results of CNN, our algorithm achieves effective contour constraints on the defects area in EBSI imaging results. Especially when the number of sensors is reduced to 6, the original distorted defective shape is greatly improved.

Furthermore, as shown in Fig. 8, It can be found that, no matter the number of sensors is 12, 10, 8 or 6, CNN once again accurately locates the regions of the double-circle defects at the same time. Especially when the number of sensors is reduced to 6, the two defective areas which were originally bonded together can be separated greatly based on contour constraint. The positive influence of detection results on tomography is fully reflected again.

4 Conclusion

In this paper, we present a novel stress wave tomography method based on deep learning and contour constraint. The method can produce high quality reconstructed image of wood internal defects. To quantitatively evaluate the imaging capability of our method, we extracted reconstructed defective contours. Then, the overlap rate between the reconstructed image shape and the real defects shape can be evaluated using the imaging accuracy based on the visualized confused matrix [15]. The results show that the imaging accuracy of the proposed method for all samples can always be maintained at more than 90%, when the number of used sensors is decreased from 12 to 6. The average imaging accuracy of our method is 93.6%, which shows its high imaging performance under sparse sampling.

Acknowledgments. This work is jointly supported by National Natural Science Foundation of China (U1809208), and Public Welfare Technology Research Project of Zhejiang Province, China (LGG19F020019).

References

1. Yamasaki, M., Tsuzuki, C.: Influence of moisture content on estimating young's modulus of full-scale timber using stress wave velocity. J. Wood Sci. **63**(3), 1–11 (2017)
2. Wang, X., Allison, R.: Decay detection in red oak trees using a combination of visual inspection, acoustic testing, and resistance microdrilling. Arboric. Urban For. **34**(1), 1–4 (2008)
3. Johnstone, D., Moore, G., Tausz, M., Nicolas, M.: The measurement of wood decay in landscape trees. Arboric. Urban For. **36**(3), 121–127 (2010)
4. Ross, R., Brashaw, B., Pellerin, R.: Nondestructive evaluation of wood. For. Prod. J. **48**(1), 14–19 (1998)
5. Feng, H., Li, G., Fu, S., Wang, X.: Tomographic image reconstruction using an interpolation method for tree decay detection. Bioresources **9**(2), 3248–3263 (2014)
6. Lei, L., Li, G.: Acoustic tomography based on hybrid wave propagation model for tree decay detection. Comput. Electron. Agric. **151**, 276–285 (2018)
7. Qiu, Q., Qin, R., Lam, J.: An innovative tomographic technique integrated with acoustic-laser approach for detecting defects in tree trunk. Comput. Electron. Agric. **156**, 129–137 (2019)
8. Du, X., Li, S., Li, G., Feng, H., Chen, S.: Stress wave tomography of wood internal defects using ellipse-based spatial interpolation and velocity compensation. Bioresources **10**(3), 3948C–3962 (2015)

9. Hettler, J., Tabatabaeipour, M., Delrue, S.: Linear and nonlinear guided wave imaging of impact damage in CFRP using a probabilistic approach. Materials **9**(11), 901 (2016)
10. Zeng, L., Jing, L., Huang, L.: A modified lamb wave time-reversal method for health monitoring of composite structures. Sensors **17**(5), 955 (2017)
11. Huang, L., Zeng, L., Lin, J., Luo, Z.: An improved time reversal method for diagnostics of composite plates using Lamb waves. Compos. Struct. **190**, 10–19 (2018)
12. Wang, X.: Acoustic measurements on trees and logs: a review and analysis. Wood Fiber Sci. **47**(5), 965–975 (2013)
13. Howard, A., Zhu, M., Chen, B.: MobileNets: Efficient Convolutional Neural Networks for Mobile Vision Applications. https://arxiv.org/abs/1704.04861
14. He, X., Peng, Y., Zhao, J.: Which and how many regions to gaze: focus discriminative regions for fine-grained visual categorization. IJCV **127**, 1235–1255 (2019)
15. Du, X., Li, J., Feng, H., Chen, S.: Image reconstruction of internal defects in wood based on segmented propagation rays of stress waves. Appl. Sci. **8**(10), 1778 (2018)

Robustness of Network Controllability Against Cascading Failure

Lv-lin Hou[1(✉)], Yan-dong Xiao[2], and Liang Lu[1]

[1] Academy of Joint Logistics, National Defense University,
Beijing 100858, People's Republic of China
`houlvlin@163.com`
[2] Science and Technology on Information Systems Engineering Laboratory,
National University of Defense Technology, Changsha 410073,
People's Republic of China

Abstract. Controllability of networks widely existing in real-life systems have been a critical and attractive research subject for both network science and control systems communities. Research in network controllability has mostly focused on the effects of the network structure on its controllability, and some studies have begun to investigate the controllability robustness of complex networks. Cascading failure is common phenomenon in many infrastructure networks, which largely affect normal operation of networks, and sometimes even lead to collapse, resulting in considerable economic losses. The robustness of network controllability against the cascading failure is studied by a linear load-capacity model with a breakdown probability in this paper. The controllability of canonical model networks under different node attack strategies is investigated, random failure and malicious attack. It is shown by numerical simulations that the tolerant parameter of load-capacity model has an important role in the emergence of cascading failure, independent to the types of network. The networks with moderate average degree are more vulnerable to the cascading failure while these with high average degree are very robust. In particular, betweenness attack strategy is more harmful to the network controllability than degree attack one, especially for the scale-free networks.

Keywords: Controllability · Robustness · Cascading failure · Complex networks

1 Introduction

The relationship between the functionality and structure of network is always the main focus in the network science. In the last decade, as controllability of complex networks become the center of much current interest, many studies have focused on investigating the influence of network structure on controllability within various fields [1–4]. The networks in real world often confront node or edge failure, and sometimes malicious attacks. When these failure or attacks happen, network functions are damaged, such as

---

Supported by the National Natural Science Foundation of China under Grant No. 61603408.

Z. Cui et al. (Eds.): IScIDE 2019, LNCS 11936, pp. 347–355, 2019.
https://doi.org/10.1007/978-3-030-36204-1_29

controllability. Therefore, robustness of network controllability under different attack strategies become a natural question.

Controllability robustness of complex networks has attracted much attention [5–7], including several different types of networks under node and edge attacks. Different attack strategies for nodes and edges have been studied, both of which can be classified into random attack and malicious attack [6]. Random attack is also called random failure, where nodes or edges are removed one by one randomly. Malicious attacks mainly includes two types: based on degree and betweenness centrality, which means that nodes or edges are removed in the descending order of degree or betweenness centrality. The node-based attack strategies often do more harm to the network controllability than the edge-based strategies [7]. To study controllability robustness of a network against malicious attack and random failure, controllability robustness index has been proposed based on global connectivity and controllability [8, 9]. The controllability robustness index is used to approach the issue of characterizing and comparing the controllability robustness of networks. Based on the controllability robustness index, some studies have proposed a few optimization methods to enhance the robustness of network controllability against malicious attack, most of them rely on adding edges, rewiring edges [9, 10] and changing edge direction [11–13]. The controllability robustness for canonical model networks as well as some real-world networks have been surveyed, and it is found that most real networks have high controllability robustness to random node failures [7].

Recently, some scholars have begun to investigate controllability robustness of networks under cascading failures, and found some valuable results [13–17]. Compared to single node attack in the above studies, cascading failures are more harmful to the network controllability. The cascading failures are very common in real-life networks, which usually cause great losses in power grid [18], internet [19] and so on. In power grids [18], some substation failure could lead to large-scale blackout of power grid because of cascading failures. In the internet [19], breakdown of some server or cutting-off of communication cable by malicious hacker increase congestion and further cause network paralysis. Pu et al. [15] investigated the robustness of controllability under cascading failures triggered by the removal of the node with the largest load, and found that cascading failures can cause extensive damage to the network controllability. Wang and Fu et al. [16] proposed a nonlinear cascading model to analyze the controllability robustness of scale-free networks against edge-based cascading failures. Nie et al. [17] studied the network controllability robustness against cascade failures under different edge attacks, and showed that the edges amount in strong connected components has a marked impact on cascading failures. Different to previous studies of network controllability, considering not all overload nodes are removed for some protection measures in real networks, a linear load-capacity model with a breakdown probability is introduced in this paper to study the robustness of network controllability under node attacks.

In this paper, we analyze the robustness of network controllability under the cascading failures with node attacks. A linear load-capacity model with a breakdown probability is introduced, and a simple measure for the controllability robustness is defined. Then, the network controllability under different attack strategies is investigated on Erdös-Rényi (ER) [20, 21] and scale-free networks [22, 23]. The rest of this paper is organized as follows: In Sect. 2, the model of network controllability and cascading failure is described. In Sect. 3, Numerical results and discussions are presented. Section 4 is the conclusion.

2 Model

2.1 Controllability of Complex Networks

Generally, a network of N nodes with linear invariant dynamics is described by the following form

$$x(\text{t}) = \text{A}x(t) + \text{B}u(t) \tag{1}$$

where $x(t) \in \text{R}^\text{N}$ is the state vector, $u(t) \in \text{R}^\text{M}$ is the input vector, A is the $N \times N$ state matrix, B is the $N \times M$ input matrix. Kalman's controllability matrix [24] is defined as follows.

$$\text{W} = \left[\text{B, AB}, \cdots, \text{A}^{\text{N}-1}\text{B}\right] \tag{2}$$

The network described by Eq. (1) is said to be controllable when $rank(\text{W}) = N$. The concept of structural controllability is defined by Lin [25] in 1974, which points out that the matrix is called a structured matrix if its elements are either independent non-zeros or constant zeros. The network is said to a structurally controllable network if non-zeros is fixed by some certain values in the matrix A and B so that $rank(\text{W}) = N$. The network controllability that we discuss in this paper is structural controllability. The nodes controlled directly by different input signals are called driver nodes. If two or more nodes share the same input signal, anyone of them can be the driver node but one input signal has only one driver node. Controllability of a network is characterized by the minimum number of driver nodes N_D, which can entirely control the network. Let the proportion of the driver nodes is denoted as n_D, i.e., $n_D = N_D/N$, where N is the total number of nodes. The minimum number of driver nodes N_D can be calculated by a practical framework [1] based on the maximum matching, which avoids a brute-force search in the controllability matrix to identify the minimum number of driver nodes. In the framework, the network is represented by its bipartite graph, and then Hopcroft-Karp algorithm [26] is used to find a maximum matching in the network. A node is unmatched if there is no link pointing at it in the maximum matching. The minimum number of driver nodes N_D equals to the number of unmatched nodes.

2.2 Cascading Failure Load-Capacity Model

The load-capacity model is the most common in real networks, which usually gives a certain initial load and capacity (security threshold) to each node. When the load of one node exceed its security threshold for some reason, the node fails. The failure will inevitably affect the local connection of the node, causing the load of nodes in the network to be redistributed, which results in the failure of some other nodes. This process is repeated, and the failed nodes gradually increase and cause the cascading failure. Until the load of every node is less than its capacity, the network achieves a stable state.

The Betweenness Centrality model is proposed by Motter et al. [27, 28], where the load of each node is determined by its Betweenness Centrality (BC) numbers [29]. It is assumed that information is always transmitted along the shortest path between nodes. The BC number of node i is defined as the number of the shortest paths between all nodes through node i in the network. Therefore, the load of node i at time t is the BC number of the node at this moment, and the maximum capacity threshold C_i of each node is proportional to its initial load $L_i(0)$, represented as:

$$C_i = (1 + \alpha)L_i(0), \quad i = 1, 2, \cdots N \tag{3}$$

In the above formulation, α is called the tolerance parameter, which determines the maximum capacity of nodes. When some nodes are removed from network after failure, it results in changes of the shortest path in the network so that the BC of nodes change, which will cause the cascading failure. Therefore, some other normal nodes become failed, and the process of cascading failures continue. Until every node satisfies that its load is not more than its maximum capacity, the network become stabilized.

Considering that not all overload nodes will be removed in real networks, the breakdown probability of overload nodes is introduced, which represents the probability that overload node fails and is removed. The model of breakdown probability in Ref. [30, 31] can be described as:

$$p(L_i(t)) = \begin{cases} 0, & L_i(t) < C_i \\ \frac{L_i(t) - C_i}{\beta C_i}, & C_i \le L_i(t) < (1 + \beta)C_i \\ 1, & (1 + \beta)C_i \le L_i(t) \end{cases} \tag{4}$$

where $P(Li(t))$ is the breakdown probability of node i, $(1 + \beta)C_i$ is the removal threshold and β is the tunable parameter.

2.3   Attack Strategies

The controllability robustness of complex networks against cascading failure is analyzed by comparing the effect of different node attack strategies. Although there are a number of destructive strategies to a network or system in real world, which can be basically classified into random failures and malicious attacks. Therefore, Both random attacks and malicious attacks strategies are used.

Random attack strategy means that nodes are randomly chosen to fail. Malicious attack strategy means that the most important node is preferred to fail by means of measures of node importance.

According to the previous studies, the popular calculation methods of node importance are based on node degree or betweenness centrality, so we mainly consider degree and betweenness measures to choose the most important node. In the degree attack strategy (DAS), the node with the highest degree will be firstly remove.

Similarly, in the betweenness attack strategy (BAS), the node degree is replaced by the node betweenness.

2.4 Robustness Metric for Network Controllability

To assess the ability of the network to maintain controllability, a standard metric is introduced in our work, used in the related literature. The metric Δn_D specifies the addition number of driver nodes after node attack in the network. We denote the minimum number of driver nodes in initial network as N_D, and the minimum number of driver nodes in the network after attack as $N_D(t)$, so the addition number of driver nodes Δn_D is represented as:

$$\Delta n_D(t) = (N_D(t) - N_D)/N, t = 1, 2, \ldots, N \qquad (5)$$

Where N is the number of nodes in the network, t is the number of attacks. The minimum number of driver nodes can be calculated by the above method of maximum matching.

3 Simulation Experiments and Results

To investigate the controllability robustness of BC model in networks with different structure property, two typical network models are chosen as the simulation experiments, the Erdös–Rényi (ER) model [20, 21] and the scale-free model (SF) [22, 23]. The ER networks are generated as follows: start from N isolated nodes, and then connect each pair of nodes with probability p, $0 \le p \le 1$, the direction of edge is assigned with even probability pointing at each node. The SF networks are generated from the static model in Ref. [23]. In this study, the networks with different average degree $<k>$ and heterogeneity are used, all of which have 1000 nodes, i.e., $N = 1000$.

3.1 Random Attack Experiments

We first studied the variation of Δn_D in the networks with different tolerance coefficients under random attack strategy, and experimental results are shown in Fig. 1. In the experiment, the number of failure nodes is 10% of total nodes in the network, which means $t = 100$.

It can be seen from Fig. 1 that the networks with lower tolerance coefficient α have the larger addition number of driver nodes Δn_D, which shows large-scale cascading failure is more likely to occur in the networks with low α. In other words, the networks with low α reduce the control robustness. When the tolerance coefficient α is larger than 0.5, the value of Δn_D reaches stable, addition number of driver nodes Δn_D approaches 0.1, which indicates that the increase of maximum security threshold in the networks with high tolerance coefficient α greatly reduce the possibility of cascading failure, enhancing the robustness of networks.

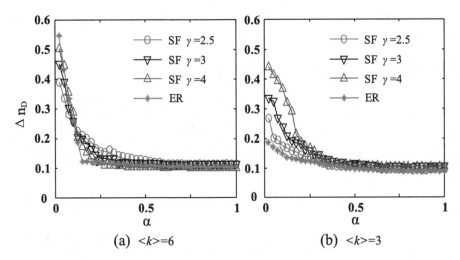

Fig. 1. Addition of driver nodes Δn_D as a function of tolerance parameter α in ER and SF networks with different power law parameter γ. (a) the average degree is $<k> = 6$, (b) the average degree is $<k> = 3$.

As shown in Fig. 1(a) and (b), when the tolerance parameter α is small, there are significant differences between two ER networks with different average degrees. On one hand, the networks with $<k> = 3$ are relatively sparse, where the load distribution of nodes are more uniform. Therefore, after removing the failure nodes the load distribution has not significantly change, and the cascading cascade is not obvious. On the other hand, n_D in initial networks is already high, so Δn_D is relative low. Comparatively, the networks with $<k> = 6$ are more intensive, which is more likely to come to cascading failure, and Δn_D is high. However, SF networks have obvious power-law degree distribution, so even if the average degree is low, there are still some nodes with higher BC because of the existence of hub nodes. The failure nodes still have a larger impact on the load distribution of SF networks, so that after nodes fail, Δn_D is high. At the same time, SF networks with different γ are comparatively analyzed. SF networks with different γ have very similar results, especially the SF networks have larger average degrees.

For ER networks, as average degree $<k>$ increases, the addition of driver nodes Δn_D increases at first and reach the peak when average degree $<k>=5$, and then shows an obvious downward trend. When average degree $<k>$ is larger than 10, Δn_D is getting closer to 0.1, which indicates that the failure of nodes do not cause cascading failure of network, and the increase of Δn_D mostly comes from the initial failure of nodes. The main cause is that there are more links in the network with high average degree $<k>$, where the failure of nodes is least likely to cause cascading failure, so the network has better control robustness. As shown in Fig. 2, Δn_D in SF networks shows a similar trend. However, until average degree of network is higher than 20, Δn_D tend towards stability. Comparatively, SF networks need more edges to reach the robustness of controllability against cascading failure attacks.

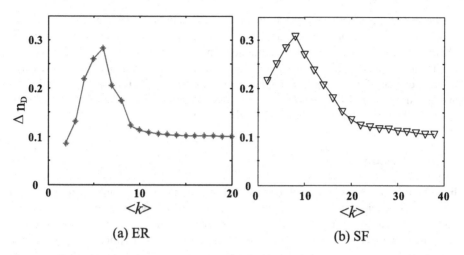

Fig. 2. Addition of driver nodes Δn_D as a function of average degree $<k>$ under random attack strategy. (a) ER networks and (b) SF networks, the power law parameter is $\gamma = 3$ in SF networks. The tolerance parameter $\alpha = 0.1$, the tunable parameter is $\beta = 0.5$.

3.2   Malicious Attack Experiments

The controllability robustness of networks against cascading failure is also investigated under malicious attack strategies, DAS and BAS. Since malicious attack strategies are more likely to cause the cascading failure, only one node with the highest degree or betweenness is chosen to attack, which means $t = 1$. For comparison, the same ER and SF networks are chosen as experiment subject. The variation of driver nodes Δn_D in the networks with different average degree $<k>$ are shown in Fig. 3.

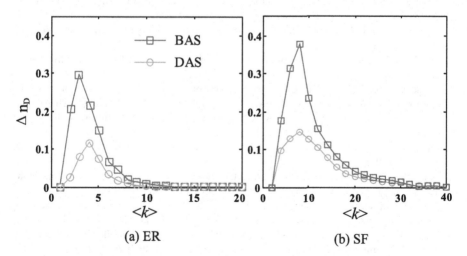

Fig. 3. Addition number of driver nodes Δn_D as a function of average degree $<k>$ under BAS and DAS. (a) ER networks and (b) SF networks, the power law parameter is $\gamma = 3$ in SF networks. The tolerance parameter $\alpha = 0.1$, the tunable parameter is $\beta = 0.5$.

From Fig. 3, we can see that Both ER and SF networks are fragile for malicious attack strategies when average degree of networks are low, especially for BAS. Comparatively, SF networks is more fragile than ER networks. The trend of curves are similar for DAS and BAS, raising up and dropping down dramatically, so that when average degree of networks $<k>$ is high enough they are quite robust. Specifically, the average degree of ER networks $<k>$ is bigger than 10 when the controllability of networks keep steady, and that of SF is almost 30.

4   Conclusion

In summary, we investigated the controllability robustness of complex networks with different topologies against cascading failure. The breakdown probability of overload nodes is introduced to deal with that not all overload nodes are removed in some real networks. The tolerance parameter of node capacity has a significant impact on the robustness of network controllability. When the node capacity is bigger than 150% of its initial load, cascading failure is very light in both ER and SF networks. The damage of BAS is heavier than that of DAS in networks with low average degree. Moreover, SF networks are more vulnerable to malicious attacks because of the power-law distribution. Optimization of controllability robustness of network under cascading failure appears to be a promising direction for future work in this area.

References

1. Liu, Y.Y., Slotine, J.J., Barabási, A.L.: Controllability of complex networks. Nature **473**(7346), 167–173 (2011)
2. Yuan, Z., Zhao, C., Di, Z.: Exact controllability of complex networks. Nat. Commun. 63–73 (2013)
3. Jia, T., Pósfai, M.: Connecting core percolation and controllability of complex networks. Sci. Rep. **4**, 5379 (2014)
4. Liu, Y.Y., Barabási, A.L.: Control principles of complex systems. Rev. Mod. Phys. **88**(3), 035006 (2016)
5. Menichetti, G., Dall'Asta, L., Bianconi, G.: Network controllability is determined by the density of low in-degree and out-degree nodes. Phys. Rev. Lett. **113**, 078701 (2014)
6. Chen, G.R., Lou, Y., Wang, L.: A comparative robustness study on controllability of complex networks. IEEE Trans. Circ. Syst. **66**(5), 828–832 (2019)
7. Lu, Z.-M., Li, X.-F.: Attack vulnerability of network controllability. PLoS ONE **11**(9), e0162289 (2016)
8. Wang, B., Gao, L., Gao, Y., Deng, Y.: Maintain the structural controllability under malicious attacks on directed networks. EPL (Europhys. Lett.) **101**, 58003 (2013)
9. Xiao, Y.-D., Lao, S.-Y., Hou, L.-L., Bai, L.: Optimization of robustness of network controllability against malicious attacks. Chin. Phys. B **121**(11), 678–686 (2014)
10. Wang, W.-X., Ni, X., Lai, Y.-C., Grebogi, C.: Optimizing controllability of complex networks by minimum Structural perturbations. Phys. Rev. E **85**, 026115 (2012)
11. Hou, L.-L., Lao, S.-Y, Liu, G., Bai, L.: Controllability and Directionality in Complex Networks. Chin. Phys. Lett. **29**, 108901 (2012)

12. Xiao, Y.-D., Lao, S.-Y., Hou, L.-L., Bai, L.: Edge orientation for optimizing controllability of complex networks. Phys. Rev. E **90**, 042804 (2014)
13. Liang, M., Jin, S.-Q., Wang, D.-J., Zou, X.-F.: Optimization of controllability and robustness of complex networks by edge directionality. Eur. Phys. J. B **89**, 186 (2016)
14. Ruths, J., Ruths, D.: Robustness of network controllability under edge removal. In: Ghoshal, G., Poncela-Casasnovas, J., Tolksdorf, R. (eds.) Complex Networks IV. SCI, vol. 476, pp. 185–193. Springer, Heidelberg (2013). https://doi.org/10.1007/978-3-642-36844-8_18
15. Pu, C.-L., Pei, W.-J., Michaelson, A.: Robustness analysis of network controllability. Phys. A: Stat. Mech. Appl. **391**(18), 4420–4425 (2012)
16. Wang, L., Fu, Y.-B., Chen, M.Z.-Q., Yang, X.-H.: Controllability robustness for scale-free networks based on nonlinear load-capacity. Neurocomputing **251**, 99–105 (2017)
17. Nie, S., Wang, X., Zhang, H., Li, Q., Wang, B.: Robustness of controllability for networks based on edge-attack. PLoS ONE **9**(2), e89066 (2014)
18. Sold, R.V., Rosas-Casals, M., Corominas-Murtre, B., Valverde, S.: Robustness of the European power grids under intentional attack. Phys. Rev. E **77**(2), 026102 (2008)
19. Barabási, A.L., Albert, R., Jeong, H.: Mean-field theory for scale-free random networks. Phys. A: Stat. Mech. Appl. **272**(1), 173–187 (1999)
20. Erdős, P., Rényi, A.: On random graphs. Publicationes Mathematicae Debrecen **6**, 290–297 (1959)
21. Erdős, P., Rényi, A.: On the evolution of random graphs. Publ. Math. Inst. Hung. Acad. Sci. **5**, 17–61 (1960)
22. Barabási, A.L., Albert, R.: Emergence of scaling in random networks. Science **286**, 509–512 (1999)
23. Goh, K.I., Kahng, B., Kim, D.: Universal behavior of load distribution in scale-free networks. Phys. Rev. Lett. **87**, 287701 (2001)
24. Kalman, R.E.: Mathematical description of linear dynamical systems, J. Soc. Ind. Appl. Math. Series A: Control **1**(2), 152–192 (1963)
25. Lin, C.T.: Structural controllability. IEEE Trans. Autom. Control **19**(3), 201–208 (1974)
26. Hopcroft, J.E., Karp, R.M.: An n^5/2 algorithm for maximum matchings in bipartite graphs. SIAM J. Comput. **2**(4), 225–231 (1973)
27. Motter, A.E., Lai, Y.-C.: Cascade-based attacks on complex networks. Phys. Rev. E **66**(6), 065102 (2002)
28. Motter, A.E.: Cascade control and defense in complex networks. Phys. Rev. E **93**(9), 098701 (2004)
29. Dou, B.-L., Wang, X.-G., Zhang, S.-Y.: Robustness of networks against cascading failures. Phys. A: Stat. Mech. Appl. **389**(11), 2310–2317 (2010)
30. Wang, J.-W., Rong, L.-L.: A model for cascading failures in scale-free networks with a breakdown probability. Phys. A **388**, 1289–1298 (2009)
31. Liu, J., Xiong, Q.Y., Shi, X., Wang, K., Shi, W.R.: Robustness of complex networks with an improved breakdown probability against cascading failures. Phys. A **456**, 302–309 (2016)

Multi-modality Low-Rank Learning Fused First-Order and Second-Order Information for Computer-Aided Diagnosis of Schizophrenia

Huijie Li[1,2], Qi Zhu[1,2(✉)], Rui Zhang[1], and Daoqiang Zhang[1]

[1] College of Computer Science and Technology,
Nanjing University of Aeronautics and Astronautics, Nanjing 211106, China
zhuqi@nuaa.edu.cn
[2] Collaborative Innovation Center of Novel Software Technology
and Industrialization, Nanjing 210093, China

Abstract. The brain functional connectivity network (BFCN) based methods for diagnosing brain diseases have shown great advantages. At present, most BFCN construction strategies only calculate the first-order correlation between brain areas, such as the Pearson correlation coefficient method. Although the work of the low-order and high-order BFCN construction methods exists, there is very little work to integrate them, that is, to design a multi-modal BFCN feature selection and classification method to combine low-order and high-order information. This may affect the performance of brain disease diagnosis. To this end, we propose a multi-modality low-rank learning framework jointly learning first-order and second-order BFCN information and apply it to the diagnosis of schizophrenia. The proposed method not only embeds the correlation information of multi-modality data in the learning model, but also encourages the cooperation between the first-order and the second-order BFCN by combining the ideal representation term. The experimental results of the three schizophrenia datasets (totally including 168 patients and 163 normal controls) show that our proposed method achieves promising classification results in the diagnosis of schizophrenia.

Keywords: Multi-modality learning · Low rank representation · Brain functional connection network

1 Introduction

Studies of resting-state functional magnetic resonance imaging (rs-fMRI) have shown that BFCN has disease-related changes [1]. A number of BFCN strategies have been proposed for the diagnosis and analysis of mental illness. Most BFCN disease diagnosis strategies first construct a brain network based on the correlation between paired-wise brain areas, and then perform feature selection and classification algorithms.

Currently, most BFCN construction methods only consider the association between pairs of brain areas or voxels, i.e. the first-order BFCN construction methods. For

© Springer Nature Switzerland AG 2019
Z. Cui et al. (Eds.): IScIDE 2019, LNCS 11936, pp. 356–368, 2019.
https://doi.org/10.1007/978-3-030-36204-1_30

example, method based on Pearson correlation (PC) coefficients [2]. The first-order BFCN is robust, but not sensitive to small signal changes. There are also some BFCN construction methods that consider the association between multiple brain areas, namely the high-order BFCN construction method. For example, Guo et al. [3] take for that the correlation between pairs of brain areas may be affected by the third brain area, and proposes a method to eliminate the effects through partial correlation. High-order BFCN construction methods can capture small changes between brain networks, but it lacks robustness. Zhu et al. [4] proposed a BFCN construction method for hybrid first-order second-order brain networks, performing a simple weighted combination of first-order and second-order brain networks. However, before the classification step, no machine learning method such as feature selection is used for brain networks to process the brain network data.

In the field of machine learning, researchers have proposed various methods for processing multi-modality data. These methods may provide a theoretical basis for combining first-order and second-order BFCN information. In fact, although there seems no multi-modality work for integrating first-order and second-order BFCNs, some multi-modality methods have been proposed and applied to the diagnosis of brain diseases. For example, Huang et al. [5] proposed a sparse composite linear descriptive analysis model to jointly identify disease-related brain features from multi-modality data. Zhang et al. [6] proposed a multi-modality multi-task method for joint feature selection and classification of Alzheimer's disease data. However, these methods ignore the structural information of multi-modality data and the data noise problem is not considered.

In view of the above problems, in this paper, we proposed a multi-modality low rank representation learning framework to fuse first and second order BFCN information and applied it to the diagnosis of schizophrenia. Our contributions mainly include the following two points:

(1) We extract the intrinsic structure information through low-rank constraint, embed the correlation of multi-modality data into the learning model, and encourage the cooperation between first-order and second-order BFCN by combining ideal representation term to obtain better diagnostic performance.
(2) To the best of our knowledge, this is the first work to combine first-order and second-order BFCN information through multi-modality learning strategy.

2 Background

2.1 First-Order and Second-Order Brain Functional Connection Network

At present, BFCN analysis has been widely used in the diagnosis of mental diseases, and brain function network construction is the core step of BFCN analysis. The BFCN construction methods can be divided into low-order methods and high-order methods. The low-order methods are highly robust, while the high-order methods are usually more sensitive to subtle changes in signals.

The most common low-order method is the BFCN construction method based on the PC coefficient, which reveals the first-order relationship of the brain interval by calculating the correlation coefficient of the paired brain areas. Let x_i and x_j represent a pair of brain areas, PC can be calculated by the following formula:

$$C_{ij}^1 = \frac{Cov(x_i, x_j)}{\sqrt{Var(x_i)Var(x_j)}} \tag{1}$$

where $Cov(\cdot, \cdot)$ is a function for calculating the covariance, and $Var(\cdot)$ is a function for calculating the variance.

In our previous work, we proposed a second-order BFCN strategy based on triplet [8] for extracting high-order information from brain areas. Specifically, let triplet (x_i, x_u, x_v) consists of x_i and its neighbors x_u and x_v. S_{uv}^i defines the distance between x_i and x_v relative to x_u:

$$S_{uv}^i = dist(x_i, x_v) - dist(x_i, x_u) \tag{2}$$

where $dist(\cdot, \cdot)$ calculates the squared Euclidean distance.

The triplet-based second-order BFCN strategy takes into account that a brain area usually interacts with its neighbors rather than distant brain areas, so it only considers second-order information among neighbors. Let N_i be a set of sequence numbers indicating the k nearest neighbors of x_i, and then, relative to all k nearest neighbors of x_i, the distance between x_i and x_v can be expressed as $\sum_{u \in N_i} S_{uv}^i$. Thus, the triplet-based second-order BFCN can be expressed as:

$$C_{ij}^2 = \begin{cases} norm\left(-\sum_{u \in N_i} S_{uv}^i\right), (j \in N_i) \\ 0, (j \notin N_i) \end{cases} \tag{3}$$

where $norm(\cdot)$ is a function that that normalizes the data.

2.2 Low-Rank Representation

Recently, LRR has performed very well in feature extraction and subspace learning [7]. For a given set of samples, the LRR looks for the low-rank component of all samples as the basis in the dictionary so that the data can be represented as a linear combination of bases [9].

Let $X = [x_1, x_2, \ldots, x_n] \in R^{d \times n}$ denote a set of data vectors, and each column of which can be represented by a linear combination of the bases in dictionary $A = [a_1, a_2, \ldots, a_m]$:

$$X = AZ \tag{4}$$

where $Z = [z_1, z_2, \ldots, z_n]$ is a coefficient matrix, and z_i is the representation coefficient vector of x_i.

Considering that dictionary A is usually over-complete, LRR seeks a low-order solution to solve the problem of possible multiple solutions by solving the following problems [10]:

$$\min_{Z} \|Z\|_* \quad s.t. \ X = AZ \tag{5}$$

where $\|\cdot\|_*$ represents the nuclear norm of the matrix [12].

2.3 Materials

In this study, three schizophrenia rs-fMRI datasets were used, including The Center for Biomedical Research Excellence (COBRE) dataset (53 patients and 67 normal controls), Nottingham dataset (32 patients and 36 normal controls), and Xiangya dataset (83 patients and 60 normal controls). Detailed information such as subjects, image acquistions, data preprocessing, and anatomical parcellation please refer to [4].

3 Method

3.1 Proposed Method

Let $X_k = [x_{k,1}, x_{k,2}, \ldots, x_{k,n}] \in R^{m \times n}$ denotes the k-th modality of the training data. Assuming that X_k contains c classes, then X_k can be divided into c subsets, expressed as $X_k = \{X_k^1, X_k^2, \ldots, X_k^c\}$. In the context of this study, X_1 represents the first-order BFCN, and X_2 represents the second-order BFCN. Considering X_k itself as the dictionary, X_k can be re-represented as $X_k = X_k Z_k + E_k$, where $Z_k = [z_{k,1}, z_{k,2}, \ldots, z_{k,n}] \in R^{n \times n}$ represents the LRR of X_k and E_k denotes the sparse noise matrix. In order to include the block diagonal structure information in the learning process and enhance the cooperation between the two modalities, we introduced the regular ideal representation term $\|Z_2 Z_2^T - P\|_F^2$, where $P = block(p_1, p_2, \ldots, p_c)$, $p_i = ones(n_i, n_i)$ is the code for X^i, n_l is the number of samples in X^i. That means, if $x_{k,j}$ belongs to class i, then the coefficients in p_i are all 1 s, whiles in others are all 0 s. Thus, we have the follow function:

$$\min_Z \sum_{k=1}^{2} \left(\|Z_k\|_* + \beta \|Z_k\|_1 + \gamma \|E_k\|_{2,1} \right) + \lambda \|Z_1 Z_2^T - P\|_F^2 \quad s.t. \ X_k = X_k Z_k + E_k \tag{6}$$

where $\|\cdot\|_1$ denotes the l_1-norm, $\|\cdot\|_{2,1}$ denotes the $l_{2,1}$-norm, $\|\cdot\|_F$ denotes the Frobenius norm, and β, γ, λ are the hyperparameters used to balance the differents parts of the function.

Combining Eq. (6) with the multimodal supervised feature selection framework [6, 13], the following objective functions can be obtained:

$$\min_{W,Z} \sum_{k=1}^{2} \left(\left\| H - Z_k^T W_k \right\|_F^2 + \alpha \|Z_k\|_* + \beta \|Z_k\|_1 + \gamma \|E_k\|_{2,1} + \xi \|W_k\|_F^2 \right) \\ + \lambda \left\| Z_1 Z_2^T - P \right\|_F^2 \quad \text{s.t. } X_k = X_k Z_k + E_k \tag{7}$$

where H denotes the ground truth of X, $W = [w_1, w_2, \ldots, w_m] \in R^{c \times m}$ is the weight matrix, and α, ξ denote the hyperparameters.

3.2   Optimization and Solution

Since the problem (7) is non-convex, we first introduce the auxiliary variables Z_k' and Z_k'' to make the problem separable:

$$\min_{W} \sum_{k=1}^{2} \left(\left\| H - Z_k^T W_k \right\|_F^2 + \alpha \|Z_k'\|_* + \beta \|Z_k''\|_1 + \gamma \|E_k\|_{2,1} + \xi \|W_k\|_F^2 \right) \\ + \lambda \left\| Z_1 Z_2^T - P \right\|_F^2 \quad \text{s.t. } X_k = X_k Z_k + E_k, Z_k' = Z_k, Z_k'' = Z_k \tag{8}$$

The problem (8) can be solved by minimizing the following augmented Lagrangian multiplier (ALM) function L:

$$L = \left\| H - Z_k^T W_k \right\|_F^2 + \alpha \|Z_k'\|_* + \beta \|Z_k''\|_1 + \gamma \|E_k\|_{2,1} + \lambda \left\| Z_1 Z_2^T - P \right\|_F^2 + \xi \|W_k\|_F^2 \\ + \langle Y_{1,k}, X_k - X_k Z_k - E_k \rangle + \langle Y_{2,k}, Z_k - Z_k' \rangle + \langle Y_{3,k}, Z_k - Z_k'' \rangle \\ + \frac{\mu}{2} \left(\|X_k - X_k Z_k - E_k\|_F^2 + \|Z_k - Z_k'\|_F^2 + \|Z_k - Z_k''\|_F^2 \right) \quad k = 1, 2 \tag{9}$$

where $\langle A, B \rangle = tr(A^T B)$, $Y_{1,k}$, $Y_{2,k}$ and $Y_{3,k}$ are Lagrange multipliers and μ is a balance parameter.

Further, the augmented Lagrangian function (9) would reduce to:

$$L = \left\| H - Z_k^T W_k \right\|_F^2 + \alpha \|Z_k'\|_* + \beta \|Z_k''\|_1 + \gamma \|E_k\|_{2,1} + \lambda \left\| Z_1 Z_2^T - P \right\|_F^2 + \xi \|W_k\|_F^2 \\ + \frac{\mu}{2} \left(\left\| X_k - X_k Z_k - E_k + \frac{Y_{1,k}}{\mu} \right\|_F^2 + \left\| Z_k - Z_k' + \frac{Y_{2,k}}{\mu} \right\|_F^2 + \left\| Z_k - Z_k'' + \frac{Y_{3,k}}{\mu} \right\|_F^2 \right) \\ - \frac{1}{2\mu} \left(\|Y_{1,k}\|_F^2 + \|Y_{2,k}\|_F^2 + \|Y_{3,k}\|_F^2 \right) \quad k = 1, 2 \tag{10}$$

The above problem can be solved by inexact ALM (IALM) algorithm, which is an iterative method that solves each variable in a decreasing coordinate manner [14–18]. The stopping criteria are $\|X_k - X_k Z_k - E_k\|_\infty < \varepsilon$, $\|Z_k - Z_k'\|_\infty < \varepsilon$ and $\|Z_k - Z_k''\|_\infty < \varepsilon$, where $\|\cdot\|_\infty$ denotes the l_∞-norm.

The solution processes are as follows, and each step has a closed solution:
Step 1 (Update Z'_k):

$$Z'_k = \underset{Z'_k}{argmin}\, \alpha\left\|Z'_k\right\|_* + \frac{\mu}{2}\left\|Z_k - Z'_k + \frac{Y_{2,k}}{\mu}\right\|_F^2 \tag{11}$$

Problem (11) can be solved by the singular value threshold (SVT) operator [11]:

$$Z'_k = US_\theta[S]V^T \tag{12}$$

where $\theta = \frac{\alpha}{\mu}$, USV^T is the SVD decomposition of $Z_k + \frac{Y_{2,k}}{\mu}$, and $S_\theta[x]$ is the soft-thresholding (shrinkage) operator [17], which defined as follows:

$$S_\theta[x] = \begin{cases} x - \theta, & if\ x > \theta \\ x + \theta, & if\ x < -\theta \\ 0, & otherwise \end{cases} \tag{13}$$

Step 2 (Update Z''_k):

$$Z''_k = \underset{Z''_k}{argmin}\, \beta\left\|Z''_k\right\|_1 + \frac{\mu}{2}\left\|Z_k - Z''_k + \frac{Y_{3,k}}{\mu}\right\|_F^2 \tag{14}$$

According to [14], the above problem has the following closed form solution:

$$Z''_k = shrink\left(Z_k + \frac{Y_{3,k}}{\mu}, \frac{\beta}{\mu}\right) \tag{15}$$

Step 3 (Update Z_k):
Z_k is updated by solving optimization problem (16):

$$Z_k = \underset{Z_k}{argmin}\left\|H - Z_k^T W_k\right\|_F^2 + \lambda\left\|Z_1 Z_2^T - P\right\|_F^2$$
$$+ \frac{\mu}{2}\left(\left\|X_k - X_k Z_k - E_k + \frac{Y_{1,k}}{\mu}\right\|_F^2 + \left\|Z_k - Z'_k + \frac{Y_{2,k}}{\mu}\right\|_F^2 + \left\|Z_k - Z''_k + \frac{Y_{3,k}}{\mu}\right\|_F^2\right) \tag{16}$$

It is easy to solve for the closed solution of Z_k:

$$Z_k = \left(2(1 + \lambda + \mu)I + \mu X_k^T X_k\right)^{-1}\left(2HW_k^T + 2\lambda PZ_l + \mu X_k^T X_k + \mu X_k^T E_k - X_k^T Y_{1,k} + \mu Z'_k - Y_{2,k} + \mu Z''_k - Y_{3,k}\right)$$
$$\left(W_k W_k^T + Z_l^T Z_l + 3I\right)^{-1} \tag{17}$$

where $l \neq k$, $Z_k^T = Z_k$ and $P^T = P$.

Step 4 (Update W_k):
W_k can be updated by solving optimization problem (18):

$$W_k = \underset{W_k}{argmin} \left\| H - Z_k^T W_k \right\|_F^2 + \xi \| W_k \|_F^2 \tag{18}$$

Similar to the previous step, it is easy to get the solution:

$$W_k = \left(Z_k Z_k^T + \xi I \right)^{-1} Z_k H \tag{19}$$

Step 5 (Update E_k):

$$E_k = \underset{E_k}{argmin}\, \gamma \| E_k \|_{2,1} + \frac{\mu}{2} \left\| X_k - X_k Z_k - E_k + \frac{Y_{1,k}}{\mu} \right\|_F^2 \tag{20}$$

In order to solve the problem (20), the following lemma is required:

Lemma 1: Let Q be a given matrix. If the optimal solution to

$$\underset{E}{min}\, \alpha \| B \|_{2,1} + \frac{1}{2} \| B - Q \|_F^2 \tag{21}$$

is U^*, the i-th column of U^* is as follows [19]:

$$[B^*]_{:,i} = \begin{cases} \frac{\| Q_{:,i} \|_2 - \alpha}{\| Q_{:,i} \|_2} Q_{:,i}, & if \| Q_{:,i} \|_2 > \alpha \\ 0, & otherwise \end{cases} \tag{22}$$

Therefore, it is clear that the optimal solution of problem (20) is:

$$[E]_{:,i} = \begin{cases} \frac{\| J_{:,i} \|_2 - \frac{\gamma}{\mu}}{\| J_{:,i} \|_2} J_{:,i}, & if \| J_{:,i} \|_2 > \frac{\gamma}{\mu} \\ 0, & otherwise \end{cases} \tag{23}$$

where $J = X_k - X_k Z_k + \frac{Y_{1,k}}{\mu}$.

Step 6 (Update multiplies):
Multipliers $Y_{1,k}$, $Y_{2,k}$, $Y_{3,k}$ and iteration step-size ρ ($\rho > 1$) are updated by Eq. (24):

$$\begin{cases} Y_{1,k} = Y_{1,k} + \mu(X_k - X_k Z_k - E_k) \\ Y_{2,k} = Y_{2,k} + \mu(Z_k - Z_k') \\ Y_{3,k} = Y_{3,k} + \mu(Z_k - Z_k'') \\ \mu = min(\rho\mu, \mu_{max}) \end{cases} \tag{24}$$

In short, we summed up the process of solving the objective function (7) in Algorithm 1.

Algorithm 1: Solving Problem x by IALM

Input: Data X_k, Lable H, Parameters $\alpha, \beta, \gamma, \xi$ and λ;

Initialization: $W_k = 0$; $Z_k = Z'_k = Z''_k = 0$; $E = 0$; $Y_{1,k} = Y_{2,k} = Y_{3,k} = 0$;
$\mu_{max} = 10^7$; $\mu = 0.1$; $\rho = 1.01$; $\varepsilon = 10^{-7}$;

While not converged do

1. Fix other variables and update Z'_k by eq. (12);
2. Fix other variables and update Z''_k by eq. (15);
3. Fix other variables and update Z_k by eq. (17);
4. Fix other variables and update W_k by eq. (19);
5. Fix other variables and update E_k by eq. (23);
6. Update the multipliers and parameters by eq. (24);
7. Check the convergence conditions $\|X_k - X_k Z_k - E_k\|_\infty < \varepsilon$, $\|Z_k - Z'_k\|_\infty < \varepsilon$ and $\|Z_k - Z''_k\|_\infty < \varepsilon$;

End while

Output: Z_k, W_k

---

4 Experiments and Discussion

4.1 Comparison of Our Method and Baseline Classification Methods

We first compare our approach to some baseline feature selection and classification methods. Comparison methods include Nearest neighbor (NN) classifier without feature selection, linear discriminant analysis (LDA) [20], support machine vector (SVM) [21], and kernel discriminant analysis (KDA) [22].

We used the 10-fold cross-validation strategy in this experiment. Specifically, each dataset is equally divided into 10 subsets, and then each subset is taken as test set in turn, with all remaining subsets as training set. This process is repeated 20 times to avoid deviations caused by the sample segmentation process. Accuracy (ACC), sensitivity (SEN), specificity (SPE), the area under the receiver operating characteristic (ROC) curve (AUC) and their respective standard deviations (STD) are used to measure performance in the classification. For their calculation process, please refer to [4].

For fair comparison, the first-order and second-order BFCN construction steps are the same for all methods, that is, the method in [4]. We set a threshold of $\{0, 0.05, 0.1, \cdots, 0.4\}$ for the first-order and second-order BFCN on all datasets. For NN, LDA, SVM, and KDA, we refer to [4], combining the first-order and second-order BFCN feature information for feature selection and classification steps. For SVM, we use a linear kernel. For KDA, we use a Gaussian kernel function. For the method we proposed, for simplicity, we use greedy strategy to select α and ξ from $\{10^{-3}, 10^{-2}, \cdots, 10^3\}$ and set the other hyperparams to 1.

Table 1 reports the comparison of our method and baseline feature selection and classification methods on three schizophrenic datasets. It can be seen that our proposed method has the best ACC, SEN and AUC performance on the three datasets compared to the baseline feature selection and classification algorithm. As can be seen from the Table 1, the SEN and SPE of these baseline methods show unbalanced results, so their SEN or SPE is very high, but the AUC results are not ideal. In contrast, the proposed method shows good performance, probably because the proposed method uses a multi-modality learning strategy to better combine the first-order and second-order BFCN information.

Table 1. The classification results (ACC/SEN/SPE/AUC ± STD%) of our method and several comparison baseline classification algorithms. The best results are shown in bold.

		ACC ± STD	SEN ± STD	SPE ± STD	AUC ± STD
COBRE	NN	66.67 ± 2.17	38.64 ± 1.89	88.55 ± 2.52	59.84 ± 5.33
	LDA	77.00 ± 1.94	70.72 ± 1.20	83.30 ± 3.22	71.99 ± 0.64
	KDA	81.00 ± 0.62	74.51 ± 4.73	**87.53 ± 3.70**	76.39 ± 2.01
	SVM	77.17 ± 1.25	68.67 ± 2.41	84.62 ± 2.07	72.66 ± 1.11
	Ours	**83.91 ± 0.50**	**79.29 ± 0.99**	**87.53 ± 1.55**	**81.75 ± 0.50**
Nottingham	NN	68.58 ± 2.53	56.62 ± 12.24	80.06 ± 7.84	66.79 ± 4.48
	LDA	68.71 ± 2.87	62.92 ± 7.48	74.83 ± 4.50	65.52 ± 2.31
	KDA	75.88 ± 1.54	67.05 ± 5.53	**84.63 ± 3.79**	73.25 ± 3.47
	SVM	75.73 ± 2.85	70.28 ± 3.18	82.11 ± 5.55	71.16 ± 3.89
	Ours	**78.08 ± 1.50**	**71.33 ± 2.27**	84.29 ± 2.15	**74.57 ± 2.01**
Xiangya	NN	65.79 ± 1.37	76.99 ± 8.68	51.05 ± 7.96	62.80 ± 3.34
	LDA	77.64 ± 1.27	84.90 ± 2.51	68.07 ± 2.96	75.71 ± 2.46
	KDA	80.83 ± 1.02	85.98 ± 3.17	74.47 ± 3.35	79.20 ± 2.89
	SVM	78.23 ± 1.63	86.16 ± 1.95	67.63 ± 2.48	75.78 ± 3.05
	Ours	**81.84 ± 0.87**	**88.55 ± 2.09**	**74.56 ± 2.69**	**79.24 ± 0.83**

4.2 Comparison of Our Method and State of Art Multi-modality Based Methods

In this experiment, we compare the proposed method with some start of art multi-modality based methods. Comparison methods include multi-kernel learning (MKL) SVM method [23], multitask feature selection (MTFS) model [6], manifold regularized MTFS (M2TFS) model [13], and multi-modal structured low-rank dictionary learning (MM-SLDL) method [24].

The experimental setup and the metrics for measuring classification performance are the same as in the previous experiment. For fair comparison, the MKL parameters are chosen from $\{0, 0.1, 0.2, \cdots, 1\}$ using greedy strategy. The parameters selection method of MTFS and MT2FS is the same as that in [13], and the parameter selection range of MM-SLDL is the same as that in [24].

The experimental results are reported in Table 2. In addition, we plot the ROC curve of the experiment and show it in Fig. 1. It can be seen that the proposed method performs best on all indicators, and the ROC curve of the comparison method is almost at the bottom right of the ROC curve of our method. This may be because strategies based on multi-modality low-rank representation can better learn and fuse first-order and second-order BFCN information.

Table 2. The classification results (ACC/SEN/SPE/AUC ± STD%) of our method and state of art multi-modality based algorithms. The best results are shown in bold.

		ACC ± STD	SEN ± STD	SPE ± STD	AUC ± STD
COBRE	MKL	78.53 ± 3.38	73.26 ± 4.11	82.69 ± 3.60	75.25 ± 3.80
	MTFS	74.60 ± 2.29	64.09 ± 3.05	82.95 ± 2.97	70.56 ± 2.13
	MT2FS	76.85 ± 1.72	63.42 ± 4.24	87.45 ± 1.48	72.22 ± 2.22
	MM-SLDL	78.61 ± 1.86	73.38 ± 4.39	82.69 ± 1.88	75.55 ± 2.96
	Ours	**83.91 ± 0.50**	**79.29 ± 0.99**	**87.53 ± 1.55**	**81.75 ± 0.50**
Nottingham	MKL	67.35 ± 2.45	56.83 ± 3.82	76.93 ± 4.07	63.35 ± 3.38
	MTFS	69.02 ± 2.89	62.17 ± 1.94	75.29 ± 3.95	65.87 ± 3.61
	MT2FS	73.88 ± 3.58	65.33 ± 2.61	81.57 ± 5.17	70.66 ± 3.87
	MM-SLDL	67.40 ± 2.36	54.33 ± 7.88	78.93 ± 3.76	61.81 ± 3.15
	Ours	**78.08 ± 1.50**	**71.33 ± 2.27**	**84.29 ± 2.15**	**74.57 ± 2.01**
Xiangya	MKL	77.43 ± 1.97	84.80 ± 2.36	67.33 ± 4.78	72.91 ± 2.51
	MTFS	76.46 ± 2.25	85.22 ± 2.34	64.33 ± 2.91	71.21 ± 2.94
	MT2FS	77.60 ± 2.34	87.97 ± 3.22	63.33 ± 4.35	71.49 ± 3.25
	MM-SLDL	77.74 ± 0.49	84.43 ± 2.22	68.33 ± 2.98	72.91 ± 1.68
	Ours	**81.84 ± 0.87**	**88.55 ± 2.09**	**74.56 ± 2.69**	**79.24 ± 0.83**

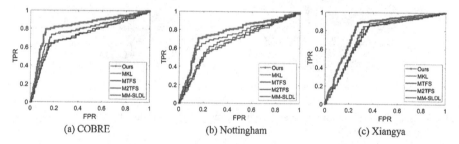

(a) COBRE   (b) Nottingham   (c) Xiangya

Fig. 1. ROC curves for our proposed method and all state of art multi-modality based methods on all datasets.

4.3 Analysis of Convergence and Parameter Sensitivity

In this section, we first show the convergence iteration of the proposed method. Figure 2 shows the change in the value of the objective function as the number of iterations increases. It can be found that our method can basically converge within 300 iterations.

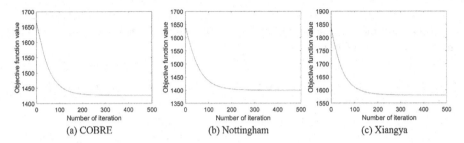

Fig. 2. The convergence property of the proposed algorithm on different datasets

We then evaluated the effect of two hyperparameters (α and ξ) on the performance of our method. The results are shown in Fig. 3. We mainly evaluated the effects of these two parameters on ACC and AUC. α and ξ have a value range of $\{10^{-3}, 10^{-2}, \cdots, 10^3\}$. For the sake of simplicity, we only show the effect of these two parameters on the results of the Xiangya dataset. Figure 3 shows that our method achieves a steady state when ξ takes a value of 10, 10^2, or 10^3. Similar results were obtained on the other two datasets. In addition, we can see that parameter ξ has a more critical impact on the results than parameter α.

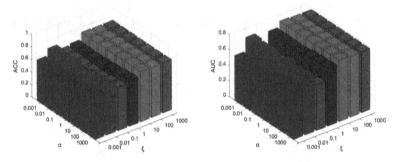

Fig. 3. The effect of parameters α and ξ on our method. Left subgraph is the result ACC, and right subgraph is result of AUC.

5   Conclusion

In summary, we propose a multi-modality feature learning method based on low-rank representation to fuse first-order and second-order BFCN information and apply it to the classification of schizophrenia. Specifically, we combine the low-rank representation method with the multi-modality learning framework and add an ideal representation term to effectively learn the first-order and second-order BFCN feature information. Experiments on three schizophrenia datasets show that our approach is superior to existing multimodal feature learning methods.

Acknowledgments. This work was supported in part by National Natural Science Foundation of China (Nos. 61501230, 61732006, 61876082 and 81771444), National Science and Technology Major Project (No. 2018ZX10201002), and the Fundamental Research Funds for the Central Universities (No. NJ2019010).

References

1. Bluhm, R.L., et al.: Spontaneous low-frequency fluctuations in the BOLD signal in schizophrenic patients: anomalies in the default network. Schizophr. Bull. **33**, 1004–1012 (2007)
2. Richiardi, J., Achard, S., Bunke, H., Van De Ville, D.: Machine learning with brain graphs: predictive modeling approaches for functional imaging in systems neuroscience. IEEE Signal Process. Mag. **30**, 58–70 (2013). https://doi.org/10.1109/MSP.2012.2233865
3. Guo, S., Kendrick, K.M., Yu, R., Wang, H.L.S., Feng, J.: Key functional circuitry altered in schizophrenia involves parietal regions associated with sense of self. Hum. Brain Mapp. **35**, 123–139 (2014). https://doi.org/10.1002/hbm.22162
4. Zhu, Q., Li, H., Huang, J., Xu, X., Guan, D., Zhang, D.: Hybrid functional brain network with first-order and second-order information for computer-aided diagnosis of schizophrenia. Front. Neurosci. **13**, 603 (2019). https://doi.org/10.3389/fnins.2019.00603
5. Huang, S., et al.: Identifying Alzheimer's disease-related brain areas from multi-modality neuroimaging data using sparse composite linear discrimination analysis. In: Advances in Neural Information Processing Systems, vol. 1431–1439 (2011)
6. Zhang, D., Shen, D.: Multi-modal multi-task learning for joint prediction of multiple regression and classification variables in Alzheimer's disease. Neuroimage **59**, 895–907 (2012). https://doi.org/10.1016/j.neuroimage.2011.09.069
7. Candès, E.J., Tao, T.: The power of convex relaxation: near-optimal matrix completion. IEEE Trans. Inf. Theory **56**, 2053–2080 (2010)
8. Zhu, Q., Li, H., Huang, J., Xu, X., Guan, D., Zhang, D.: Hybrid functional brain network with first-order and second-order information for computer-aided diagnosis of schizophrenia. Front. Neurosci. **13**, 603 (2019). https://doi.org/10.3389/fnins.2019.00603
9. Liu, G., Lin, Z., Yan, S., Sun, J., Yu, Y., Ma, Y.: Robust recovery of subspace structures by low-rank representation. IEEE Trans. Pattern Anal. Mach. Intell. **35**, 171–184 (2013). https://doi.org/10.1109/TPAMI.2012.88
10. Zhang, N., Yang, J.: Low-rank representation based discriminative projection for robust feature extraction. Neurocomputing **111**, 13–20 (2013). https://doi.org/10.1016/j.neucom.2012.12.012
11. Candès, E.J., Recht, B.: Exact matrix completion via convex optimization. Found. Comput. Math. **9**, 717–772 (2009). https://doi.org/10.1007/s10208-009-9045-5
12. Fazel, M.: Matrix Rank Minimization with Applications. Dissertation (2002)
13. Jie, B., Zhang, D., Cheng, B., Shen, D.: Manifold regularized multitask feature learning for multimodality disease classification. Hum. Brain Mapp. **36**, 489–507 (2015)
14. Liu, G., Lin, Z., Yan, S., Sun, J., Yu, Y., Ma, Y.: Robust recovery of subspace structures by low-rank representation. IEEE Trans. Pattern Anal. Mach. Intell. **35**, 171–184 (2013). https://doi.org/10.1109/TPAMI.2012.88
15. Wright, J., Ganesh, A., Rao, S., Ma, Y.: Robust principal component analysis: exact recovery of corrupted low-rank matrices, vol. 1, pp. 289–298 (2009). 58

16. Zhu, C., Wei, L., Zhou, R., Wang, X., Wu, A.: Robust subspace segmentation by self-representation constrained low-rank representation. In: Neural Processing Letters, pp. 1671–1691 (2018) https://doi.org/10.1007/s11063-018-9783-y

17. Lin, Z., Chen, M., Wu, L., Ma, Y.: The Augmented Lagrange Multiplier Method for Exact Recovery of Corrupted Low-Rank Matrices. Eprint Arxiv. 9 (2010)

18. Zhang, Z., Liu, L., Shen, F., Shen, H.T., Shao, L.: Binary multi-view clustering. IEEE Trans. Pattern Anal. Mach. Intell. **41**, 1 (2018)

19. Yang, J., Yin, W., Zhang, Y., Wang, Y.: A fast algorithm for edge-preserving variational multichannel image restoration. SIAM J. Imaging Sci. **2**, 569–592 (2009)

20. Zhang, X., Jia, Y.: A linear discriminant analysis framework based on random subspace for face recognition. Pattern Recogn. **40**, 2585–2591 (2007). https://doi.org/10.1016/j.patcog.2006.12.002

21. Chang, C., Lin, C.: LIBSVM. ACM Trans. Intell. Syst. Technol. **2**, 1–27 (2011). https://doi.org/10.1145/1961189.1961199

22. Cai, D., He, X., Han, J.: Speed up kernel discriminant analysis. VLDB J. **20**, 21–33 (2011)

23. Zhang, D., Wang, Y., Zhou, L., Yuan, H., Shen, D.: Multimodal classification of Alzheimer's disease and mild cognitive impairment. Neuroimage **55**, 856–867 (2011). https://doi.org/10.1016/j.neuroimage.2011.01.008

24. Foroughi, H., Shakeri, M., Ray, N., Zhang, H.: Face recognition using multi-modal low-rank dictionary learning. In: 2017 IEEE International Conference on Image Processing (ICIP), pp. 1082–1086. IEEE (2017). https://doi.org/10.1109/ICIP.2017.8296448

A Joint Bitrate and Buffer Control Scheme for Low-Latency Live Streaming

Si Chen⬤, Yuan Zhang$^{(\boxtimes)}$⬤, Huan Peng⬤, and Jinyao Yan⬤

Communication University of China, Beijing, China
{chinsi,yzhang,hpeng,jyan}@cuc.edu.cn

Abstract. Live video streaming has experienced explosive growth on the mobile Internet. Unlike on-demand streaming, live video streaming faces more challenges due to the strong requirement of low latency. To balance several inherently conflicting performance metrics and improve the overall quality of experience (QoE), the adaptive bitrate algorithm is widely used under time-varying network conditions. However, it does not perform well at low latency. In this paper, we present a joint bitrate and buffer control scheme (JBBC) for low-latency live streaming based on latency-constrained bitrate adaptation and playback rate adaptation. Experiments demonstrate that the proposed algorithm has better performance on overall QoE than most existing adaptive schemes, achieving a more stable bitrate selection and relatively lower delay on the premise of almost no rebuffering.

Keywords: Live video streaming · ABR (adaptive bitrate algorithm) · AMP (adaptive media playout) · QoE (quality of experience)

1 Introduction

Recent years have witnessed a rapid increase in the requirement of live video services [1]. Unlike on-demand streaming, live streaming over HTTP, such as DASH (Dynamic Adaptive Streaming over HTTP), faces more challenges for it has a strong requirement of low latency to enable real-time interaction between content providers and users. It is necessary to reduce the latency while maintaining high video quality and smooth playback.

At present, the adaptive bitrate (ABR) algorithm is the primary choice of content providers for providing high-quality video streaming services to users under fluctuating network conditions. Existing algorithms mainly choose download bitrate for each video segment based on network resources and playback buffer conditions, such as BOLA [2], MPC [3] and Pensieve [4]. However, these ABR algorithms rely on reserving a large buffer to absorb the risk of bandwidth fluctuation, without consideration of the impact of latency on QoE.

Many recent studies have focused on live streaming services. Some researches optimize the network transmission module, try to use HTTP2.0 to build a low latency live broadcast system, but they ignore the impact of video download

© Springer Nature Switzerland AG 2019
Z. Cui et al. (Eds.): IScIDE 2019, LNCS 11936, pp. 369–380, 2019.
https://doi.org/10.1007/978-3-030-36204-1_31

bitrate, thus cannot provide high-quality video service [6,7]. There are also some researches focus on the client-side, the main idea is to control the client-buffer occupancy for it directly affects end-to-end latency. [8] designs an adaptive bitrate algorithm based on PID (Proportional-Integral-Derivative) controller to control the buffer occupancy to a target value so as to maintain low latency without rebuffering. DTBB [9] proposes a buffer-based rate adaptation algorithm with a dynamic threshold, which can decrease the rate transitions and provide a seamless playback under a low latency. [10] tests the performance of PID, DTBB, OSMF [11] under various network conditions, and they set video framerate to 25 fps and segment duration to 2 s. The results show that the mean latency of these algorithms is between 4 s–9 s, which indicates that although these ABR algorithms take latency into consideration, they still don't perform well at low latency.

In addition to ABR algorithms, adaptive media playback (AMP) is also widely studied to provide smooth video services [12–14]. It mainly determines the playback rate of the video at the client based on the buffer status or buffer fluctuations to prevent buffer from overflow and underflow. In most AMP researches, it is generally accepted that the playout speed variations are not noticeable within 25% [15–17]. In [18], subjective experiments were carried out for six types of video, each at different frame rates. The results show that 15 fps is a critical value and the subjective experience will drop sharply if the frame rate is below this value. [19] examined the relationship between video bitrate, frame rate, and video motion quality. The results show that at various video bitrates, the frame rate changes have the least impact on the video motion quality when the frame rate fluctuates within 15%. An example is that motion pictures that shot at a frame rate of 24 fps are shown on European PAL/SECAM broadcast television at 25 fps [15], which indicates that the QoE deterioration might not occur if the playback rate changes within the range of 5%. Therefore, it is worth noting that since AMP can dynamically adjust the buffer status by adjusting the playback rate while no damaging to QoE, it could be a good way to further reduce latency on the basis of ABR.

In this paper, we propose a joint bitrate and client-buffer control scheme in DASH client for low-latency live streaming. Our main contributions are summarized as follows:

- We design a latency-constrained bitrate adaptation algorithm to improve the overall QoE;
- In addition to controlling the download bitrate of video segments, we also design a playback rate adaptation algorithm to further reduce the end-to-end latency.

The remainder of the paper is organized as follows. Section 2 introduces the framework of the live streaming system. Section 3 describes the bitrate adaptation and client-side playback adaptation method. Section 4 tests the performance of our algorithm. Section 5 concludes this paper.

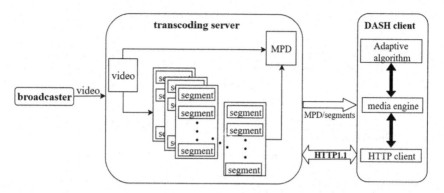

Fig. 1. Universal framework for live streaming system.

2   System Overview

The framework of a general live streaming system is shown in Fig. 1. Video is captured, encoded, and then uploaded to a transcoding server, which divides the video into multiple video segments and each segment is encoded at several different bitrates. The client requests video segments from the transcoding server, and the server responds with a manifest file MPD (Media Presentation Description) that provides a list of available bitrates to the client and directs the client to CDN (content delivery network) to download the requested segment. The ABR algorithm is deployed at the client to select the download bitrate for each video segment based on current network conditions. To absorb network instabilities, the client uses a buffer to store several video segments before playing out. Time delay occurs in every stage of the video transporting process, including encoding, transcoding, downloading, client buffering, and so on. In this work, since the network status of the uploader is uncontrollable, we focus on the download and playback process at the client-side.

3   Joint Bitrate and Buffer Control Scheme

The goal of our algorithm is to improve the QoE, which is mainly related to the following factors:

- Higher bitrate: improve the average bitrate of all video segments, but high bitrate may cause rebuffering.
- Less rebuffering: avoid the interruption of playback caused by buffer underflow, which is proved to be the most serious QoE impairment; we use rebuffering time to measure it.
- fewer oscillations: avoid frequent bitrate switching; we measure this metric using the change of download bitrate.
- Lower latency: reduce the end-to-end latency, which means that client buffer size must be small enough to meet low-latency requirements.

Table 1. Inputs of existing DASH-client algorithm vs JBBC

Influence factors		Dash	Ours
Network conditions	Bandwidth	✓	✓
	Throughput	✓	✓
	RTT		✓
	Jitter	✓	✓
Video content	Coding bitrare	✓	✓
	Frame rate		✓
	I/B/P frame	✓	✓
	Resolution		
	Frame size	✓	✓
	Total video time	✓	
	The number of video segments	✓	
	Switch	✓	✓
	Download time	✓	✓
Client	Buffer size	✓	✓
	Target buffer		✓
	Rebuffer	✓	✓
	Delay		✓

In this work, we intend to jointly address the bitrate adaptation and the playback adaptation problem at the client, which means that we need to determine the cliend-side playback rate while determining the download bitrate. We use reinforcement learning to generate the proposed adaptation algorithm, i.e. JBBC, in which we represent our control strategy as a neural network that maps observations (e.g., video content information, network bandwidth conditions, client buffer status) to the bitrate and playback rate decisions.

3.1   Inputs

Table 1 lists the inputs of the existing dash-client algorithm and ours. Compared with ABR algorithms in the on-demand scenario, we focus more on the factors related to delay. Besides, some information cannot be obtained in the live streaming such as the video frame sizes, the number of remaining video frames to be downloaded, and so on, while these can be obtained in the on-demand streaming.

Through the above analysis, the inputs of our ABR algorithm can be shown as $s_t = (\bar{x}_t, \bar{b}_t, \bar{\tau}_t, \bar{d}_t, \bar{q}_t, \bar{n}_t)$. $\bar{x}_t$ is the predicted network throughput for the past k video segments. We use the bandwidth estimation algorithm in dash.js [21] to estimate the available bandwidth x_t, which uses the network information of the past three segments to estimate the current network throughput. $\bar{b}_t$ is a vector of the client-buffer occupancy when downloading past k video segments; $\bar{\tau}_t$ is a vector of the rebuffing time during downloading past k video segments; $\bar{d}_t$ is a vector of the end-to-end delay when playing past k video segments; $\bar{q}_t$ is a vector of bitrates of past k video segments; $\bar{n}_t$ is a vector of available bitrates for the next video segment. It should be mentioned that all the selected features need to be normalized before training.

Besides, we treat video download bitrate and bandwidth as a form of signal, so that we can use FFT (Fast Fourier Transform) to decompose signals into a complex-valued function of frequency. We add the FFT of $\bar{x}_t$ and FFT of $\bar{q}_t$ to the input. In this way, the trend of bitrate changes and bandwidth fluctuation can be characterized, which can facilitate decision making.

3.2   Outputs

In this work, we need to decide the clied-side playback rate while determining the download bitrate. The mechanism of playback rate adjustment is described as follows. We set a pair of buffer thresholds: quick-play-threshold B_up and slow-play-threshold B_low. Our adaptive algorithm controls the playback rate at the client by adjusting the buffer thresholds. To simplify the decision mechanism of our algorithm, we introduce target-buffer to calculate B_up and B_low in (1). In this way, we can adjust the playback rate by deciding the value of the target buffer.

$$
\begin{aligned}
B\_up &= target\_buffer + 0.1 \\
B\_low &= target\_buffer * 2/3
\end{aligned}
\tag{1}
$$

The playback rate φ can be determined based on the buffer threshold and the current buffer size B in (2), where ρ is a parameter to control the aggressiveness of playback rate adaptation, the upper limit of the playback rate adjustment is set to 5%.

$$
\varphi = \begin{cases}
1, & \text{if } B \in [B\_low, B\_up] \\
1 + min(\frac{B - B\_up}{\rho}, 0.05), & \text{if } B > B\_up \\
1 + max(\frac{B - B\_low}{\rho}, -0.05), & \text{if } B < B\_low
\end{cases}
\tag{2}
$$

In a word, in order to implement a joint bitrate and buffer control scheme, our adaptive algorithm need to output the download bitrate and the value of target buffer.

3.3   Actor-Critic Network

In this work, we tend to learn a mapping from inputs to actions, where action $a(t)$ contains the download bitrate and the target buffer value. We formulate the adaptive bitrate decision and target buffer decision problem within an asynchronous actor-critic [22] framework. As shown in Fig. 2, we use a 1D convolution layer (CNN) to extract features of inputs. The actor-network learns the adaptive strategy of bitrate and target buffer. The critic network is used to guide the training of the actor-network.

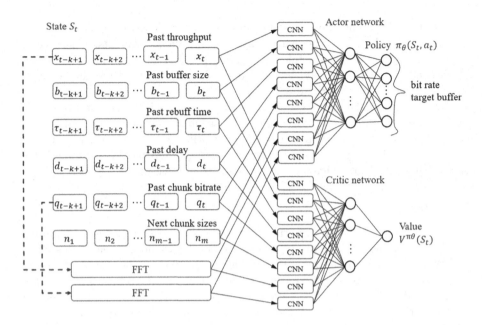

Fig. 2. Actor-critic algorithm.

In actor-network, the key idea is to maximise the cumulative discounted reward. The gradient of cumulative discounted reward can be computed as:

$$\nabla_\theta E_{\pi_\theta} \left[\sum_{t=0}^{\infty} \gamma^t r_t \right] = E_{\pi_\theta} [\nabla_\theta \log \pi_\theta(s, a) A^{\pi_\theta}(s, a)] \qquad (3)$$

where r_t is the reward when taking action a_t at state s_t, $\gamma \in [0, 1]$ is a discount parameter reflecting the impact of the current action on the subsequent reward, $\pi_\theta(s, a)$ represents the probability of selecting action a in state s; $A^{\pi_\theta}(s, a)$ represents the difference between the expected reward and the reward generated by the action we choose.

$$A^{\pi_\theta}(s, a) = Q^{\pi_\theta}(s, a) - V^{\pi_\theta}(s) \qquad (4)$$

where $Q^{\pi_\theta}(s, a)$ represent the value when choosing action a in state s, $V^{\pi_\theta}(s)$ represents the value of state s. Thus, the update of actor-network parameter θ follows:

$$\theta \leftarrow \theta + \sigma \sum_t \nabla_t \log \pi_\theta(s_t, a_t) A(s_t, a_t) \qquad (5)$$

Critic network is used to estimate the value of $A^{\pi_\theta}(s, a)$, training the critic network parameter w follows the Temporal Difference method [23].

$$w \leftarrow w - \sigma' \sum_t \nabla_w (Q^{\pi_\theta}(s, a; w) - V^{\pi_\theta}(s_t; w))^2 \qquad (6)$$

3.4   QoE Metrics

The goal of our algorithm is to improve the QoE, which means higher bitrate, less rebuffering, few oscillations, lower latency. The QoE after playing n video segments is shown in (7).

$$QoE = \sum_{1}^{n} \left(\alpha \cdot \varphi \cdot q_t - \beta \cdot \tau_t - \eta \cdot d_t - \mu \cdot |q_t - q_{t-1}| \right) \tag{7}$$

where φ represent the playback rate at the client, q_t is the download bitrate, τ_t is the rebuffing time, d_t is the end-to-end latency, α, β, η, μ are weight coefficients. We set weight coefficients as in AITrans[1] competition, where α is 1, β is 1.5, η is 1, μ is 1.

3.5   Training

We use the simulator provided in AITrans to simulate the download process. We train our algorithm using network traces provided by HSDPA [24] and AITrans and video traces provided by AITrans competition which contain the information of video frame size, I/P frame and each frame arriving time to CDN. To speed up training, we train 16 agents in parallel. Each agent continually sends [state, action, reward] to the central agent. The central agent compute the gradient and update network parameters of the 16 agents.

In the process of training, we find that the model is difficult to converge to the optimal value. The main reason is that the model does not fully explore various possibilities. When the model explores a locally optimal direction, it keeps training in this direction, which may deviate from the global optimal value. So we intend to add an entropy regularization term to the actor's update rule (5) to encourage exploration. At the beginning of training, the distribution of all actions should be equal, so the entropy of policy is the largest at this time. With the training going on, our model gradually learns a better policy $\pi_\theta(s, a)$, which means that the probability of one of the actions will be higher and the entropy of policy will gradually decrease, even tending to 0. However, the current strategy may not be the optimal solution, so we pushing θ to the direction of higher entropy to encourage exploring more possibilities. Therefore, the update of actor-network parameter θ can be modified to (8).

$$\theta \leftarrow \theta + \sigma \sum_{t} \nabla_t \log \pi_\theta(s_t, a_t) A(s_t, a_t) + \delta \nabla_\theta H(\pi_\theta(\cdot|s_t)) \tag{8}$$

In the early stage of training, we set a larger entropy weight δ to encourage the agent to explore more possibilities. The entropy weight δ decreases gradually with training. Adjusting the weight of the entropy can help the agent to converge to a better policy.

---

[1] https://www.aitrans.online/.

Table 2. Performance of JBBC under various bandwidth

	Average bitrate (kbps)	Average delay (s)	Rebuffering ratio
1500 kbps	1200	0.35	0
1000 kbps	987	1.77	0.01
600 kbps	593	1.51	0.03

4   Experiments and Results

In this section, we evaluate the proposed adaptation algorithm JBBC using public mobile dataset HSDPA collected in Norway and compare the performance of JBBC with DTBB [9] and LAPAS [10]. The test video is from a live game with a duration of 25 min, which is encoded with the H.264 format at the bitrate of 500 kbps and 1200 kbps, respectively. Each version of the video is divided into equal-length video segments of 1 s. The frame rate is 25 fps.

We first evaluate the performance of JBBC on the constant bandwidth of 600 kbps, 1000 kbps, and 1500 kbps, respectively. Average bitrate, average delay, and rebuff rate are shown in Table 2. It shows that in the case of sufficient bandwidth, JBBC can continue to select high bitrate and reduce the delay to 0.35 s without rebuffering. If the bandwidth is not sufficient enough, JBBC tries to choose a higher bitrate while almost no rebuffering, and eventually the delay is reduced to about 1.5 s–2.5 s.

Then we compare JBBC with DTBB and LAPAS on several fluctuation bandwidths, as shown in Fig. 3. If the rebuff flag is true in Fig. 3, it means that rebuffering has occurred. In Fig. 3(a), we test with a fluctuation bandwidth of 0.5 Mbps to 2 Mbps. The delay can be reduced to less than 1 s if the bandwidth is higher than the maximum available download bitrate. However, the delay in LAPAS cannot continue to drop even if the bandwidth is sufficient. Figure 3(b) shows the bitrate selection, latency, and rebuff time of each algorithm over a fluctuating bandwidth of 0.5 Mbps–1.5 Mbps. When the bandwidth is in the range of the highest and lowest available download bitrates, JBBC and LAPAS can reduce the delay to around 2 s with almost no stagnation, while DTBB frequently results in rebuffering with a delay of more than 6 s. In Fig. 3(c), the bandwidth between 0–1 Mbps is used for testing, wherein the bandwidth at some moments is smaller than the lowest available download bitrate of the video. The results show that JBBC can maintain the buffer stability for about 2 s, and the bitrate selection is more stable than the LAPAS.

Figure 4 shows the CDF curve of these algorithms over 200 network traces extracted from HSDPA. The abscissa is the normalized QoE value, and the ordinate is the cumulative probability distribution of QoE. Since DTBB has higher delay and rebuff rate under various network, its QoE is less than zero in 20% of the network traces. For JBBC and LAPAS, rebuffering rarely occurs and a relatively stable delay status can be achieved. However, the delay of JBBC is lower than LAPAS, especially in the case of better network conditions. Also,

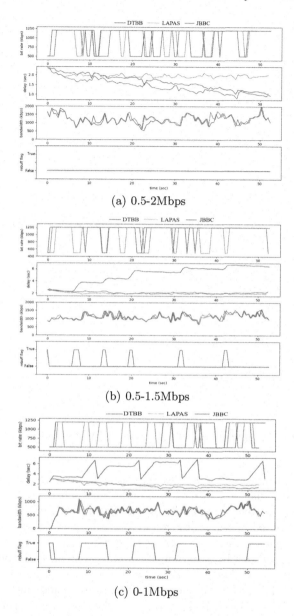

(a) 0.5-2Mbps

(b) 0.5-1.5Mbps

(c) 0-1Mbps

Fig. 3. Bitrate and delay under various fluctuation bandwidth

JBBC has a more stable rate selection and relatively low bitrate switching, so it has a higher QoE than LAPAS.

Finally, we test the effect of adjusting the playback rate in the end-to-end delay control. The blue curve in Fig. 5 shows the results of bitrate selection, delay conditions, and rebuffering under fluctuation bandwidth when we use playback

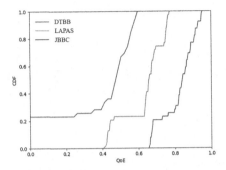

Fig. 4. Comparison of QoE for each algorithm

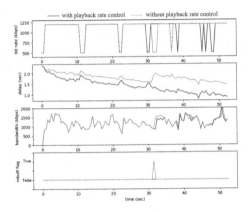

Fig. 5. Bitrate and delay comparison with and without playback control

rate control in our proposed algorithm. And the orange curve shows the results when we don't use playback rate control under fluctuation bandwidth. The initial delays of the two methods are almost the same. However, by controlling the playback rate, the delay in the orange curve begins to decrease to less than 2 s gradually and remains stable for a long time while the delay in orange curve fluctuates dramatically. The experiments show that the average delay of JBBC is 1.90 s without controlling playback rate while controlling the playback rate can reduce the delay to 1.44 s under 0 to 2M variable bandwidth. More importantly, adjusting the client's playback rate can also avoid rebuffering as shown in Fig. 5. Therefore, we can conclude that controlling the playback rate at the client can reduce the end-to-end delay without QoE deterioration, and can ensure the stability of the transmission process at the same time.

5   Conclusion

In this paper, we have presented a joint bitrate and buffer control scheme for low-latency live streaming. We use the reinforcement learning method to generate an adaptive algorithm, which consists of two parts: one is the latency-contrained adaptive bitrate selection, which is used to balance several inherently conflicting performance metrics and improve the QoE, and the other is the client-side adaptive playback rate control, which is used to further reduce the end-to-end latency and avoid rebuffering. The experiments have shown that our proposed algorithm has better performance on overall QoE than most existing adaptive schemes, achieving a more stable bitrate selection and relatively lower delay on the premise of almost no rebuffering.

Acknowledgments. This work was supported by National Natural Science Foundation of China (61472389).

References

1. Cisco: Cisco Visual Networking Index: Global Mobile Data Traffic Forecast Update, 2016–2021 White Paper (2017)
2. Li, Z., Zhu, X., Gahm, J., et al.: Probe and adapt: rate adaptation for HTTP video streaming at scale. IEEE J. Sel. Areas Commun. **32**(4), 719–733 (2014)
3. Yin, X., Jindal, A., Sekar, V., et al.: A control-theoretic approach for dynamic adaptive video streaming over HTTP. In: ACM Conference on Special Interest Group on Data Communication, pp. 325–338. ACM (2015)
4. Mao, H., Netravali, R., Alizadeh, M.: Neural adaptive video streaming with pensieve. In: Proceedings of the Conference of the ACM Special Interest Group on Data Communication, pp. 197–210. ACM (2017)
5. Huang, T.Y., et al.: A buffer-based approach to rate adaptation: evidence from a large video streaming service. In: SIGCOMM. ACM (2014)
6. Wei, S., Swaminathan, V.: Low latency live video streaming over HTTP 2.0. In: Proceedings of the Network and Operating System Support on Digital Audio and Video Workshop, p. 37. ACM (2014)
7. van der Hooft, J., De Boom, C., Petrangeli, S., Wauters, T., De Turck, F.: An HTTP/2 push-based framework for low-latency adaptive streaming through user profiling. In: NOMS 2018 - 2018 IEEE/IFIP Network Operations and Management Symposium, Taipei, pp. 1–5 (2018)
8. Wang, J., Meng, S., Sun, J., Quo, Z.: A general PID-based rate adaptation approach for TCP-based live streaming over mobile networks. In: Proceedings of the IEEE ICME, Seattle, USA, pp. 1–6, July 2016
9. Xie, L., Zhou, C., Zhang, X.: Dynamic threshold based rate adaptation for HTTP live streaming. In: Proceedings of the IEEE ISCAS, Baltimore, MD, USA, pp. 1–4, May 2017
10. Zhang, G., Lee, J.Y.B.: LAPAS: latency-aware playback-adaptive streaming (2019, in press)
11. Adobe OSMF. https://sourceforge.net/projects/osmf.adobe/
12. Chen, Y., Liu, G.: Adaptive media playout assisted rate adaptation scheme for HTTP adaptive streaming over lte system. In: 2016 IEEE International Conference on Multimedia & Expo Workshops (ICMEW), pp. 1–6. IEEE (2016)
13. Su, Y.F., Yang, Y.H., Lu, M.T., et al.: Smooth control of adaptive media playout for video streaming. IEEE Trans. Multimed. **11**(7), 1331–1339 (2009)
14. Li, M., Yeh, C.L., Lu, S.Y.: Real-time QoE monitoring system for video streaming services with adaptive media playout. Int. J. Digit. Multimed. Broadcast. (2018)

15. Kalman, M., Steinbach, E., Girod, B.: Adaptive media playout for low-delay video streaming over error-prone channels. IEEE Trans. Circuits Syst. Video Technol. **14**(6), 841–851 (2004)

16. Fan, M., Yang, J., Zhao, Y.: Probability estimation based adaptive media playout algorithm. In: 2010 2nd International Conference on Advanced Computer Control, vol. 4, pp. 267–271. IEEE (2010)

17. Chuang, H.C., Huang, C.Y., Chiang, T.: Content-aware adaptive media playout controls for wireless video streaming. IEEE Trans. Multimed. **9**(6), 1273–1283 (2007)

18. Ou, Y.F., Liu, T., Zhao, Z., et al.: Modeling the impact of frame rate on perceptual quality of video. In: 2008 15th IEEE International Conference on Image Processing, pp. 689–692. IEEE (2008)

19. Alsrehin, N.O., Klaib, A.F.: VMQ: an algorithm for measuring the video motion quality. Bull. Electr. Eng. Inform. **8**(1), 231–238 (2019)

20. Mok, R.K.P., Luo, X., Chan, E.W.W., Chang, R.K.C.: QDASH: a QoE-aware DASH system. In: Proceedings of the Annual ACM SIGMM Conference on Multimedia Systems (MMSys 2012), pp. 11–22 (2012)

21. DASH reference player. http://mediapm.edgesuite.net/dash/public/nightly/samples/dash-if-reference-player/index.html

22. Mnih, V., Badia, A.P., Mirza, M., et al.: Asynchronous methods for deep reinforcement learning. In: International Conference on Machine Learning, pp. 1928–1937 (2016)

23. Sutton, R.S., Barto, A.G.: Reinforcement Learning: An Introduction. MIT Press, Cambridge (2018)

24. Riiser, H., Endestad, T., Vigmostad, P., et al.: Video streaming using a location-based bandwidth-lookup service for bitrate planning. ACM Trans. Multimed. Comput. Commun. Appl. (TOMM) **8**(3), 24 (2012)

Causal Discovery of Linear Non-Gaussian Acyclic Model with Small Samples

Feng Xie[1], Ruichu Cai[1(✉)], Yan Zeng[1], and Zhifeng Hao[1,2]

[1] School of Computer Science,
Guangdong University of Technology, Guangzhou, China
`cairuichu@gmail.com`
[2] School of Mathematics and Big Data, Foshan University, Foshan, China

Abstract. Linear non-Gaussian Acyclic Model (LiNGAM) is a well-known model for causal discovery from observational data. Existing estimation methods are usually based on infinite sample theory and often fail to obtain an ideal result in the small samples. However, it is commonplace to encounter non-Gaussian data with small or medium sample sizes in practice. In this paper, we propose a Minimal Set-based LiNGAM algorithm (MiS-LiNGAM) to address the LiNGAM with small samples. MiS-LiNGAM is a two-phase and greedy search algorithm. Specifically, in the first phase, we find the skeleton of the network using the regression-based conditional independence test, which helps us reduce the complexity in finding the minimal LiNGAM set of the second phase. Further, this independence test we applied guarantees the reliability when the number of conditioning variables increases. In the second phase, we give an efficient method to iteratively select the minimal LiNGAM set with the skeleton and learn the causal network. We also present the corresponding theoretical derivation. The experimental results on simulated networks and real networks are presented to demonstrate the efficacy of our method.

Keywords: LiNGAM · Non-Gaussian · Small samples · Causal discovery

1   Introduction

In recent years, causal discovery algorithm has drawn much more attention in machine learning [1,2] and has been widely used in Neuroscience [3,4], Economics [5,6], Human Action Sequences [7], Epidemiology [8], and other fields. However, with the absence of prior knowledge, the constraint-based methods can only find a set of Markov equivalence classes, such as PC [9], IC [10] and their variants. To solve the above problem, the Linear non-Gaussian Acyclic Model (LiNGAM) [11,12] was proposed by introducing disturbance variables with non-Gaussian distribution in the structural equation modeling (SEM). The LiNGAM can learn a complete causal network from purely observational data.

At present, there are two main approaches for solving the LiNGAM [12,13]. The first approach converts the discovery process into a function-optimization

© Springer Nature Switzerland AG 2019
Z. Cui et al. (Eds.): IScIDE 2019, LNCS 11936, pp. 381–393, 2019.
https://doi.org/10.1007/978-3-030-36204-1_32

problem, such as ICA-LiNGAM algorithm [11] and Bayes-LiNGAM algorithm [14] et al. The second approach is a direct estimation method, i.e., select the exogenous variables one by one using the independence between variables and residuals in a finite number of steps, such as DirectLiNGAM algorithm [15] and its improved version Pairwise-LiNGAM algorithm [16] et al. However, the theory of the above algorithms is based on sufficient samples. When the samples are insufficient, especially the number of samples is less than 100, the results of the network are ineffective (see the discussion in Sect. 3.2). Further, in real life, we are often faced with limited samples. If we want to get enough samples, it may take a lot of time or money, and sometimes it's impossible. To this situation, the existing estimation methods may fail to discover the causal networks.

Therefore, the development of a practical method with small samples is necessary. With the same sample size, the smaller the number of nodes, the higher the accuracy of the results of the causal network (see the detailed discussion in Sect. 3.2). An intuitive strategy is to learn the causal network iteratively with the minimal number of nodes' set, which helps us make the most of valid information to learn the causal network.

Following the above analysis, we propose a Minimal Set-based LiNGAM algorithm to settle this problem, MiS-LiNGAM in short. MiS-LiNGAM is a greedy search algorithm, i.e., it iteratively selects the minimal LiNGAM set to learn the causal network. In order to reduce the complexity of finding the minimal LiNGAM set, we utilize a two-phase method to conduct this process. Specifically, in the first phase, we find neighbor nodes for every node in pairs using the d-separation, i.e., find the skeleton of the network. For the traditional conditional independence test of the d-separation, the result tends to be unreliable when the number of conditioning variables increases. With Darmois-Skitvitch theorem [17] and regression method, this test can transform the conditional independence test into the independence test between residuals [18]. In the second phase, we define the minimal LiNGAM Set according to the blocking property of the collider nodes and give an efficient method to iteratively find this set from data in finite steps.

The remainder of this paper is organized as follows. In Sect. 2, we review the related work of LiNGAM and its variants. Next, we introduce the LiNGAM and some important definitions. Besides, we discuss the effects of the small samples on the LiNGAM. In Sect. 4, we present our algorithm and its basic theory. In Sect. 5, we present preliminary experimental results using artificial and real networks. Finally, we present conclusions in Sect. 6.

2   Related Work

Causal discovery from the observational data is part of the theoretical background of this work. Typical constraint-based algorithms have been exploited to discover the underlying causal network [19,20]. However, these methods can only find a set of Markov equivalence classes but not a unique DAG, such as PC [9], IC [10] and their variants. It is worth noting that Shimizu et al. [11,12]

addressed the above issue with the proposed LiNGAM (Linear non-Gaussian Acyclic Model). LiNGAM can distinguish these structures in the same Markov equivalence classes and get a unique causal graph.

Various methods have been conducted to estimate the basic LiNGAM model [12], which can roughly be divided into two categories. The first category converts the discovery process into a function-optimization problem. For example, Shimizu et al. [11] cleverly transformed the problem to blind source separation and used Independent Component Analysis (ICA) to estimate the model, which is called ICA-LiNGAM. Later, Hoyer and Hyttinen [14] developed a Bayesian method and constructed a scoring function in terms of likelihood estimation, which is called Bayes-LiNGAM. However, this strategy easily falls into local optima and relies on parameters. The second category is a direct estimation method and eliminates parameters, i.e., selects the exogenous variables one by one in finite steps and gets a causal order. For example, Shimizu et al. [15] utilized the independence between variables and residuals to identify the exogenous variable, which is called DirectLiNGAM. In order to reduce the complexity of the DirectLiNGAM algorithm, Hyvärinen and Smith [16] calculated the one-dimensional differential entropies of variables and residuals instead of calculating their pairwise independence in identifying exogenous variables, which is called Pairwise-LiNGAM. Pairwise-LiNGAM algorithm is an improved version of DirectLiNGAM algorithm.

Apart from estimating the basic model, researchers have improved the model for different scenarios. (1) When the number of variables was greater than the number of observations, Sogawa et al. [21] utilized negative entropy to select exogenous variables, and applied this method to gene-expression data. Cai et al. [22] employed the idea of divide-and-conquer and proposed the SADA framework for estimating sparse networks. (2) When the data were time series, Hyvärinen et al. [23] combined the non-Gaussian instantaneous model and proposed the SVAR (Structural Vector AutoRegressive) model. (3) For LiNGAM with latent confounders, Hoyer et al. [24] proposed LvLiNGAM to address the LiNGAM with latent confounder. Shimizu and Bollen [25] adopted the Bayesian model selection method to estimate the direction with individual-specific confounder variables, which is called mixed-LiNGAM. (4) For the high-dimensional dataset, Loh and Bühlmann [26] proposed a scoring algorithm based on a moralized graph for linear SEM. (5) When the noise variables are weak non-Gaussian distribution, Cai et al. [27] recently proposed a KFCI algorithm based on the kurtosis of variables. (6) When the noise variables are arbitrary distribution, Hoyer et al. [28] proposed the PClingam algorithm to address this problem. Note that PClingam algorithm also firstly estimates the skeleton and then learns the causal network. But, in their paper, the algorithm is only the basic combination and the authors focus on the basic theory of identifying the directed acyclic graphs when the sample size is sufficient. To the best of our knowledge, there is no existing work by minimal LiNGAM set to address the problem of small samples under LiNGAM.

3   Background

3.1   LiNGAM and Preliminaries

Assume that the observational data $\mathbf{X} = \{x_1, x_2, ..., x_n\}$ generation process satisfies a linear SEM process, such that, without loss of generality, each observational variable x_i is assumed to have a zero mean. Now, we add the assumption that disturbance variables follow non-Gaussian distribution, then the Linear SEM is LiNGAM [12]. The model can be expressed as

$$x_i = \sum_{k(j)<k(i)} b_{ij}x_j + e_i \qquad (1)$$

where $k(i)$ is the causal order of x_i and b_{ij} is the connection strength from x_j to x_i. All disturbance variables e_i are independent of each other, with zero means and non-zero variances. That is $p(e_1, e_2, ...e_n) = \prod_i p_i(e_i)$, where $p(e_i)$ is the probability density of e_i.

Note that the generating process of LiNGAM is recursive. So it can be represented as a directed acyclic graph (DAG) $\mathcal{G}(\mathbf{X})$ by drawing directed edges from the causal order $k(i)$ of each variable. Let $\bar{\mathcal{G}}(\mathbf{X})$ denote the undirected graph, which is said to be the **skeleton** of $\mathcal{G}(\mathbf{X})$. Next, we briefly review the basic concepts of the Graph [19].

Definition 1 (path). *In a $\bar{\mathcal{G}}(\mathbf{X})$, any sequence of edges connecting two variables is a path.*

Definition 2 (collider). *In a $\mathcal{G}(\mathbf{X})$, if x_i and x_j both point to a node x_k (as in $x_i \rightarrow x_k \leftarrow x_j$), then x_k is a collider on that path between x_i and x_j.*

We say a path is an active path if there is no collider.

Definition 3 (d-separation). *A path p is said to be d-separated by a set of nodes Z if and only if*

a. *p contains a chain $i \rightarrow m \rightarrow j$ or a collider $i \rightarrow m \leftarrow j$ such that the middle node m is in Z, or*
b. *p contains a collider $i \rightarrow m \leftarrow j$ such that the middle node m is not in Z and such that no descendant of m is in Z.*

A set Z is said to d-separate X from Y if and only if Z blocks every path from a node in X to a node in Y.

3.2   Effects of Small Samples on the LiNGAM

Next, we will show how sample size affects the result of a causal network using the existing methods. More precisely, we design some experiments from different sample sizes and different dimensions. Here, we use the ratio of Number of Sample/ Number of Dimension (N.S/N.D) to analyze the effects of small samples and use the F1 score as our metric. The F1 score is calculated as $2(P \times R)/(P + R)$, where P and R are precision and recall, respectively. The result is reported in Fig. 1.

From Fig. 1, we find the following results, (a) The F1 scores of three algorithms have less than 50% when the sample size is 100. This demonstrates the existing method fails to solve the LiNGAM with small samples. (b) With decreasing N.S/N.D, the F1 score will decrease. In other words, if N.S/N.D increases, then the F1 score will increase at the same time. This implicitly tells us that locally learning the causal network will be a better strategy compared with the globally learning.

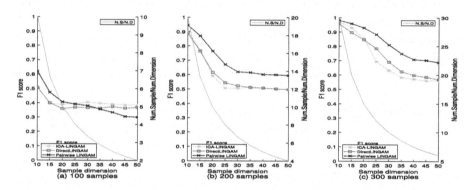

Fig. 1. Results of three algorithms over various samples and dimensions, where the disturbance variables follow Uniform distribution. For each setting, we run 100 trials and take the average as the final result

4   MiS-LiNGAM: Minimal Set-Based LiNGAM Algorithm

We begin with the definition of LiNGAM set.

Definition 4 (LiNGAM set/minimal LiNGAM set). *Let* $\mathbf{X} = \{x_1, x_2, ..., x_n\}$ *are generated according to the LiNGAM. If all elements in* $\mathbf{S} \subseteq \mathbf{X}$ *satisfy LiNGAM, then* $\mathbf{S}$ *is a LiNGAM set. Further, If there is no another* $\mathbf{T} \subseteq \mathbf{X}$ *LiNGAM set such that* $|\mathbf{T}| < |\mathbf{S}|$, *then* $\mathbf{S}$ *is a minimal LiNGAM set.*

where $|\mathbf{T}|$ and $|\mathbf{S}|$ denote the number of dimensions in $\mathbf{T}$ and $\mathbf{S}$ respectively.

In order to improve precision for small samples, an intuitive idea is to divide variables into small LiNGAM sets and locally learn the causal network. However, finding the LiNGAM set from the observational data is a difficult problem. This is because it is easy to introduce latent confounders, which would impede the learning of the causal network.

Fortunately, the non-Gaussianity property of the variables is helpful in addressing such difficulty, and it heavily relies on the following Theorem.

Darmois-Skitovitch Theorem (D-S Theorem) [17]: *Define two random variables x_1 and x_2 as linear combinations of independent random variables $s_i, (i = 1, ...n)$:*

$$x_1 = \sum_{i=1}^{n} \alpha_i s_i, x_2 = \sum_{i=1}^{n} \beta_i s_i \tag{2}$$

Then, if x_1 and x_2 are independent, all variables s_i for which $\alpha_i \beta_i \neq 0$ are Gaussian. In other words, if there exists a non-Gaussian s_j for which $\alpha_j \beta_j \neq 0$, x_1 and x_2 are dependent.

Next, we give two theorems to show the properties of the LiNGAM set, which are further explored to facilitate the finding of the LiNGAM set from observational data

Theorem 1. *Let $x_i, x_j \in \mathbf{X}$, if $[x_i - f(x_j)] \perp\!\!\!\perp x_j$ or $[x_j - g(x_i)] \perp\!\!\!\perp x_i$, where f and g are regression functions, then $\{x_i, x_j\}$ is a LiNGAM set. Otherwise, x_i and x_j have a common ancestor, i.e., require other variables to be a LiNGAM set.*

Proof. According to the linearity assumption, any variable x_i can be regarded as a linear sum of a set of noise variables. Therefore, we can express x_i and x_j as

$$x_i = \sum_{p=1}^{m} a_p e_p, x_j = \sum_{q=1}^{m} a_q e_q \tag{3}$$

where there is at least one $a_p \neq 0$ and $a_q \neq 0$.

Here, we firstly analyze the case $[x_i - f(x_j)] \perp\!\!\!\perp x_j$. By the linearity assumption,

$$[x_i - f(x_j)] = X_i - \frac{\mathrm{Cov}(X_i, X_j)}{\mathrm{Var}(X_j)} \cdot X_j = \sum_{p=1}^{m} a_p e_p - \frac{\mathrm{Cov}(X_i, X_j)}{\mathrm{Var}(X_j)} \sum_{q=1}^{m} a_q e_q. \tag{4}$$

Due to $[x_i - f(x_j)] \perp\!\!\!\perp x_j$ and D-S Theorem, we know $[x_i - f(x_j)]$ and x_j have no common noise variables. Combining Eqs. 3 and 4, x_j can be expressed as $x_i + \sum_{r=1}^{m} a_r e_r$, where there is at least one $a_r \neq 0$. That is to say, $\{x_i, x_j\}$ is a LiNGAM set.

Similarly, for the case $[x_j - g(x_i)] \perp\!\!\!\perp x_i$, we can obtain

$$[x_j - g(x_i)] = X_j - \frac{\mathrm{Cov}(X_i, X_j)}{\mathrm{Var}(X_i)} \cdot X_i = \sum_{q=1}^{m} a_q e_q - \frac{\mathrm{Cov}(X_i, X_j)}{\mathrm{Var}(X_i)} \sum_{p=1}^{m} a_p e_p. \tag{5}$$

Combining Eqs. 3 and 5, x_i can be expressed as $x_j + \sum_{r=1}^{m} a_r e_r$, where there is at least one $a_r \neq 0$. That is to say, $\{x_i, x_j\}$ is a LiNGAM set.

Next, we analyze the case $[x_i - f(x_j)] \not\perp\!\!\!\perp x_j$ and $[x_j - g(x_i)] \not\perp\!\!\!\perp x_i$. Based on the D-S Theorem, we know $[x_i - f(x_j)]$ and x_j have at least one common noise variable, e.g., e_k. According to the linearity assumption, x_i and x_j must have at least one common ancestor that contains e_k. Theorem 1 is proven.

Theorem 2. *Let* $\mathbf{S} \subseteq \mathbf{X}$. *For any* $x_i, x_j \in \mathbf{S}$, *if all variables on their active path are in set* $\mathbf{S}$, *then* $\mathbf{S}$ *is a LiNGAM set.*

Proof. According to the definition of path and graph theory, we can obtain there is no confounder in $\mathbf{S}$, that is to say, $\mathbf{S}$ is a LiNGAM set.

In fact, for a dataset $\mathbf{X}$, there are many possible LiNGAM sets and it is very complex to find all LiNGAM sets by Theorems 1 and 2. Therefore, we design a two-phase greedy search algorithm to address it, i.e., firstly estimate the skeleton and iteratively select the minimal LiNGAM set to learn the causal network. More precisely, in the first phase, we find all neighbor nodes using the d-separation. Considering the traditional conditional independence test of the d-separation, the result tends to be unreliable when the number of conditioning variables increases. So, we adopt the Regression-based d-separation [18], which with Darmois-Skitvitch theorem [17] and regression method, the conditional independence test can be transformed into the independence test between residual and variable [18]. In the second phase, we iteratively select the minimal LiNGAM set with the resulting graph of the previous iteration.

Theorem 3 (Regression-based d-separation[18]). *Let* $x_i, x_j \in \mathbf{X}$. *If there exists* $Z \subset \mathbf{X} - \{x_i, x_j\}$ *such that* $[x_i - f(Z)] \perp\!\!\!\perp [x_j - g(Z)]$, *where* f *and* g *are regression functions, then* Z *d-separates* x_i *and* x_j. *In other words,* x_i *and* x_j *have no direct edge.*

Now, based on the Theorem 3, one can determine whether there is a directed edge or not for any pair variables and estimate the skeleton of graph. Then, one can iteratively find the minimal LiNGAM set based on the Theorems 1 and 2, and learning the subgraph of this set using the existing methods, i.e., Pairwise-LiNGAM. Based on the above analyses, we propose a complete algorithm, termed the MiS-LiNGAM algorithm to estimate the LiNGAM with small samples, given in Algorithm 1.

5   Experiments and Discussion

In this section, we evaluate our algorithm on simulated causal networks and real causal networks. We employ Fisher's z-transform to test the independence for non-Gaussian data in our method, with threshold 95%. We compare the proposed MiS-LiNGAM algorithm with Pairwise-LiNGAM [16][1], DirectLiNGAM [15][2] and ICA-LiNGAM [11][3]. The experiments were compiled and ran using MAT-LAB 2017b on a Windows PC equipped with a dual-core 3.40 GHz CPU and 8 GB RAM. Since data are generated randomly, we take an average of 100

---

[1] Matlab package available at https://www.cs.helsinki.fi/u/ahyvarin/code/pwcausal/, Here, we adopt mxnt (maximum entropy approximations) to estimate the likelihood ratios in the Pairwise-LiNGAM algorithm.

[2] Matlab package available at https://sites.google.com/site/sshimizu06/Dlingamcode.

[3] Matlab package available at https://sites.google.com/site/sshimizu06/lingam.

Algorithm 1. MiS-LiNGAM

Input: Data set $\mathbf{X} = \{x_1, ..., x_n\}$
Output: Causal network $\mathcal{G}$

1: Initialize complete undirected graph $\mathcal{G}$, and $s = 2$;
2: //Phase 1: Estimate the skeleton of network
3: **for** each pair $\{x_i, x_j\} \in \mathbf{X}$ and **do**
4:     **if** $\exists Z \in Ne(x_i, x_j)$ such that $[x_i - f(Z)] \perp\!\!\!\perp [x_j - g(Z)]$, where f and g are regression functions. **then**
5:         delete the edge $x_i - x_j$ from $\mathcal{G}$.
6:     **end if**
7: **end for**
8: //Phase 2: Iteratively find minimal LiNGAM set and learn the causal network

9: **repeat**
10:     **for** each subset $\mathbf{S} \subset \mathbf{X}$, where $|\mathbf{S}| = s$ **do**
11:         **if** $\mathbf{S}$ is a LiNGAM set in existing $\mathcal{G}$ by Theorem 1 and 2 **then**
12:             $\mathcal{G} \leftarrow$ Learning the causal network($\mathbf{S}$).//One may use the existing method
13:         **end if**
14:     **end for**
15:     s=s+1;
16: **until** there is no edge to estimate
17: **Return:** $\mathcal{G}$

---

trials as the final result. To quantify the accuracy of algorithms, we use the Recall, Precision and F1 score as our metrics. The F1 score is calculated as $2(P \times R)/(P + R)$, where P and R are precision and recall, respectively.

5.1   Results on Simulated Networks

In this subsection, the data were randomly generated by LiNGAM, where the average indegree was 1.5, and the disturbance variables followed a Uniform distribution or Laplace distribution with zero mean and unit variance. The number of samples is set to be 2V, 3V, 4V, and 5V, where V is the number of dimensions, i.e., $N.S/N.D = 2, 3, 4, 5$. Here, V is set to 10, 20, 30, 40, 50, and 60.

The results are visualized in Figs. 2 and 3. We find that the three metrics of the proposed algorithm is superior to those of the other three algorithms, indicating that it can perform causal discovery in various sample sizes. On the contrary, the Recalls of the other algorithms are very low, especially for ICA-LiNGAM, indicating that many ground-truth edges have not been discovered. Besides, the three metrics of Pairwise-LiNGAM are the smallest when the dimension is small, not even up to 20% for 10 dimensions. The reason is that Pairwise-LiNGAM is not reliable when removing the effects of exogenous variables for small samples.

5.2   Results on Real-World Networks

Here, we select six groups of real-world Bayesian networks in different fields (http://www.cs.huji.ac.il/~galel/Repository/). The statistics and description of these networks are listed in Table 1. The data are generated by LiNGAM according to these networks, where the disturbance variables follow a Uniform distribution or Laplace distribution with zero mean and unit variance, so that we can evaluate the performance of our algorithm on both sub-Gaussian and super-Gaussian setting. We also set the number of samples from 2 V to 5V, i.e., $N.S/N.D = 2, 3, 4, 5$.

Table 1. Statistics on the networks.

Networks	Num. variables	Avg degree	Max in-degree	Description
Sachs	11	3.09	3	Medicine data
Child	20	1.25	2	Disease data
Water	32	4.12	5	Wastewater treatment
Alarm	37	2.49	4	Medicine data
Haildfinder	56	2.36	4	Weather forecasting
Win95pts	76	2.95	7	Printer troubleshooting

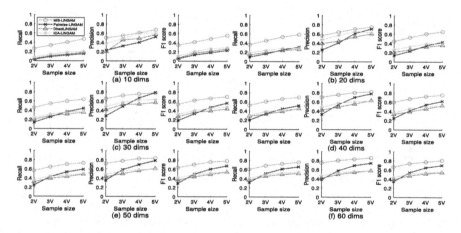

Fig. 2. Results of four algorithms over various sample sizes, where the noise variables follow Uniform distribution.

Results are illustrated by Figs. 4 and 5. Our proposed algorithm achieves satisfactory metrics compared with the other three algorithms for sub-Gaussian and super-Gaussian datasets, indicating that it can perform causal discovery in various networks. Especially, the F1 score of our algorithm reaches more than 50%, even up to 80% when the $N.S/N.D = 5$. We find that Pairwise-LiNGAM almost performs better than ICA-LiNGAM and DirectLiNGAM. But, for Sachs and Water datasets, ICA-LiNGAM achieves better than Pairwise-LiNGAM. The main reason is that the causal network of these two datasets is relatively dense,

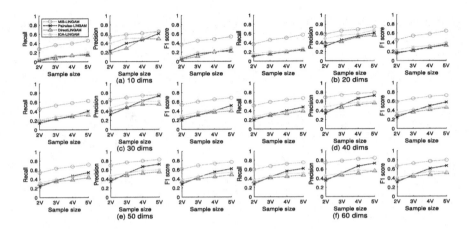

Fig. 3. Results of four algorithms over various sample sizes, where the noise variables follow Laplace distribution.

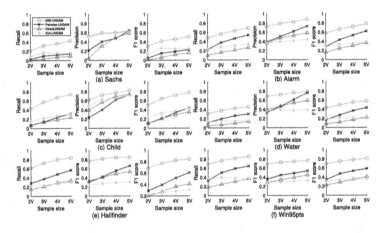

Fig. 4. Results of three algorithms in six real-world networks over various sample sizes, where the noise variables follow Uniform distribution.

in other words, the Num.Arcs/Num.Variables of them are larger than other networks.

In conclusion, not only in simulation networks but also in real-world networks, MiS-LiNGAM shows more excellent performance in different distributions. Even when the sample size is less than 40, the three metrics of MiS-LiNGAM can still perform better than other algorithms. These analyses verify that MiS-LiNGAM algorithm is more stable and efficient with small samples.

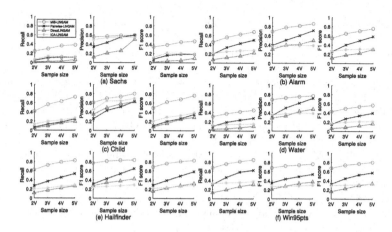

Fig. 5. Results of three algorithms in six real-world networks over various sample sizes, where the noise variables follow Laplace distribution.

6   Conclusion

In this paper, we proposed the MiS-LiNGAM algorithm to solve the LiNGAM with small samples. In order to maximize the use of valid information to learn causal network, we define the minimal LiNGAM set, adopting a greedy strategy to iteratively find this set, and learn the causal network. Theoretical analysis proves the correctness of MiS-LiNGAM. Further, experimental results reveal that the proposed algorithm can improve the accuracy of causal networks when the sample size is small.

Acknowledgement. This work was supported in part by the NSFC-Guangdong Joint Fund under Grant U1501254, in part by the Natural Science Foundation of China under Grant 61876043 and Grant 61472089, in part by the Natural Science Foundation of Guangdong under Grant 2014A030306004 and Grant 2014A030308008, in part by the Science and Technology Planning Project of Guangdong under Grant 2013B051000076, Grant 2015B010108006, and Grant 2015B010131015, in part by the Guangdong High-Level Personnel of Special Support Pro- gram under Grant 2015TQ01X140, in part by the Pearl River S&T Nova Program of Guangzhou under Grant 201610010101, and in part by the Science and Technology Planning Project of Guangzhou under Grant 201902010058.

References

1. Spirtes, P., Glymour, C., Scheines, R., Tillman, R.: Automated search for causal relations: theory and practice (2010)
2. Pearl, J.: The seven tools of causal inference, with reflections on machine learning. Commun. ACM **62**(3), 54–60 (2019)
3. Lele, X., et al.: A pooling-lingam algorithm for effective connectivity analysis of fMRI data. Front. Comput. Neurosci. **8**, 125 (2014)

4. Sanchez-Romero, R., Ramsey, J.D., Zhang, K., Glymour, M.R.K., Huang, B., Glymour, C.: Estimating feedforward and feedback effective connections from fMRI time series: assessments of statistical methods. Netw. Neurosci. **3**(2), 274–306 (2019)

5. Zhang, K., Chan, L.: Minimal nonlinear distortion principle for nonlinear independent component analysis. J. Mach. Learn. Res. **9**(Nov), 2455–2487 (2008)

6. Al-Yahyaee, K.H., Mensi, W., Al-Jarrah, I.M.W., Tiwari, A.K.: Testing for the Granger-causality between returns in the U.S and GIPSI stock markets. Phys. A: Stat. Mech. Appl. **531**, 120950 (2019)

7. Cai, R., Zhang, Z., Hao, Z., Winslett, M.: Understanding social causalities behind human action sequences. IEEE Trans. Neural Netw. Learn. Syst. **28**(8), 1801–1813 (2016)

8. Helajärvi, H., Rosenström, T., et al.: Exploring causality between TV viewing and weight change in young and middle-aged adults. The cardiovascular risk in young finns study. PLoS One **9**(7), e101860 (2014)

9. Spirtes, P., Glymour, C.: An algorithm for fast recovery of sparse causal graphs. Soc. Sci. Comput. Rev. **9**(1), 62–72 (1991)

10. Pearl, J., Verma, T.: A theory of inferred causation. In: Proceedings of the Second International Conference on Principles of Knowledge Representation and Reasoning, KR 1991, pp. 441–452 (1991)

11. Shimizu, S., Hoyer, P.O., Hyvärinen, A., Kerminen, A.: A linear non-Gaussian acyclic model for causal discovery. J. Mach. Learn. Res. **7**(Oct), 2003–2030 (2006)

12. Shimizu, S.: LiNGAM: non-Gaussian methods for estimating causal structures. Behaviormetrika **41**(1), 65–98 (2014)

13. Shimizu, S.: Non-gaussian methods for causal structure learning. Prev. Sci. **20**, 1–11 (2018)

14. Hoyer, P.O., Hyttinen, A.: Bayesian discovery of linear acyclic causal models. In: Proceedings of the Twenty-Fifth Conference on Uncertainty in Artificial Intelligence, pp. 240–248. AUAI Press (2009)

15. Shimizu, S., et al.: DirectLiNGAM: a direct method for learning a linear non-Gaussian structural equation model. J. Mach. Learn. Res. **12**(Apr), 1225–1248 (2011)

16. Hyvärinen, A., Smith, S.M.: Pairwise likelihood ratios for estimation of non-Gaussian structural equation models. J. Mach. Learn. Res. **14**(Jan), 111–152 (2013)

17. Kagan, A.M., Rao, C.R., Linnik, Y.V.: Characterization problems in mathematical statistics (1973)

18. Zhang, H., Zhou, S., Zhang, K., Guan, J.: Causal discovery using regression-based conditional independence tests. In: Thirty-First AAAI Conference on Artificial Intelligence (2017)

19. Spirtes, P., Glymour, C.N., Scheines, R.: Causation, Prediction, and Search. MIT Press, Cambridge (2000)

20. Pearl, J.: Causality: Models, Reasoning and Inference, vol. 29. Springer, Heidelberg (2000)

21. Sogawa, Y., Shimizu, S., Shimamura, T., HyväRinen, A., Washio, T., Imoto, S.: Estimating exogenous variables in data with more variables than observations. Neural Netw. **24**(8), 875–880 (2011)

22. Cai, R., Zhang, Z., Hao, Z.: SADA: a general framework to support robust causation discovery. In: International Conference on Machine Learning, pp. 208–216 (2013)

23. Hyvärinen, A., Zhang, K., Shimizu, S., Hoyer, P.O.: Estimation of a structural vector autoregression model using non-gaussianity. J. Mach. Learn. Res. **11**(May), 1709–1731 (2010)
24. Hoyer, P.O., Shimizu, S., Kerminen, A.J., Palviainen, M.: Estimation of causal effects using linear non-gaussian causal models with hidden variables. Int. J. Approx. Reason. **49**(2), 362–378 (2008)
25. Shimizu, S., Bollen, K.: Bayesian estimation of causal direction in acyclic structural equation models with individual-specific confounder variables and non-Gaussian distributions. J. Mach. Learn. Res. **15**(1), 2629–2652 (2014)
26. Loh, P.-L., Bühlmann, P.: High-dimensional learning of linear causal networks via inverse covariance estimation. J. Mach. Learn. Res. **15**(1), 3065–3105 (2014)
27. Cai, R., Xie, F., Chen, W., Hao, Z.: An efficient kurtosis-based causal discovery method for linear non-Gaussian acyclic data. In 2017 IEEE/ACM 25th International Symposium on Quality of Service, pp. 1–6. IEEE (2017)
28. Hoyer, P.O., et al.: Causal discovery of linear acyclic models with arbitrary distributions. In: Proceedings of the Twenty-Fourth Conference on Uncertainty in Artificial Intelligence, pp. 282–289. AUAI Press (2008)

Accelerate Black-Box Attack
with White-Box Prior Knowledge

Jinghui Cai[1], Boyang Wang[2], Xiangfeng Wang[1(✉)], and Bo Jin[1]

[1] Shanghai Key Lab for Trustworthy Computing, School of Computer Science
and Technology, East China Normal University, Shanghai 200062, China
51174500002@stu.ecnu.edu.cn, {xfwang,bjin}@sei.ecnu.edu.cn
[2] School of Mechanical Engineering, Shanghai Jiao Tong University, Shanghai, China
wby920422@sjtu.edu.cn

Abstract. We propose an efficient adversarial attack method in the
black-box setting. Our Multi-model Efficient Query Attack (MEQA)
method takes advantage of the prior knowledge on different models' relationship to guide the construction of black-box adversarial instances. The
MEQA method employs several gradients from different white-box attack
models and further the "best" one is selected to replace the gradient of
black-box model in each step. The gradient composed by different model
gradients will lead a significant loss to the black-box model on these
adversarial pictures and then cause misclassification. Our key motivation
is to estimate the black-box model with several existing white-box models, which can significantly increase the efficiency from the perspectives
of both query sampling and calculating. Compared with gradient estimation based black-box adversarial attack methods, our MEQA method
reduces the number of queries from 10000 to 40, which greatly accelerates
the black-box adversarial attack. Compared with the zero query black-
box adversarial attack method, which also called transfer attack method,
MEQA boosts the attack success rate by 30%. We evaluate our method
on several black-box models and achieve remarkable performance which
proves that MEQA can serve as a baseline method for fast and effective
black-box adversarial attacks.

Keywords: Efficient black-box attack · Gradient estimation · Transfer
attack · Model robustness

1 Introduction

With the development of deep neural network, machine learning plays an
extremely important role in wide application domains. For instance, many applications such as face recognition and flaws detection already enter daily life. We
believe machine learning will become a fundamental technique and be used to
solve more difficult tasks in future. To emphasize, the robustness and security of
machine learning models are important but usually be overlooked. Szegedy et al.
[16] found that the convolution networks can be easily confused with even tiny

© Springer Nature Switzerland AG 2019
Z. Cui et al. (Eds.): IScIDE 2019, LNCS 11936, pp. 394–405, 2019.
https://doi.org/10.1007/978-3-030-36204-1_33

perturbation. Furthermore, recent work proposed by Athalye et al. [1] showed that the adversarial 3D objects can also be misclassified by neural network. In practical applications, these vulnerable machine learning methods are considered as latent threats to the security of the self-driven car or other machine learning systems.

Many powerful adversarial attacks, such as BIM [10] and C&W [2], were proposed after L-BFS [16] and FGSM [5], while achieved quite high attack success rate. However all these attacks focus on white-box technique which indicates that the adversary can get all the information of model including structure and parameters. In the white-box setting, the gradient backward from the model loss can effectively guide the update of adversarial images by maximizing target neural network loss, which leads to misclassification or incorrect segmentation. It only costs about 10 iteration of backward and gets a good attack success rate.

However, it is impossible for us to get all the model information in reality. The black-box setting adversarial attack is more concerned in recent studies because of its practicality. Black-box adversarial attack means that the information of model is limited to the adversary. We can only get the prediction confidence vector of the model instead of the gradient because we know nothing about the model parameters. In hard-label black-box setting, we can only get the prediction label instead of confidence vector, i.e., the change of loss is difficult to observe. As the gradient based white-box attack algorithms achieve great success, the gradient estimation attack method ZOO ATTACK (Zeroth Order Optimization Attack [3]) is first proposed to attack the black-box model in white-box attack similar way. However the ZOO Attack is powerful but costly on the amount of query times. How to decrease the query times is one of the most important research task in query-based black-box adversarial attack domain. There are lots of recent works decrease query times by simplifying gradient estimation while our method can efficiently solve this task by replacing the gradient with our prior knowledge.

In this paper, we proposed a black-box adversarial attack method called Multi-model Efficient Query Attack (MEQA) method. We use the known white-box models to guide the adversarial attack on black-box model in order to employ relationship between these white-box models and the target black-box model. The previous work [11] shows that transfer attacks from white-box model to black-box model have different success rate and the level of rate is related to the similarity between white-box model and black-box model. Because of this conclusion, we hope to use several models to estimate the target model. Our work is similar with ensemble learning [17] which can improve the transfer attack. Compared with the ensemble transfer attack, we query the prediction confidence from the target model to guide the attack instead of using the estimated prediction confidence from white-box models. Great improvement will be guaranteed on success rate with negligible increased cost. Compared with the gradient estimation method, our method decrease the query count to 40 every picture and have an approximate success result on lots black-box models. Figure 1 displays how MEQA method procedures adversarial images from original images.

Contributions. We summarize our main contribution about adversarial attack and relevance between models as follows:

1. We proposed a new query-based black-box adversarial attack called MEQA. The MEQA Method needs only 40 queries to the target model per image and achieve a high attack success rate, which decrease 99% query times than the state-of-art methods. To the best of our knowledge, MEQA Method is the first work to combine the model knowledge with the query-based black-box attack.

2. We evaluate MEQA attack in different setting and discuss the effectiveness of MEQA method. MEQA method is based on our white-box model zoo, which means that more models we have, stronger the MMEQ is. It has its practical meaning because it's hard to avoid using some regular model or module such as ResNet, VGG, shortcut or inception. Because these models or modules are the guarantee of performance in the deep learning architecture.

3. We construct a new way to measure the correlation between models. The correlation between models can be confirmed by its transfer attack success rate before. The higher transfer attack success rate means the more similar twos models are. It builds the relation on one model to one model. However, we think the relation should be built on multi models to one model. If one model is out of our model zoo and MEQA attack success rate decreases a lot, this model is critical to estimate the target model and has a high similarity with the target model. In this new rules, we can get the conclusion such like the model zoo containing ResNet50, AlexNet is similarity to target model ResNet101. This relation graph can help us understand the deep model better or be used in other tasks.

2   Related Work

To solve the problem that there is only limited information in black-box setting, several methods are proposed to attack black-box models from different perspectives. There are three common ways to apply black-box adversarial attack. First, Papernot et al. [12] found that the adversarial attack can transfer across models and the transfer attack success rate is related to the structure similarity between models. These transfer attacks have same complexity with the white-box attacks while their highest success rate is about 63% if original model and target model are similar such as ResNet50 and ResNet101 [7]. This study reflects that adversarial pictures are transferable but weak across the models.

Papernot et al. [13] proposed a method to estimate the black-box model by training a substitute model. They produce the adversarial picture by applying white-box adversarial attack on substitute model and querying the label from the target model and updating their substitute model. They alternately expand their training set and update the model. The well trained substitute model has a similar output with the black-box model. This method needs a lot time to train the substitute model, but it is efficient to procedure the high success rate

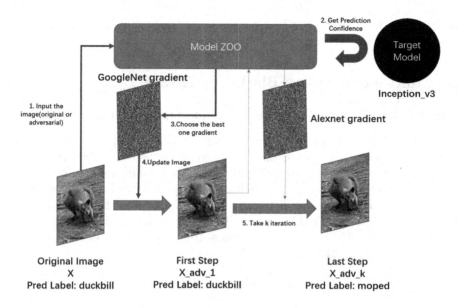

Fig. 1. MEQA method framework

adversarial picture to the target model after the training has been finished. Furthermore, it's one solution of the hard-label black-box setting task.

Another way is called query based black-box attack. Its key point is to estimate the gradient which maximize the target model loss with query result from the model. ZOO (Zeroth Order Optimization) Attack [3] achieve remarkable attack effectiveness by solving the gradient estimation with zeroth order optimization, while it needs almost one million queries for one picture. To solve the problem that too many queries is needed, AutoZOOM (autoencoder zeroth order optimization method) [18] encoder the picture to a low dimension vector and use the coarse to fine method to decrease the directions of the gradient estimation. Beyond zeroth order optimization way, the authors in [8] proposed to use a natural evolution strategy (NES) to enhance query efficiency. NES is a efficient way to select the gradient estimation directions. This work is parallel to AutoZOOM because they accelerate the gradient estimation in different way. In 2019, the authors in Guo et al. [6] proposed a simple method to black-box setting. They randomly choose a direction and add a constant perturbation on this direction. If the tiny perturbation confuses the target model and loss rise, they add the perturbation. If not, they add it inversely and check if the loss rise. They achieve a good attack performance and decrease the query count to ten thousand. It's still expensive on query count and time cost compared with the transfer attack. We think that gradient estimation method dose not concern about the similarity between white-box model and black-box model is the main reason why it needs so many queries. Our work use both the white-box models we have and the response of black-box model to produce adversarial images.

It combines the advantage of transfer attack methods and query based attack methods.

3   Multi-model Query Black-Box Attack

3.1   Black-Box Attack Loss Function and Optimization

In this paper, we focus on combining the advantage of transfer attack methods and gradient estimation based black-box attack methods. We find the image gradient in transfer attack is meaningful and can be used in gradient estimation. MEQA method improves the query efficiency of gradient estimation greatly. In our black-box setting, the prediction confidence is known to the adversary. The black-box model can be treat as a classification function $F : [0,1]^d \Rightarrow \mathbb{R}^k$. The input picture is a d dimension vector and output is prediction of k classes. We can modeling our untarget black-box task as

$$\max \quad \ell(F(x_{adv}), y), \tag{1}$$
$$\text{s.t.} \quad \|x_{adv} - x\|_p \le \epsilon,$$

where ℓ means the loss function of DNN, which is cross entropy in classification task and $\|x_{adv} - x\|_p \le \epsilon$ means the constraints on the perturbation of the adversarial picture. p will change for different tasks. For example, we set $p = 1$ if we want to change as least as pixel. In our setting, we choose $p = 2$ which means Euclidean distance from adversarial picture to clear picture. It is a common way to measure the perturbation. If we hope to confused the deep neural network to classify the adversarial picture as a special class. The loss function can be built as

$$\min \quad \ell(F(x_{adv}), t), \tag{2}$$
$$\text{s.t.} \quad \|x_{adv} - x\|_p \le \epsilon,$$

where t is the target class label. For the reason that solutions of target attack problems and untarget attack problems are similar, we will focus on the solution in the untarget attack setting. If the parameters of F are known to us, the BIM can solve the problem 1 well. It changes the problem with constraints to the problem in Lagrange form. The Lagrange dual problem can be solved with the iteration gradient sign method. Every step takes

$$x_{i+1} = x_i + sign(\nabla F(x_i)), \tag{3}$$

where $\nabla F(x_i)$ comes from the backward of the model.

In black-box setting, the gradient estimation method is try to estimate the $\nabla F(x_i)$ by zeroth order optimization method, while we want to use several F_{known} to estimate F_{black} instead of estimating $\nabla F(x_i)$ directly. The loss function of our method changes to

$$\max \quad \ell(F_{black}(X_{adv}), Y)$$
$$\text{s.t.} \quad \|X_{adv} - X\|_2 \leq \epsilon$$
$$|\sum_j \lambda_i \ell(F_j(X_{adv}), Y) - \ell(F_{black}(X_{adv}), Y)|^2 \leq \kappa$$
$$F_j \in \text{ModelZoo} \tag{4}$$

where F_j means the jth model in our model zoo and λ_j means the weight of the jth model in the loss function. λ_j is determined by the similarity between the model. λ_j is not a default value by our prior knowledge. It's calculated by model according to the response of model. If one white model's gradient guidance is better than another white model, it will have a larger λ_j. We hope our method change the λ_j for different scenarios to meet the constraint $|\sum_j \lambda_j \ell(F_j(X_{adv}), Y) - \ell(F_{black}(X_{adv}), Y)|^2 \leq \kappa$ and after that we can solve the black-box problem by the white-box substitute. In other words, we get the $\nabla F_{black}(X)$ from backwarding $\sum_j \lambda_j \ell(F_j(X_{adv}), Y)$. It's apparent to us that more models in warehouse, more powerful our attack is. Fortunately, the phenomenon that adversarial attack can transfer between models has shown that there are common heat map between different models. In reality, it is common to use the pretrained model or some popular model or module. So there exists λ_j which can meet the constraints. Algorithm 1 summarizes the MEQA framework to solve the problem 4. We use the basic iterative method (BIM) framework to optimize the loss function, and its complexity on an imagenet picture decreased from $O(iteration * 3 * 224 * 224)$ to $O(iteration * len(modelzoo))$. In every step, we choose the best performance adversarial picture. This greedy strategy is effective and efficient. The reason why we choose the BIM as our basic method is that it's powerful and easy to understand. If we use the FGSM as our basic method, we can average the gradient of different white-box models with different λ_j weights.

Algorithm 1. Multi Model Efficiently Query Attack

Input: Model Zoo M, $F_{white}^i \in M$, F_{black}, X, Y, ϵ
Output: Adversarial Picture X_{adv}

1: $X_{now} = X$
2: **while** $\arg\max F_{black}(X) = Y$ **do**
3:     **for** each $F_{white}^i \in M$ **do**
4:         $X_{temp}^i = X_{now} + \epsilon * \nabla F_{white}^i(X_{now})$
5:         $LossList^i = \ell(F_{black}(X_{temp}^i), Y)$
6:     $maxindex = \arg\max_i LossList$
7:     Update X_{now} with $X_{now} = X_{temp}^{maxindex}$
8: $X_{adv} = X_{now}$

---

4   Experimental Evaluation

In this section, we evaluate our attack against a list of traditional transfer attack algorithms and state-of-art black-box attack algorithms. Because there are too many models and white-box attack algorithms to produce the transfer attack, we choose the best transfer attack on target model. White-box attack algorithms contain FGSM [5], BIM [10], MI-FGSM [4], C&W [2]. The state-of-art black-box attack algorithms contain AutoZOOM, SimBA [6]. We focus on three dimension to evaluate the black-box adversarial attack: success rate, query count and perturbation norm. Success rate means the probability our attack confuses the target model. Query count is related to the time cost and the risk of being detected, and it's the main target of the gradient estimation research. In our experiments, we use the iteration times instead of query counts and time cost. There are almost 4 queries in one iteration and one iteration of a picture will only cost one second. Perturbation norm means the risk of being detected by human eye. Besides, we evaluate our method on a totally different structure from our model zoo and show how it works on black-box model.

4.1   Setup

We evaluate our method on 800 images from ImageNet Validation Set [14]. We evaluate our method in the untarget setting, and our method is easily change to the target setting and have a good result in theory.

Then we check the black-box loss and model chosen in every step. These analysis help us understand the relation between models and how to build our model zoo.

4.2   Attack Performance

Success Rate and Perturbation Norm. In Fig. 2, we plot the change of attack success rate with the iteration increasing. In this experiment, we attack the black-box model resnet101 with our model zoo containing alexnet [9], vgg11_bn [15], resnet50. When the original picture is updated nine times, it achieves 94% success rate with a tiny pertubation. Nine times iteration means that we just query 27 times at most for every picture. The 99% reduction of the number of queries is a great improvement compared to the gradient estimation method.

Evaluating on Different Black-Box Model. To prove that our attack method is generally working on different black-box model as soon as our model zoo contains base models, we evaluate our attack method from different model zoos to different target black-box models. We confirm that our model zoo in every experiment doesn't contain the target model. Table 1 shows the attack performance on different target model. We compare our result with the best transfer attack and SIMBA (10000 Queries). From the result, our method has a similar attack performance with the SIMBA on several models and decreases

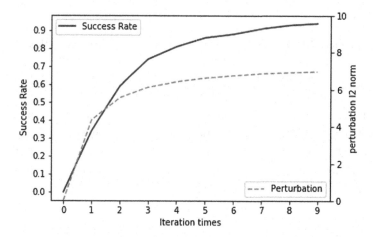

Fig. 2. MEQA performance on Resnet101

almost 99% query times. However, this is still a gap when target model has a different module from all models in our model zoo. The perturbation norm of MEQA is larger than SIMBA because of the value of epsilon and the early stop mechanism in generating adversarial picture. Perturbation of MEQA adversarial images still hard to detect by human. This is only 61.48% success rate to attack AlexNet which is the simplest model in deep neural networks. It is an interesting phenomenon which we will discuss later. In theory, the worst result of our attack equals to the best transfer attack when this is no model or only one model has relation with the target model. This hypothesis is impossible in reality. The adversarial attack has the transfer ability is the proof that there exists relation between the models we usually used. The best result of our attack equals to the white-box BIM attack when the target model is just in our model zoo. Considering model's defense to the transfer attack, our attack method has higher success rate in this condition because we can get help from other models to jump out of the gradient mask trap.

The Procedure of MEQA Method. To analyze the procedure of our attack, we check the mean loss in every step in different scenario. In a similar black-box model setting, we choose AlexNet, VGG11_bn, ResNet50, DenseNet121 in our model zoo and ResNet101 as target model. The loss of target model is shown in left of Fig. 3. The loss rises apparently which means our model zoo has a well representation of the target model. The blue line means the attack performance on pictures which are correctly classified before the attack and the red line means the attack performance on pictures which are wrong classified before attack. These pictures of red line are meaningless for evaluating our attack method, but they can also give us some clue to analyze our attack. We can find the blue line grows slower at first iteration compared with other iteration, but we doesn't find this phenomenon in red line. This phenomenon looks like gradient mask mentioned by authors [17]. The gap between target model and white-box

Table 1. Black-box attack performance on different target model

Target model	Transfer attack	SIMBA (10000 queries)		MEQA method (40 queries)	
	Success rate	Success rate	L2 norm	Success rate	L2 norm
AlexNet	27.35%	97.90%	5.59	61.48%	9.74
VGG11_bn	50.92%	98.9%	3.51	95.33%	6.92
ResNet50	63.29%	98.9%	4.59	96.95%	6.54
ResNet101	56.03%	100%	4.41	97.65%	6.36
DenseNet121	60.03%	100%	4.01	97.38%	6.59
Inception_v3	41.29%	92.18%	6.85	84.98%	8.13

models looks like a gradient mask at the saddle point and avoids the loss rises fast at first. The first step of our attack looks like a random walk to jump out the gradient trap and helps to get good result in next steps. Then we build a dissimilar black-box model setting. The combination of AlexNet, VGG11_bn, ResNet101, GoogleNet becomes our model zoo and Inception_v3 is chosen as target model. The loss of target model is shown in Fig. 3. It's obvious that the loss in the dissimilar setting is lower than the loss in the similar setting. It proves that the performance of MEQA is highly related to the content of model zoo. In Fig. 4, we show which model of model zoo is selected to produce the gradient in every step. We can conclude that models with similar structure is likely to be selected in every step, and the probability of selecting simple structure model rises with iteration increases because simple structure models are good at fine-tuning the adversarial images.

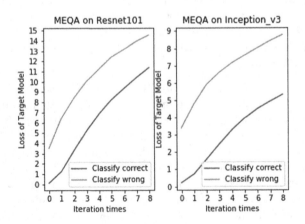

Fig. 3. MEQA performance on Resnet101 (Color figure online)

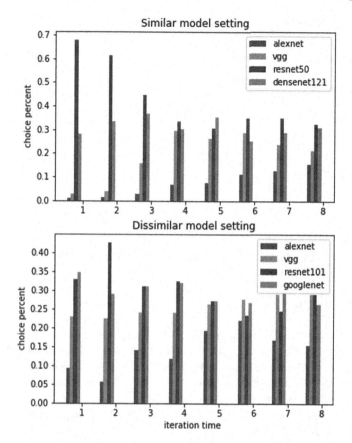

Fig. 4. Every step model choice in different setting

Model Zoo's Models. In Fig. 2, this is a strange phenomenon that the model zoo without the AlexNet has a low attack success rate on AlexNet. It's an interesting finding that even complicate networks such as VGG or ResNet are impossible to replace simple models such as AlexNet. To evaluate that models with simple structure have good transferability and they are important to our model zoo, we add the SqueezeNet into our model zoo and the attack success rate increases greatly as Fig. 5 shows. This phenomenon shows the shortcoming of the MEQA method that it will have a bad performance if the relation between model zoo and black-box model is weak. However, this phenomenon can easily be avoided if the base model structure is contained in model zoo.

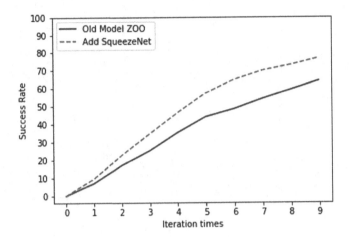

Fig. 5. MEQA on AlexNet with different model zoo

5   Conclusion

MEQA method is an efficient black-box adversarial attack method that achieves a high-level success rate with only 40 queries per image. It combines the advantage of transfer attack methods and gradient estimation based attack methods. Besides, it offers a new way to solve the black-box adversarial attack tasks. From the attack procedure, we can get the hints that models combination without ensemble learning can also estimate the black-box model. Greedy strategy is fast but not robust when the relation between model zoo and target model is hard to describe in linear space. MEQA method framework can be changed to a reinforcement learning framework which needs training but still effective when standard MEQA fails. We leave this method as our future work.

Acknowledgement. This work is supported by NSFC 61702188 and U1509219.

References

1. Athalye, A., Engstrom, L., Ilyas, A., Kwok, K.: Synthesizing robust adversarial examples. arXiv preprint arXiv:1707.07397 (2017)
2. Carlini, N., Wagner, D.: Towards evaluating the robustness of neural networks. In: 2017 IEEE Symposium on Security and Privacy (SP), pp. 39–57. IEEE (2017)
3. Chen, P.Y., Zhang, H., Sharma, Y., Yi, J., Hsieh, C.J.: ZOO: zeroth order optimization based black-box attacks to deep neural networks without training substitute models. In: Proceedings of the 10th ACM Workshop on Artificial Intelligence and Security, pp. 15–26. ACM (2017)
4. Dong, Y., et al.: Boosting adversarial attacks with momentum. In: Proceedings of the IEEE Conference on Computer Vision and Pattern Recognition, pp. 9185–9193 (2018)

5. Goodfellow, I.J., Shlens, J., Szegedy, C.: Explaining and harnessing adversarial examples. arXiv preprint arXiv:1412.6572 (2014)
6. Guo, C., Gardner, J.R., You, Y., Wilson, A.G., Weinberger, K.Q.: Simple black-box adversarial attacks. arXiv preprint arXiv:1905.07121 (2019)
7. He, K., Zhang, X., Ren, S., Sun, J.: Deep residual learning for image recognition. In: Proceedings of the IEEE Conference on Computer Vision and Pattern Recognition, pp. 770–778 (2016)
8. Ilyas, A., Engstrom, L., Athalye, A., Lin, J.: Black-box adversarial attacks with limited queries and information. arXiv preprint arXiv:1804.08598 (2018)
9. Krizhevsky, A., Sutskever, I., Hinton, G.E.: ImageNet classification with deep convolutional neural networks. In: Advances in Neural Information Processing Systems, pp. 1097–1105 (2012)
10. Kurakin, A., Goodfellow, I., Bengio, S.: Adversarial examples in the physical world. arXiv preprint arXiv:1607.02533 (2016)
11. Liu, Y., Chen, X., Liu, C., Song, D.: Delving into transferable adversarial examples and black-box attacks. arXiv preprint arXiv:1611.02770 (2016)
12. Papernot, N., McDaniel, P., Goodfellow, I.: Transferability in machine learning: from phenomena to black-box attacks using adversarial samples. arXiv preprint arXiv:1605.07277 (2016)
13. Papernot, N., McDaniel, P., Goodfellow, I., Jha, S., Celik, Z.B., Swami, A.: Practical black-box attacks against machine learning. In: Proceedings of the 2017 ACM on Asia Conference on Computer and Communications Security, pp. 506–519. ACM (2017)
14. Russakovsky, O., et al.: ImageNet large scale visual recognition challenge. Int. J. Comput. Vis. **115**(3), 211–252 (2015)
15. Simonyan, K., Zisserman, A.: Very deep convolutional networks for large-scale image recognition. arXiv preprint arXiv:1409.1556 (2014)
16. Szegedy, C., et al.: Intriguing properties of neural networks. arXiv preprint arXiv:1312.6199 (2013)
17. Tramèr, F., Kurakin, A., Papernot, N., Goodfellow, I., Boneh, D., McDaniel, P.: Ensemble adversarial training: attacks and defenses. arXiv preprint arXiv:1705.07204 (2017)
18. Tu, C.C., et al.: AutoZOOM: autoencoder-based zeroth order optimization method for attacking black-box neural networks. arXiv preprint arXiv:1805.11770 (2018)

A Dynamic Model + BFR Algorithm for Streaming Data Sorting

Yongwei Tan[1,2], Ling Huang[1,2(✉)], and Chang-Dong Wang[1,2]

[1] School of Data and Computer Science,
Sun Yat-sen University, Guangzhou, China
`tanyw8@mail2.sysu.edu.cn`, `huanglinghl@hotmail.com`,
`changdongwang@hotmail.com`
[2] Guangdong Province Key Laboratory of Computational Science,
Guangzhou, China

Abstract. Streaming data is widely generated in our lives. This has promoted a lot of research on streaming data mining, such as streaming data clustering and filtering. In our work, we present a problem about data stream processing, namely, streaming data sorting. There are some important characteristics of streaming data. Firstly, streaming data comes in the form of streams. It is usually assumed that streaming data is infinite, so it cannot be stored completely in memory. Secondly, we must process the streaming data in real time, otherwise we may lose the opportunity to deal with it forever. Based on these characteristics, we propose a dynamic algorithm that can make full use of memory and minimize error to solve the problem of streaming data sorting, which is further combined with the BFR algorithm to sort a particular type of streaming data. Some experiments are conducted to confirm the effectiveness of the proposed algorithms.

Keywords: Streaming data · Sorting · Dynamic model · Hash ·
Real-time processing · Stream management system

1 Introduction

Nowadays, streaming data is constantly being generated, and it contains a lot of valuable information, which is infinite but fleeting [1–5]. Streaming data comes in the form of continuous flows [6]. Most of the time, the computer can not meet the requirement of storing all the data. When the memory required to store the data exceeds the storage limit of the computer, we have to discard the redundant data. In fact, in streaming data mining processes, only a small portion of data can be retained. What we need to do is to use limited memory to store the most efficient information while removing the low-value data.

Although many efforts have been made in streaming data mining [7], there are few studies on the problem of streaming data sorting. Because to some extent the streaming data sorting problem can be regarded as a large-scale data sorting problem [8–10]. It is feasible to apply the large-scale data sorting algorithm

© Springer Nature Switzerland AG 2019
Z. Cui et al. (Eds.): IScIDE 2019, LNCS 11936, pp. 406–417, 2019.
https://doi.org/10.1007/978-3-030-36204-1_34

directly to streaming data sorting, but the general large-scale data sorting algorithms consume a lot of memory and time [11–14]. For ever-changing streaming data processing, time and space consumption is immeasurable. Based on the characteristics of streaming data, it is impossible to directly sort it all. Obtaining the approximate solution of a problem is much more efficient than the exact solution [15], so we do not need to obtain the exact sorting results, but minimize the error of the results with limited storage space [16–18].

In this paper, we raise a streaming data sorting problem, and propose a space-efficient dynamic algorithm that can approximate the sorting result. Due to some characteristics of streaming data, we usually only store a small amount of data, and most of the data will be discarded forever, which is why sorting the streaming data is almost impossible [15]. Based on the following two general conclusions of streaming data processing: (1) When processing streaming data, finding an approximate solution is more efficient than an exact solution and (2) A series of Hash-related techniques are considered very useful [15], we build a dynamic model that can efficiently approximate the sorting results. Finally, we combine it with the BFR algorithm to process streaming data that satisfies a normal distribution.

2   A Dynamic Model for Streaming Data Sorting

Before describing the proposed model, we introduce a traditional data sorting algorithm. *QuickSort* is an improved algorithm for *BubbleSort*. The main idea is as follows. Firstly, we need to randomly select a record r_{random} from R. Then, $\forall i = 1, \ldots, n$, if $k_i \leq k_{random}$, r_i is moved before r_{random}, else r_i is moved after r_{random}. The above process is repeated till R is sorted. The time complexity of the *QuickSort* algorithm is $O(nlogn)$ [19,20].

Streaming data is constantly being updated, and using the traditional sorting algorithms requires constantly sorting new datasets. Moreover, different from limited datasets, data streams can not be stored in the memory as a whole and the main part of the data would be discarded over time. So the traditional large-scale data sorting algorithms are not suitable for streaming data processing [8,9]. In fact, we don't need to sort all the streaming data. We only need to calculate the sorting result of the current data in the streaming data that has flowed through. While the memory of a computer is much smaller than the memory required for streaming data, our approach can optimally estimate the sorting result of the current data by setting up a dynamic model and only a small amount of all streaming data needs to be sorted.

In this paper, we denote the streaming data set as S, which is assumed to come in the following form:

$$S = \{s_1, s_2, s_3, \ldots, s_k, s_{k+1}, \ldots\} \tag{1}$$

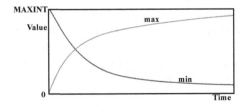

Fig. 1. The variation of *max* and *min* of streaming data over time.

Assume that the first N elements of S are stored in memory as a test dataset and will be used in step 1 and step 2.

$$\tilde{S} = S_{test} = \underbrace{\{s_1, s_2, s_3, \ldots, s_N\}}_{N} \tag{2}$$

Our algorithm is divided into 3 steps, which are detailed below.

2.1 Step 1: Find the Range of Streaming Data

The goal of step 1 is to find the range of the streaming data. It needs to maintain the maximum *max* and minimum *min* of the streaming data S. And they should be determined in advance before sorting. Of course, they will be continuously updated during the process of data streams. But it can be proved that the maximum *max* and the minimum *min* tend to be stable over time, which is one of the important factors to ensure the correctness of our algorithm.

$$max = MAX\{\tilde{S}\} \qquad min = MIN\{\tilde{S}\} \tag{3}$$

When these two values change greatly, it will be considered that the composition and distribution of the streaming data have also changed greatly. At this time, the model can be adaptively corrected to ensure that the error is always minimal.

Algorithm 1. Determination of the range of streaming data.

Input: The partial set of streaming data, $\tilde{S}$; Initial value of the maximum, $max = 0$;
   Initial value of the minimum, $min = MAXINT$;
Output: Maximum value of stream data, max; Minimum value of stream data, min;
1: **for all** S_i such that $S_i \in \tilde{S}$ **do**
2:    compare S_i with max;
3:    compare S_i with min;
4:    if $S_i > max$ then $max = S_i$, if $S_i < min$ then $min = S_i$;
5: **end for**;

Figure 1 shows the variation of the maximum *max* and minimum *min* of streaming data over time. We can see that *max* and *min* of streaming data will rise or fall rapidly in a short time, and finally both tend to be stable.

Algorithm 2. Fitting the distribution of streaming data.

Input: The partial set of streaming data, $\tilde{S}$; The value of the maximum, max; The value of the minimum, min; Number of small intervals, ϕ; Counter set, C;

Output: The value of each counter, C_i, $i \in [0, \phi - 1]$;

1: $SizeofInterval \leftarrow \lceil (max - min)/\phi \rceil$;
2: the i-th interval $\leftarrow [min + SizeofInterval * (i - 1), min + SizeofInterval * i]$;
3: The i-th interval corresponds to the i-th counter C_i;
4: **while** S_i coming **do**
5:     if $S_i \in [min + SizeofInterval * (j - 1), min + SizeofInterval * j]$;
6:        $C_j + +$;
7: **end while**;

2.2   Step 2: Fit the Distribution of Streaming Data

Experiences have shown that specific datasets in our life tend to satisfy certain distributions. For example, human height dataset can fit a normal distribution, while the height of the sea surface is approximately evenly distributed. Determining the distribution of data provides an important basis for further analysis of the data. The distribution of the data is the most important factor in determining the algorithm model in our algorithm. From the results of step 1, we can judge the approximate interval of streaming data. Through these two values: max and min, we can divide more intervals of the same size. For example, we can divide it into ϕ intervals. ϕ is manually specified, the larger ϕ, the more accurately the distribution of streaming data is fitted. And each interval's size of range, denoted as $SizeofInterval$, is calculated as follows.

$$SizeofInterval = \lceil (max - min)/\phi \rceil \tag{4}$$

In order to fit the distribution of the streaming data, we will introduce ϕ counters, denoted as $C_i, i \in [0, \phi - 1]$, corresponding to these ϕ intervals. When the streaming data of a certain interval arrives, the corresponding counter value will be increased by one. In fact, the algorithm in this section is very effective. It just needs a simple hash function to ensure that the values of these counters are accurate, and we can roughly determine the distribution of the streaming data through them. And the value of the i-th counter is as follows:

$$C_i = \sum_{j=0}^{N-1} 1 \quad \text{if} \quad s_j \in [min + SizeofInterval * (i - 1), min + SizeofInterval * i]$$
$$\forall i \in [0, \phi - 1] \tag{5}$$

The main purpose of fitting the distribution of streaming data is to find the best program for memory allocation for sorting. This program can ensure the most efficient use of space with limited memory. Because we believe that the more concentrated range of data needs to be allocated more memory space to ensure sufficient sample size. By enlarging the sample size, the final error will be decreased. This will be shown in next step.

Figure 2 shows how to fit the distribution of an unknown dataset. For example, Fig. 2(a) shows a normal distribution dataset, where the height of the blue box indicates the value of the corresponding counter, and the horizontal diameter of the box indicates its corresponding interval. The value of $C_i(i \in [0, \phi - 1])$ in this step roughly reflects the distribution of streaming data, which will determine the allocation of the memory in next step. In fact, like the values of max and min, the values of $C_i(i \in [0, \phi - 1])$ are constantly changing during the arrival of streaming data. When the distribution of streaming data changes greatly, the $C_i(i \in [0, \phi - 1])$ value will change accordingly, and the model will be dynamically corrected to ensure the accuracy of the results.

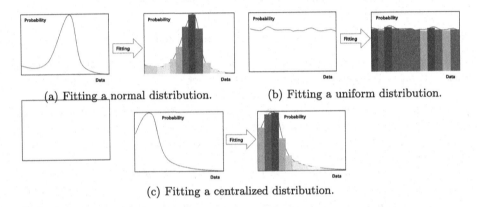

(a) Fitting a normal distribution.   (b) Fitting a uniform distribution.

(c) Fitting a centralized distribution.

Fig. 2. Fitting the distribution of streaming data. (Color figure online)

2.3   Step 3: Estimate the Sorting Result of Current Data

Firstly, similar to the operation in step 2, we divide ϕ intervals for the streaming data. Then we maintain ϕ counters and ϕ buckets for the intervals. These buckets are used to store some of the data flowing through the corresponding intervals. Then we need to determine how to allocate the memory of a computer for each bucket. Our method is to allocate more memory to the buckets corresponding to the denser intervals according to the distribution of the streaming data. According to the results of step 1 and step 2, the computer memory will be prorated. Assume that the maximum number of data that can be stored in the computer is MEM. First, the result $C_i(i \in [0, \phi - 1])$ from step 2 will be normalized to $P_i(i \in [0, \phi - 1])$.

$$P_i = C_i / \sum_{j=0}^{\phi-1} C_j \times 100\% \quad i \in [0, \phi - 1] \tag{6}$$

Then a hash bucket B_i is maintained for the i-th interval, and the bucket's size of B_i is set to $MEM * P_i$.

$$Sizeof B_i : |B_i| = MEM * P_i \quad i \in [0, \phi - 1] \tag{7}$$

As the streaming data arrives, the counter value of its corresponding interval is first increased by one. At the same time, if the corresponding hash bucket is not full, the data will be placed in it, and if it is full, a figure in the bucket will be randomly replaced and discarded forever. Assume that the current data is hashed to bucket B_j.

$$B_j = \underbrace{\{b_1, b_2, b_3, ..., b_i \circledast, b_n\}}_{P_i * MEM} \quad b \in S \quad (\circledast \quad current) \tag{8}$$

We will sort the data in bucket B_j as follows.

$$B_{sort} = \underbrace{\{b_{q_1}, b_{q_2}, b_{q_3}, ..., b_{q_k} \circledast, ...b_{q_n}\}}_{P_i * MEM} \quad (\circledast \quad current) \tag{9}$$

Assume that $SortResultinB_j$ is the sorting result of the current data in bucket B_j.

$$SortResultinB_j = q_k \tag{10}$$

It can be considered that if the current data is sorted as the $SortResultinB_j$-th in the bucket B_j, then it ranks at approximately $\frac{SortResultinB_j}{|B_j|=MEM*P_j} \times 100\%$ in all the data flowing through its interval.

$$SortResult = C_j \times \frac{SortResultinB_j}{MEM \times P_j} + \sum_{k=0}^{j-1} C_k \tag{11}$$

The estimated value $SortResult$ of the current data sorting result in all the data is its estimated sorting result in the j-th interval plus the sum of counter values of all the previous $j - 1$ intervals.

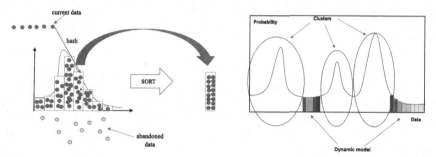

(a) Dynamic model for streaming data sorting (b) Combine BFR with dynamic model

Fig. 3. Streaming data processing model. (Color figure online)

Figure 3(a) shows the process of step 3. The red dots represent the streaming data, and the blue dot represents the current incoming data. And the yellow dots below the coordinate axis indicate the discarded data. This figure reveals the core model of our algorithm.

Algorithm 3. Determine the sorting result of the current data.

Input: The set of streaming data, S; The maximum of data stored in the computer, MEM; Number of intervals, ϕ; Hash buckets, $B_i, i \in [0, \phi - 1]$; Counter set, $C = \{C_1, C_2, ..., C_{\phi-1}\}$; Normalized result of $C_i(i \in [0, \phi - 1])$, $P_i(i \in [0, \phi - 1])$;

Output: The sorting result of the current data, $SortResult$;

1: $P_i \leftarrow C_i / \sum_{i=0}^{\phi-1} C_i$;
2: size of i-th bucket: $|B_i| = MEM * P_i$;
3: **while** S_i coming **do**
4:    **if** $S_i \in [min + SizeofInterval * (j - 1), min + SizeofInterval * j]$;
5:    $C_j + +$;
6:    $SortResultinB_j t \leftarrow quicksort(B_j)$;
7:    $SortResult = C_j * SortResultinB_j/(MEM * P_j) + \sum_{k=0}^{j-1} C_k$;
8: **end while;**

---

3   The Dynamic Model + BFR Algorithm

BFR algorithm is a scaling clustering algorithm for large-scale datasets in European space. The BFR algorithm assumes that the shape of the clusters satisfies the normal distribution with the centroid as expected value [15, 21]. As we mentioned earlier, many datasets satisfy a normal distribution. In this section, we consider combining the BFR algorithm with our proposed dynamic model to sort the streaming data that satisfies the BFR condition.

3.1   The BFR Algorithm

When clustering streaming data using the BFR algorithm, the data is read from the file into memory in chunks. After processing a chunk, the new chunk is read in to overwrite the previous data. This is very similar to the processing of streaming data, so we consider clustering the streaming data using the BFR algorithm.

BFR divides the data into three categories: *DiscardSet*, *CompressedSet* and *RetainedSet*. And the core of the BFR algorithm is to use the *N-SUM-SUMSQ* method to represent *DiscardSet* and *CompressedSet*. We only need to use $2 * d + 1$ values to represent a cluster of d-dimensional data through this method [15, 21]. So using the BFR algorithm to process streaming data saves a lot of memory, which will be allocated to some hash buckets to handle the sorting problem of *RetainedSet*.

3.2   Combine BFR Algorithm with the Dynamic Model

In order to estimate the sorting result of the current data, we first cluster the streaming data using the BFR algorithm. For each *DiscardSet* or *CompressedSet*, we only need to maintain a counter and *N-SUM-SUMSQ*. And for the *RetainedSet*, like Sect. 3, we also need to maintain some hash buckets and corresponding counters.

Suppose we get m clusters($Cluster_i, i \in [0, m-1]$) after clustering the streaming data like Fig. 3(b), and they are represented by $N_i\text{-}SUM_i\text{-}SUMSQ_i (i \in [0, m-1])$. The counters corresponding to these clusters are $C_i, i \in [0, m-1]$ and the corresponding intervals are $[a_i, b_i]$. Assume that the centroids and variances of these clusters are $Mean_i, (i \in [0, m-1])$ and $Var_i(i \in [0, m-1])$, respectively.

$$Mean_i = SUM_i/N_i \tag{12}$$

$$Var_i = SUMSQ_i/N_i - (SUM_i/N_i)^2 \tag{13}$$

Calculating the mean and variance of each cluster does not require any data in the cluster. It just need the $N\text{-}SUM\text{-}SUMSQ$ value of the cluster. And the data in $Cluster_i$ satisfies the normal distribution with a mean of $Mean_i$ and variance of Var_i [21]. So $\frac{s-Mean_i}{\sqrt{Var_i}}$ satisfies the standard normal distribution.

$$\frac{s - Mean_i}{\sqrt{Var_i}} = \frac{s - SUM_i/N_i}{\sqrt{SUMSQ_i/N_i - (SUM_i/N_i)^2}} \sim N(0,1) \quad (s \in Cluster_i) \tag{14}$$

If the current data s is assigned to $Cluster_i$, then we assume that its sort result in the cluster is $SortResultinClu_i$:

$$SortResultinClu_i = [\Phi(\frac{s - Mean_i}{\sqrt{SUMSQ_i/N_i - (SUM_i/N_i)^2}}) - \Phi(a_i)] \times C_i \tag{15}$$

where $\Phi(double)$ is the cumulative distribution function(CDF) of the standard normal distribution.

After the value of $SortResultinClu_i$ is obtained, it is combined with the algorithm in Sect. 3. And the sort result $SortResult$ of the current data is $SortResultinClu_i$ plus the sum of the counters' values before it.

$$SortResult = SortResultinClu_i + \sum_{k=0}^{i-1} C_k \tag{16}$$

$$= [\Phi(\frac{s - Mean_i}{\sqrt{SUMSQ_i/N_i - (SUM_i/N_i)^2}}) - \Phi(a_i)] \times C_i + \sum_{k=0}^{i-1} C_k$$

If the current data s is not assigned to any cluster, it will be assigned to $RetainedSet$. For the data in $RetainedSet$, we can directly use the algorithm in Sect. 3 to estimate its sorting result.

4 Experimental Study

In this section, in order to measure the error of our algorithm, we conduct some experiments by using some large datasets like $\mathbb{D}$(with size of N) to simulate the arrival of streaming data. Sorting these datasets can get accurate sorting results of the current data, $TrueSortResult$. And the estimated value($EstimateSortResult$) is obtained by our algorithm in this paper.

Assume that the number of the streaming data that has flowed through is $NumofAllData = k$, so the current data is d_k.

Sort first k elements of dataset $\mathbb{D}$.

$$\mathbb{D} = \{\underbrace{d_1, d_2, d_3, ..., d_k \circledast}_{NumofAllData=k}, d_{k+1}..., d_n\} \quad (\circledast \quad current) \tag{17}$$
$$\underbrace{\hspace{7cm}}_{N}$$

$$\mathbb{D}_{sort} = \{\underbrace{d_{q_1}, d_{q_2}, ..., d_{q_j} \circledast, ..., d_{q_k}}_{NumofAllData=k}\} \quad (TrueSortResult = q_j \circledast) \tag{18}$$

The definition $ERROR$ is the absolute value of the true sorting result $TrueSortResult$ minus the estimated value $EstimateSortResult$ divided by the total number $NumofAllData$.

Definition 1. $ERROR = \frac{|TrueSortResult - EstimateSortResult|}{NumofAllData} \times 100\%$.

4.1   Experiments Based on the Dynamic Model

This experiment is based on three different datasets: a normally distribution dataset $\mathbb{A}$, an evenly distributed dataset $\mathbb{B}$ and a partially distributed dataset $\mathbb{C}$. Where $\mathbb{A}$ satisfies a normal distribution with a mean of 10000 and a variance of 3000, and the elements in $\mathbb{B}$ is a random number between $[0, 100000000]$. And the records in $\mathbb{C}$ are mainly located in a small part of $[0, 100000000]$. The algorithm is executed on each of these three datasets, and the partial sorting results are shown as follows: Tables 1, 2 and 3. In these tables, $DATA$ represents randomly selected data, $TRUE$ represents its true ordering in the dataset, $ESTIMATE$ represents the approximate sorting result obtained by our algorithm, and $ERROR$ represents the deviation of the estimated sorting results from the true results.

Table 1 shows the results on an evenly distributed dataset. Table 2 shows the results on a normally distributed dataset. And Table 3 shows the results on a partially centralized distributed dataset. In general, Fig. 4(a) shows that the application of our algorithm on an evenly distributed dataset is better than the effect on the other two datasets. When applying this algorithm on a partially centralized dataset, although the error is kept low at most time, occasionally the error is unstable. And the error of the other two datasets is always maintained in a stable state.

4.2   Experiments Based on Dynamic Model+BFR Algorithm

This experiment used a dataset $\mathbb{D}$ of size $2,000,000$ to simulate the streaming data. It is a combination of two normally distributed datasets $(D_1 \ D_2)$. And $D_1 \sim N(300000, 3000)$, $D_2 \sim N(700000, 3000)$.

Table 4 shows the partial sorting results and errors. Figure 4(b) shows the error comparison between the Dynamic Model and the Dynamic Model+BFR

Table 1. Evenly distributed dataset $\mathbb{B}$.

DATA	49392949	37836319	64432818	62741804	17472561	75722528	33344046
TRUE	64128	144158	617045	539170	25078	519591	195403
ESTIMATE	64130	144171	616999	539169	25079	519682	195310
ERROR	$1.57867e^{-5}$	$3.50062e^{-5}$	$4.85804e^{-5}$	$1.17842e^{-6}$	$7.17458e^{-6}$	$1.33479e^{-4}$	$1.62529e^{-4}$
DATA	2400785	70787305	76100087	44446815	86875690	94612071	50652163
TRUE	22738	550130	375625	199343	659034	585848	97412
ESTIMATE	22608	550246	375703	199294	659062	585889	97397
ERROR	$1.4059e^{-4}$	$1.5054e^{-4}$	1.5895^{-4}	$1.11875e^{-4}$	$3.70379e^{-5}$	$6.63116e^{-5}$	$7.99305e^{-5}$

Table 2. Normally distributed dataset $\mathbb{A}$.

DATA	10225	9681	9421	6926	11647	13177	11998
TRUE	153515	425116	293166	97356	529115	203727	660730
ESTIMATE	152518	424387	293370	91209	525012	203524	641473
ERROR	$3.44526e^{-3}$	$7.83125e^{-4}$	$2.94467e^{-4}$	$9.65121e^{-3}$	$5.48681e^{-3}$	$8.51742e^{-4}$	$2.17498e^{-2}$
DATA	11054	8677	10137	5875	8707	8266	8334
TRUE	312264	121312	269473	75651	171292	50740	110835
ESTIMATE	312688	121497	259917	72266	178803	49406	116126
ERROR	$8.65258e^{-4}$	$5.01777e^{-4}$	$1.83748e^{-2}$	$3.77048e^{-3}$	$1.46149e^{-2}$	$7.38894e^{-3}$	$1.37993e^{-2}$

Table 3. Partially centralized distributed dataset $\mathbb{C}$.

DATA	53939391	5784	9762255	66352221	42408334	78455775	349
TRUE	224191	269153	381056	531657	535791	213142	15421
ESTIMATE	224193	266994	381113	531724	535761	213145	14753
ERROR	$6.91126e^{-6}$	$2.3193e^{-3}$	$8.22776e^{-5}$	$1.05195e^{-4}$	$4.01181e^{-5}$	$1.25873e^{-5}$	$7.54473e^{-4}$
DATA	4812876	55127985	8046	8637	4422	9965	7465
TRUE	272870	700488	206941	78091	84633	44434	170304
ESTIMATE	272868	700474	206043	76765	85384	44613	166600
ERROR	$3.84572e^{-6}$	$1.55943e^{-5}$	$1.74733e^{-3}$	$7.34463e^{-3}$	$1.95866e^{-3}$	$2.00736e^{-3}$	$8.12751e^{-3}$

Table 4. Experiments based on Dynamic Model+BFR.

DATA	704149	199681	699777	700033	701243	699872	197206
TRUE	109873	591826	117243	270842	345715	48659	1033033
ESTIMATE	109515	686857	117427	270928	345630	48344	1365000
ERROR	$1.23711e^{-3}$	$1.02087e^{-1}$	$1.08697e^{-4}$	$1.35026e^{-4}$	$4.86328e^{-5}$	$1.32167e^{-3}$	$1.76074e^{-1}$
DATA	201482	703923	204977	703616	704528	201022	202814
TRUE	1081038	54534	1081374	79210	327038	958848	1039440
ESTIMATE	921814	54475	1032068	78982	326869	1116592	902149
ERROR	$1.32426e^{-1}$	$1.20402e^{-4}$	$3.60242e^{-2}$	$4.38412e^{-4}$	$1.88246e^{-4}$	$1.04195e^{-1}$	$116295e^{-1}$

Algorithm. From Fig. 4, we can conclude that the error caused by the Dynamic Model+BFR algorithm is larger than our algorithm. However, compared with the Dynamic Model, the Dynamic Model+BFR algorithm saves much more memory space, and the calculation method of it is simpler. Even without sorting any data, the Dynamic Model+BFR algorithm can estimate the sorting result of the current data. And these advantages are lacking in the original dynamic algorithm.

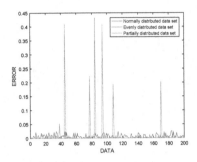
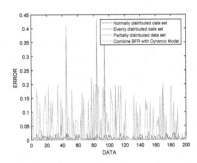

(a) Error comparison based on three different data sets.

(b) Error comparison between the Dynamic Model and the Dynamic Model+BFR.

Fig. 4. Error comparison.

5   Conclusion and Future Work

In this paper, we propose the Dynamic Model and combine it with the BFR algorithm to solve the streaming data sorting problem accordingly. Compared with some large-scale data sorting algorithms that can only sort the fixed datasets, our algorithm can process streaming data in real time, and can still estimate the sorting result of current data through existing information when a large amount of data is forced to be discarded. And we also allocate memory space efficiently by the method of fitting data distribution. By using the N-SUM-$SUMSQ$ method of the BFR algorithm, the Dynamic Model+BFR algorithm saves a lot of memory. Finally, we used a normal distributed dataset and an evenly distributed dataset to test the error range of our algorithms. The experimental results show that it is a low-error streaming data sorting algorithm which can efficiently utilize the memory. Additionally, we have shown that our algorithm is stable and effective in processing some types of streaming data (e.g. evenly distributed). But it still shows instability on some datasets (e.g. partially centralized distributed), and our future work is to find ways to reduce this instability.

Acknowledgment. This project was supported by NSFC (61876193) and Guangdong Natural Science Funds for Distinguished Young Scholar (2016A030306014).

References

1. Wang, C.D., Lai, J.H., Huang, D.: Incremental support vector clustering. In: ICDM Workshop, pp. 839–846 (2011)
2. Huang, D., Lai, J.H., Wang, C.D.: Incremental support vector clustering with outlier detection. In: ICPR, pp. 2339–2342 (2012)
3. Wang, C.D., Lai, J.H., Yu, P.S.: Dynamic community detection in weighted graph streams. In: SDM, pp. 151–161 (2013)

4. Ding, Y., Huang, L., Wang, C.D., Huang, D.: Community detection in graph streams by pruning zombie nodes. In: PAKDD, pp. 574–585 (2017)
5. Liang, W.B., Wang, C.D., Lai, J.H.: Weighted numerical and categorical attribute clustering in data streams. In: IJCNN, pp. 3066–3072 (2017)
6. Chen, J., Lin, X., Xuan, Q., Xiang, Y.: FGCH: a fast and grid based clustering algorithm for hybrid data stream. Appl. Intell. **49**(4), 1228–1244 (2019)
7. Wang, C.D., Lai, J.H., Huang, D., Zheng, W.S.: SVStream: a support vector-based algorithm for clustering data streams. IEEE Trans. Knowl. Data Eng. **25**(6), 1410–1424 (2013)
8. Laga, A., Boukhobza, J., Singhoff, F., Koskas, M.: MONTRES: merge on-the-run external sorting algorithm for large data volumes on SSD based storage systems. IEEE Trans. Comput. **66**(10), 1689–1702 (2017)
9. Shabaz, M., Kumar, A.: SA sorting: a novel sorting technique for large-scale data. J. Comput. Netw. Commun. **2019**, 3027578:1–3027578:7 (2019)
10. Sukhwani, B., et al.: Large payload streaming database sort and projection on FPGAs. In: SBAC-PAD, pp. 25–32 (2013)
11. Aggarwal, C.C., Han, J., Wang, J., Yu, P.S.: A framework for clustering evolving data streams. In: VLDB, pp. 81–92 (2003)
12. Al-Shammari, A., Zhou, R., Naseriparsa, M., Liu, C.: An effective density-based clustering and dynamic maintenance framework for evolving medical data streams. Int. J. Med. Inform. **126**, 176–186 (2019)
13. Vial, J.J.B., Devanny, W.E., Eppstein, D., Goodrich, M.T., Johnson, T.: Quadratic time algorithms appear to be optimal for sorting evolving data. In: Proceedings of the Twentieth Workshop on Algorithm Engineering and Experiments, ALENEX 2018, New Orleans, LA, USA, 7–8 January 2018, pp. 87–96 (2018)
14. Domino, K., Gawron, P.: An algorithm for arbitrary-order cumulant tensor calculation in a sliding window of data streams. Appl. Math. Comput. Sci. **29**(1), 195–206 (2019)
15. Leskovec, J., Rajaraman, A., Ullman, J.D.: Mining of Massive Datasets, 2nd edn. Cambridge University Press, Cambridge (2014)
16. Fernandez-Basso, C., Francisco-Agra, A.J., Martín-Bautista, M.J., Ruiz, M.D.: Finding tendencies in streaming data using big data frequent itemset mining. Knowl.-Based Syst. **163**, 666–674 (2019)
17. Liang, C., Li, M., Liu, B.: Online computing quantile summaries over uncertain data streams. IEEE Access **7**, 10916–10926 (2019)
18. Vial, J.J.B., Devanny, W.E., Eppstein, D., Goodrich, M.T., Johnson, T.: Optimally sorting evolving data. In: 45th International Colloquium on Automata, Languages, and Programming, ICALP 2018, 9–13 July 2018, Prague, Czech Republic, pp. 81:1–81:13 (2018)
19. Cormen, T.H., Leiserson, C.E., Rivest, R.L., Stein, C.: Introduction to Algorithms, 2nd edn. The MIT Press and McGraw-Hill Book Company, Cambridge and New York (2001)
20. Goodrich, M.T., Tamassia, R.: Algorithm Design and Applications, 1st edn. Wiley, Hoboken (2014)
21. Bradley, P.S., Fayyad, U.M., Reina, C.: Scaling clustering algorithms to large databases. In: Proceedings of the Fourth International Conference on Knowledge Discovery and Data Mining (KDD-98), New York City, New York, USA, 27–31 August 1998, pp. 9–15 (1998)

Smartphone Behavior Based Electronical Scale Validity Assessment Framework

Minqiang Yang[1], Jingsheng Tang[2], Longzhe Tang[2], and Bin Hu[1,3,4(✉)]

[1] Gansu Provincial Key Laboratory of Wearable Computing,
School of Information Science and Engineering, Lanzhou University,
Lanzhou, China
bh@lzu.edu.cn
[2] School of Life Sciences, Lanzhou University, Lanzhou, China
[3] CAS Center for Excellence in Brain Science and Intelligence Technology,
Shanghai Institutes for Biological Sciences, Chinese Academy of Sciences,
Beijing, China
[4] Beijing Institute for Brain Disorders, Capital Medical University,
Beijing, China

Abstract. In the study, we developed a smartphone-based electronical scale validity assessment framework. 374 college students are recruited to fill in Beck Depression Inventory. A total of 544 filling of scales are collected, which may be filled accordingly or concealed. Via an electronical scale based WeChat applet and backend application, temporal and spatial behavioral data of subjects during the scale-filling process are collected. We established an assessment model of the validity of the scale-filling based on the behavior data with machine learning approaches. The result shows that smartphone behavior has significant features in the dimension of time and space under different motivations. The framework achieves an valuable assessment of the effectiveness of the scale, whose key indicators such as accuracy, sensitivity and precision are over 80% under multiple dimension behavior data classification. The framework has a good application prospect in the field of psychological screening.

Keywords: Validity assessment · Smartphone · WeChat applet · Behavior data

1 Introduction

As a main-stream method for the assessment of psychological disorder or psychiatry, numbers of scales are developed. Subjects or psychologists used to fill traditional paper-and-pencil questionnaires, but past approach has been enormously time consuming so that it cannot be deployed extensively. More and more people choose scales over Internet and its effectiveness was proven [1]. As the widespread application of Mobile Internet and high holding rate of smartphone, personal computers are replaced by smartphones in most cases, as well as collecting data with scales [2, 3]. Nowadays smartphones have powerful sensing and computational capabilities, they have the potential to passively collect social and behavioral data [4]. Those behavior metrics have associations with subjects' mental condition [5] and social anxiety [6].

© Springer Nature Switzerland AG 2019
Z. Cui et al. (Eds.): IScIDE 2019, LNCS 11936, pp. 418–429, 2019.
https://doi.org/10.1007/978-3-030-36204-1_35

In the present study, we use WeChat applet to collect subjects' data, like scale options, position of tablet click and time for each question. A novelty scale collection framework is designed and implemented which evaluates the validity of the scale filling based on machine learning methods with temporal and spatial features. The result shows that the presented framework has significant assessment capability for validity of scale-filling.

The contribution of this research is threefold.

(1) We design and implement a scale data collection framework with validity assessment, which are used to collect the data for the study, already adopted as the official psychological test tools of our University.
(2) We propose a validity assessment model based the behavior data collected by the WeChat applet during the process of filling out the scales.
(3) To address the activity pattern that a subject does the test truly or mendaciously, temporal and spatial features are extracted from the original behavior data.

Section 2 introduces relevant prior research in electronical version of scales and pattern discovery. In Sect. 3, we introduced the experimental architecture. Section 4 presents the specific methods and results of electronical scale validity assessment. The results were analyzed in Sect. 5, and the shortcomings of this study were summarized.

2   Related Work

The electronic versions of many scales were well accepted and slightly quicker to complete than paper version. They are equivalent in performance and more effective than the paper version [7–10]. Schwarzer's study focused on a comparison of data collected on the Internet with data collected in the traditional paper-and-pencil manner. It reveals that the psychometric properties investigated in this study were satisfactory. It was suggested that innovative methods of data collection should be considered when developing a psychometric scale [1].

There are two main dimensions for the scale assessment, reliability and validity. Validity is the degree to which measurement tools or methods can accurately measure the things that need to be measured. While the reliability concerns the extent to which an experiment, test or any measuring procedure yields the same results on repeated trials [11, 12].

The electronic versions of scales has been tested for reliability and validity [10], but in practice many factors may affect the validity. A validity assessment method is required to test the validity of an electronical scale to improve the validity of the electronical scale.

A lot of papers were investigated on pattern discovery with statistic and machine learning methods. Some studies collect user's behavior data with the powerful sensors of smartphones. It is proved that those kind of behavior data are relevant to one's mental health and social condition [3–6]. The level of social anxiety in individuals can be manifested through athletic behaviors, including physical activity, mobility patterns, and the use of Smartphone Sensing Methods (SSMs) in behavioral science for research is a promising field [4, 6].

Smartphones and mobile Internet facilitate the long-term collection of user behavior data in multiple dimensions [13, 14], and the above-mentioned research also shows that the user's behavior on the mobile side is also a degree of user's inner portrayal. But there is no relevant work for the validity assessment of scales based on mobile terminal behavior and machine learning.

3 Methodology

We ported the traditional paper-and-pencil scales to the mobile application. A total of 374 volunteers participated in the study to fill in the scale, we used mobile terminals and applications to collect the relevant behavior data of subjects and uploaded it to the server's database during subjects fill in the scale. These data are used to train the machine learning classification model to assess the validity of the scale-filling.

3.1 Framework Design

The basic framework of this study is designed as follows. The users fill in the scale by using the WeChat applet, and the relevant data is submitted by the WeChat applet to the Django server, and the data is stored in MySQL database.

To ensure that subjects are corresponding to their behavior data, subjects who use the WeChat applet for the first time is required to authenticate their personal information. Subjects need to fill in their own name, gender, student number and short message verification.

We construct binary classification problem with truthfully completed scales (TCS) and deceivingly completed scales (DCS). And we extract spatial and temporal features from the raw behavior data and train the classification model by k-Nearest Neighbor (kNN) algorithm and Support Vector Machine (SVM) algorithm, the model will be used for the evaluation of the validity of scale-filling (Fig. 1).

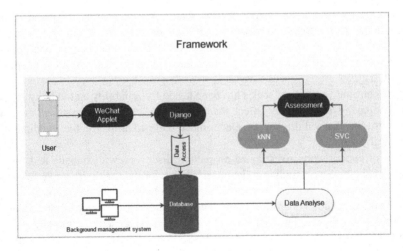

Fig. 1. Framework design

3.2   WeChat Application

WeChat applet is a new way to connect users and services. The WeChat applet in this study was developed by the team members to implement the migration of traditional paper scales to electronical scales. After completing one question, it will jump to the next question. This method of filling allows the program to record some behavioral data during the user's filling process for data analysis in the study.

Besides the options of the questions, the answer time and the coordinate of the screen click are also recorded. These functions are implemented by the following code:

Algorithm 1. Timer implemented algorithm

```
1    Start() {
2        // Timer initialization
3          this.data.d1 = new Date() },
4    // Click on the option to trigger time recording
5    radioChange(e) {
6        this.data.d2 = new Date()
7        this.data.duration = this.data.d2 - this.data.d1
8        this.data.time = this.data.time.concat(this.data.duration)
9        this.data.d1 = new Date() }
```

In Algorithm 1, the time interval between two clicks on the screen is recorded as the subjects' answer time.

Algorithm 2. Record the coordinates of each click on the screen

```
1    // coordinate data of the finger click
2    {type: "tap", timeStamp: 3222, target: {...}, currentTarget: {...}, detail: {...}, ...}
3        detail:{x: 143, y: 176}
4    getCoordinate(e){this.data.coordinate = this.data.coordinate.concat(e.detail)}
```

In Algorithm 2, we used the coordinate data recorded in the form click callback event as the behavior data of the screen click (Fig. 2).

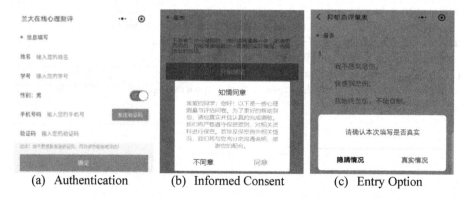

  (a) Authentication   (b) Informed Consent   (c) Entry Option

Fig. 2. Applet interface overview

3.3 Backend

The backend of this applet was developed using Django 2.0 framework. Django can quickly develop web applications with secure and powerful scalability. Nginx and uwsgi are chosen to set up the web server. Scale answers and behavior data such as time and click position are encapsulated and transferred to server via RESTful API.

We integrated data preprocessing, feature extraction and the classification algorithm into the backend application, and use the model we trained to evaluate the validity of each user's scale-filling.

3.4 Inventory

This study used Chinese version of the second edition of the Beck Depression Inventory (BDI-II-C). The Beck Depression Inventory was developed to quantitatively assess the intensity of depression and has considerable reliability and validity. The scale can effectively discriminate the different degrees of depression in patients, and can also reflect changes in the intensity of depression after a period of time [15].

4   Validity Assessment

4.1 Experiment

We distributed leaflets at Lanzhou University campus and recruited a total of 374 participants, all of whom are undergraduate or graduate students at Lanzhou University. According to the requirements, 374 people filled in the scales. Including 164 males and 210 females. The average age of participants was 20.16 years old, the minimum age was 17, and the maximum age was 24.

We design two entry points of the scale-filling, "careful filling" and "concealed filling". Some of subjects filled in the scale carefully as they would like, those read the questions carefully, fill in the scale according to their actual situation. Another group of volunteers choose "concealed filling" to finish some false scale, which was told to fill in the purpose of concealing their actual situation. This is to simulate the way in which the intention to hide their true psychological state is deliberately intended. The TCS are valid scales. And the DCS are used to simulate the invalid scales in actual scene, they will lead to reduced validity.

After each subject completes the scale, we will check the data from questionnaires that are filled truthfully, and if it is not to meet the requirements, the data will be deleted. The criteria are as follows: (1) Not filled in as required under our supervision. (2) The average filling time for each question is less than 500 ms. (3) Submit the results multiple times by the same person. After data preprocessing, we got 270 effective TCS and 274 effective DCS.

4.2   Feature Extraction and Classification

The raw behavior data we collected includes the time, the option and the click coordinate for each question. With those data, we extract corresponding three dimension features for the classification algorithms.

Temporal Dimension Feature

The original raw data is the time in milliseconds for the user to answer each question. We calculated the mean and variance of the answer time for all questions in each questionnaire:

$$\bar{t} = \frac{\sum_{i=1}^{n} t_i}{n} \tag{1}$$

$$S^2 = \frac{\sum (t - \bar{t})^2}{n - 1} \tag{2}$$

The answer time of each question and the mean value $\bar{t}$ and the variance S^2 of the questionnaire answer time are used as feature values, and the data filled in according to the real situation and the data filled in the deceived condition are respectively given to the tags 1 and 0 for training.

Option Dimension Feature

The option result of the scale is represented by four scores of 0 to 3. We define a value P_O to represent the probability of occurrence of each option in a scale. The calculation formula is as follows:

$$P_O = \frac{n(O)}{N} \tag{3}$$

N is the total number of questions in the scale, and O stands for different options, which is divided into 0, 1, 2, and 3 types. $n(O)$ represents the number of occurrences of an option in a questionnaire. We calculate the probability P_O of each option, and sort them from large to small as the features.

Spatial Dimension Feature

The data of the click coordinates is divided into x-axis data and y-axis data. Calculate the variance of the x-axis and y-axis of all click coordinates. Then extract the most selected option and calculate the mean and variance of the x-axis and y-axis of all clicks for that option. The above results are used as features to perform machine learning classification training. At the same time, calculate the center of gravity of all the click coordinates of each questionnaire, establish the polar coordinate system with the center of gravity as the origin, calculate the polar coordinates of each click position, and analyze the distribution law.

Classification

The data analysis in this paper uses two machine learning classification methods, SVM and kNN, which are suitable for small-scale data classification with labeled data.

SVM. Support vector machines (SVMs) are a set of supervised learning methods used for classification, regression and outlier detection.

In general, SVMs generate a mapping function from a set of labeled training data. They can be used for nonlinear classification, where kernel functions are used to map the input space into higher-dimensional feature spaces. This effectively maps the non-linearity of the relationship to a linear one. In this higher-dimensional space, a separating hyper plane is constructed to classify the data set [16].

kNN. The K-Nearest Neighbor is widely used to solve pattern recognition or classification problems [17, 18]. kNN is the most popular non-parametric classifier because the kNN asymptotic or infinite sample size error is less than twice the Bayesian error [19, 20]. In short, kNN finds the k instances closest to the input instance in the training set. According to the principle that the minority obeys the majority, if most of the k instances belong to a certain class, the input instance is divided into the class.

In practical applications, if the k value is too small and the model complexity is high, it is easy to learn the noise and thus determine the noise category. However, if the value of k is too large, it is equivalent to use the training data in the larger neighborhood to make prediction. At this time, the dissimilar training examples will cause interference and make prediction errors. We usually measure the distance between two points in high dimensional space by Euclidean distance.

4.3   Result

We used two machine learning classification methods, kNN and SVM, to train and calculate the classification performance metrics for comparison. We tried the three dimension respectively and also tried classification with all the three dimension features simultaneously. The decision tree algorithm is used for feature selection, and the score rank of the features are obtained. In this section, we present two parts of classifier performance evaluation and behavioral data analysis.

For kNN, the k value is given a range of 1 to 31, cross-validated with different k values, and the correct rate curve is drawn. The k value with the best fit is selected as the model parameter, the accuracy of the model is cross-validated, and the curve is drawn to determine the optimal parameters of the model. As to SVM, we applied linear kernel, polynomial and Gaussian kernel with stepper parameter iterator for C and gamma. The Gaussian kernel got the highest accuracy. The model correct rate curve and ROC curve are drawn to evaluate the performance of the classifier.

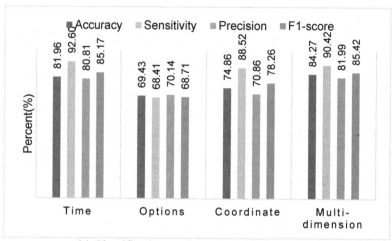

(a) Classification performance metrics of kNN

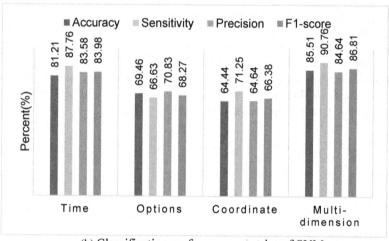

(b) Classification performance metrics of SVM

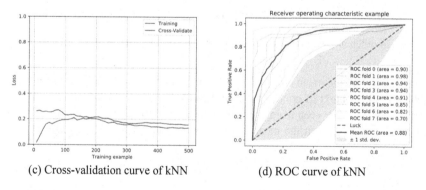

(c) Cross-validation curve of kNN   (d) ROC curve of kNN

Fig. 3. Machine learning model classification performance metrics comparison.

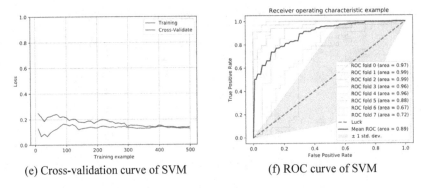

(e) Cross-validation curve of SVM   (f) ROC curve of SVM

Fig. 3. (*continued*)

Figure 3(a), (b) indicate the classification performance of kNN and SVM in different dimensions. It can be seen that the model classification performance in multiple dimensions is better than the performance of other single dimensions. However, the sensitivity metrics of time in Fig. 3(a) is higher than multi-dimension, which is the TPR (True Positive Rate). This model can be used for early large-scale screening, reducing the rate of missed diagnosis and increasing the validity of electronical scales.

The cross-validation curve and ROC (Receiver Operating Characteristic Curve) curve of kNN and SVM are shown in Fig. 3(c), (d), (e), (f), they indicate that the model fits well and has a fairly good classification effect.

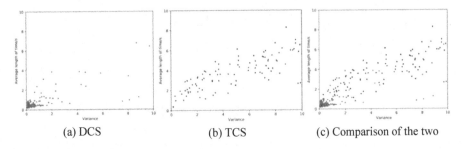

(a) DCS   (b) TCS   (c) Comparison of the two

Fig. 4. In different cases, the comparison of the mean and variance of the answer time.

Figure 4 shows the mean and variance distribution of the answer time. As shown in Fig. 4(a), in the context of concealing, the mean and variance of the answer time are generally low. It can be seen from Fig. 4(b) that under normal circumstances, there is a linear relationship between the mean and variance of the subject's response time.

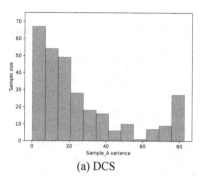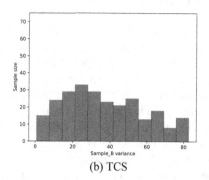

(a) DCS   (b) TCS

Fig. 5. The horizontal axis of the two histograms is the variance of the frequency of occurrences of the four options in each questionnaire, and the vertical axis is the number of samples. The red histogram represents concealed filling (a) and the blue histogram is truthful filling (b). (Color figure online)

As can be seen from Fig. 5, there is a significant difference in the distribution of the options results of the careful filling and concealed filling. Most of the concealed samples distributed at the left end or right end of the histogram, that is, most of the samples have small variance $var(P_O)$. This shows that in the concealed scene, the answers were scattered evenly in four options. In the careful filling samples, the distribution of the questionnaire answers looks like a normal distribution.

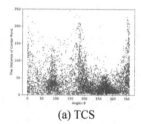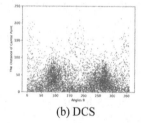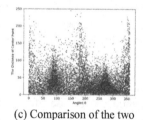

(a) TCS   (b) DCS   (c) Comparison of the two

Fig. 6. Each point represents the polar coordinates of click position relative to the center of gravity of all the click points for one scale. The horizontal axis represents the magnitude of the polar angle, and the vertical axis represents the distance of the polar axis of the point. The blue scatter plot represents the click coordinate distribution that is truthful filling (a). The red scatter plot represents the click coordinate distribution that is concealed filling (b), and (c) is the overlay comparison of the two diagrams. (Color figure online)

In Fig. 6, the distribution of scatters presents four peaks, which correspond to four different options in the scale. It can be seen that in the condition of truthful filling, the average click position is closer to an answer option. In the deceived filling, the average click position have more discrete distribution.

5 Discussion

Our results show that the TCS and the DCS are very different. From the temporal dimension, if the subject tends to conceal his or her own situation, the click time of each question will be generally longer or shorter than TCS. There may be two reasons. One is that the subject clicks on the answer without reading seriously. The other is that the subject tries to hide his or her real situation and deliberately choose an answer that is different from his or her real situation. From the perspective of the option distribution dimension, the answers by the deceived subjects are more evenly distributed among the four options, but in the TCS, the answers distributed in a normal distribution. This shows that in the DCS, the probability of occurrence of the four answers are almost the same, and for the TCS, due to the psychological state difference of the subjects, answers are likely different from each other. In the spatial dimension, the average coordinate distribution of DCS is more discrete than TCS, and the average click coordinates of TCS will be partially concentrated near an option. The reason may be that if one fills in the scale truthfully, he or she will click the option precisely, if one conceal his or her thought or does not fill in seriously, the click position may be far away from the center of the option.

There are some shortcomings in our experimental design and data, which should be solved in the future. Firstly, more samples are required for training to enhance the stability of classification model. Secondly, some subjects filled the scale online without supervision. Therefore, some data collected does not meet the requirements. In data analysis, these noises may result in a decrease in the accuracy rate of the model. Furthermore, we try to simulate the scales-filling scene to approach real situation, but there are cases that can't be simulated. For example, in different test environments, differences in people's mental state may have impacts.

We developed a new electronical scale validity assessment framework by collecting user behavior data and evaluating validity. We use machine learning to train differences in behavioral characteristics, and obtain a validity evaluation model with high accuracy rate. This work applies the psychological scale to the smartphone, extracts behavior data, and combines with machine learning to evaluate the validity of the scale. This method can be used for psychologists to quickly screen out effective scales. In the future, we will expand the scope of the subjects' occupation and increase the number of samples. We will collect more behavioral data, such as facial expressions for analysis.

Acknowledgment. This work was supported in part by the National Natural Science Foundation of China [Grant No. 61632014, No. 61627808, No. 61210010], in part by the National Basic Research Program of China (973 Program) under Grant 2014CB744600, in part by the Program of Beijing Municipal Science & Technology Commission under Grant Z171100000117005.

References

1. Schwarzer, R., Mueller, J., Greenglass, E.: Assessment of perceived general selfefficacy on the internet: data collection in cyberspace. Anxiety Stress Coping **12**(2), 145–161 (1999)
2. Statistical Report on Internet Development in China (2018)

3. Chen, J.Y., Zheng, H.T., Xiao, X., Sangaiah, A.K., Jiang, Y., Zhao, C.Z.: Tianji: implementation of an efficient tracking engine in the mobile Internet era. IEEE Access **5**, 16592–16600 (2017)
4. Harari, G.M., Müller, S.R., Aung, M.S., Rentfrow, P.J.: Smartphone sensing methods for studying behavior in everyday life. Curr. Opin. Behav. Sci. **18**, 83–90 (2017)
5. Boonstra, T.W., Nicholas, J., Wong, Q.J., Shaw, F., Townsend, S., Christensen, H.: Using mobile phone sensor technology for mental health research: integrated analysis to identify hidden challenges and potential solutions. J. Med. Internet Res. **20**(7), e10131 (2018)
6. Gong, J., et al.: Understanding behavioral dynamics of social anxiety among college students through smartphone sensors. Inf. Fusion **49**, 57–68 (2019)
7. Drummond, H.E., Ghosh, S., Ferguson, A., Brackenridge, D., Tiplady, B.: Electronic quality of life questionnaires: a comparison of pen-based electronic questionnaires with conventional paper in a gastrointestinal study. Qual. Life Res. **4**(1), 21–26 (1995)
8. Pouwer, F., Snoek, F.J., Van Der Ploeg, H.M., Heine, R.J., Brand, A.N.: A comparison of the standard and the computerized versions of the Well-being Questionnaire (WBQ) and the Diabetes Treatment Satisfaction Questionnaire (DTSQ). Qual. Life Res. **7**(1), 33–38 (1997)
9. Velikova, G., et al.: Automated collection of quality-of-life data: a comparison of paper and computer touch-screen questionnaires. J. Clin. Oncol. **17**(3), 998 (1999)
10. Ryan, J.M., Corry, J.R., Attewell, R., Smithson, M.J.: A comparison of an electronic version of the SF-36 General Health Questionnaire to the standard paper version. Qual. Life Res. **11**(1), 19–26 (2002)
11. Carmines, E.G., Zeller, R.A.: Reliability and Validity Assessment, vol. 17. Sage Publications, Thousand Oaks (1979)
12. Nieuwenhuijsen, K., De Boer, A.G.E.M., Verbeek, J.H.A.M., Blonk, R.W.B., Van Dijk, F.J. H.: The depression anxiety stress scales (DASS): detecting anxiety disorder and depression in employees absent from work because of mental health problems. Occup. Environ. Med. **60**(Suppl 1), i77–i82 (2003)
13. Guo, Y., Hu, X., Hu, B., Cheng, J., Zhou, M., Kwok, R.Y.: Mobile cyber physical systems: current challenges and future networking applications. IEEE Access **6**, 12360–12368 (2017)
14. Hu, X., et al.: Emotion-aware cognitive system in multi-channel cognitive radio ad hoc networks. IEEE Commun. Mag. **56**(4), 180–187 (2018)
15. Beck, A.T., Ward, C.H., Mendelson, M., Mock, J., Erbaugh, J.: An inventory for measuring depression. Arch. Gen. Psychiatry **4**(6), 561–571 (1961)
16. Al-Anazi, A., Gates, I.D.: A support vector machine algorithm to classify lithofacies and model permeability in heterogeneous reservoirs. Eng. Geol. **114**(3-4), 267–277 (2010)
17. Cover, T., Thomas, M., Peter, E.: Nearest neighbor pattern classification. IEEE Trans. Inf. Theory **13**(1), 21–27 (1967)
18. Zhang, B., Srihari, S.N.: Fast k-nearest neighbor classification using cluster based trees. IEEE Trans. Pattern Anal. Mach. Intell. **26**(4), 525528 (2004)
19. Hart, P.: The condensed nearest neighbor rule (Corresp.). IEEE Trans. Inf. Theory **14**(3), 515–516 (1968)
20. Yu, X.-G., Yu, X.-P.: The research on an adaptive k-nearest neighbors classifier. In: 2006 International Conference on Machine Learning and Cybernetics, pp. 1241–1246. IEEE (2006)

Discrimination Model of QAR High-Severity Events Using Machine Learning

Junchen Li[1], Haigang Zhang[2], and Jinfeng Yang[2(✉)]

[1] Tianjin Key Lab for Advanced Signal Processing,
Civil Aviation University of China, Tianjin, China
[2] Institute of Applied Artificial Intelligence of the Guangdong-Hong Kong-Macao
Greater Bay Area, Shenzhen Polytechnic, Shenzhen 518055, China
zhg2018@sina.com

Abstract. The Quick Access Recorder (QAR) is an airborne equipment designed to store raw flight data, which contains a mass amount of safety related parameters such as flap angle, airspeed, altitude, etc. The assessment of QAR data is of great significance for the safety of civil aviation and the improvement of pilots skills. The existing QAR assessment approaches mainly utilizes the exceedance detection (ED) that relies on the pre-defined parameter threshold, which could miss potential flight risks. In this paper, we perform anomaly detection on the takeoff and landing phases based on an improved random forest (RF) method. The evaluation is performed on the dataset generated by a fleet of B-737NG, which shows that the method is able to discriminate the high-severity events accurately on the high dimensional multivariate time series, which also shows that the model can identify the events with potential risk pattern on the imbalanced dataset even if the event has not been pre-defined before.

Keywords: Flight data · Aviation safety · Anomaly detection · Random forest

1 Introduction

Quick Access Recorder (QAR) is an airborne equipment that can effectively store flight data and is designed to provide easy access to a removable medium. A QAR record various kinds of aircraft parameters such as pilots' operational related parameters, altitude of the aircraft and alarm information during a whole flight. Beyond providing an important basis for the air crash investigation and aircraft maintenance, the analysis of QAR data has received extensive attention for its' significance for the improvement of flight operation technology. In addition, flight management departments has immense demands for the macro safety analysis and fine technique management. Usually they interested in the anomaly patterns which are more likely to lead to flight incidents.

© Springer Nature Switzerland AG 2019
Z. Cui et al. (Eds.): IScIDE 2019, LNCS 11936, pp. 430–441, 2019.
https://doi.org/10.1007/978-3-030-36204-1_36

Currently, the aviation industry always evaluate the QAR data through the exceedance detection (ED). By using the related regulations and standard operational manual as a guidance, airlines can choose which flight parameters need to be monitored and the threshold of those parameters will be specified by domain expert. The list of the parameters is always chosen to coincide with the airline's standard operating procedures, such as the pitch angle, landing speed, landing gear retraction, etc. If a particular flight parameter exceed the threshold under a certain condition, it will be marked as an exceedance event and airlines will investigate the events to identify and correct the irregularities.

Over the years, ED method has been continually improved and is now trusted by the industry, but there are two main drawbacks that still greatly limit the fully utilize of QAR data. Firstly, the ED method can only detected events that already exist on the predetermined list [1], if the flights have potential risk pattern but do not on the list, the anomaly will not be identified, which may cause sever flight incidents. Secondly, flight data is considered as the highest level of confidentiality of airlines as it contains business information such as flight quality and the company's operation strategy, which poses a huge challenge for the academia to acquiring the real world flight data.

In this paper we propose a detection method based on supervised learning model. By training and testing our model on the QAR high-severity events dataset constructed by a airline's flight data, we believe the discrimination method can not only identify the existed exceedance events effectively, but also have the ability to discover the potential risks patterns. Finally an inspection of domain expert is conducted to the suspected anomalous flight, which will help to identify the source of the risk and provide technical guidance for the pilot to improve the operational techniques.

2   Related Works

Using QAR data for flight anomaly detection has also received extensive attention in the academia. A number of methods have been developed to detect high-risk flights through either the probabilistic and statistical approaches or unsupervised learning approaches. The probabilistic and statistical approaches focus on the distribution of data: Qi et al. [2] and Shao et al. [3] concentrate on the exceedance events in the landing phase. After learning the common statistical distributions around the data, they established a mathematical model to analyze the risk threshold of the operating parameters. Sun et al. [4] use the difference test to analyze the flight record data. By studying three kinds of events: climb speed high, descent rate high and long touch down from the QAR data, the K-W test model can be obtained, which can be used to analyze the difference between flight parameters and give guidance to safety management. Lei et al. [5] conducted a statistical analysis to the key parameters affecting the attitude of the aircraft, such as the steering column and the throttle, and identified the reasons for the long touchdown. Unsupervised learning approaches have also achieved good results: Li et al. [6] use density-based spatial clustering of applications to analyze the multi-time series of flights in the takeoff and landing phases, which can find the fixed proportion of outliers. Zheng et al. [7] innovatively introduced the dynamic time warping (DTW) algorithm into the analysis of QAR data and finally quantified the risk level of different flight operation modes.

In order to maximize the advantages of machine learning in QAR data analysis, we collected real world flight data from a large airline and build an already labeled QAR high-severity events (QHSE) dataset. With this dataset, we can make fully use the supervised learning method for the flight data analysis. In this paper, we develop a risk discrimination model based on random forest (RF), which can discriminate the risk flight pattern and identify the pilot's risk driving behavior. Our method has made the following contributions:

1. Establishing a QAR high severity events (QHSE) dataset: We collected a large amount of real world flight data from an airline. Datasets for takeoff (TKO) and final approach to landing (FNA-LND) are built respectively. Finally we labeled the QHSE dataset according to expert's opinion.

2. Developing a risk discrimination model: By training from the predetermined QHSE on the list, our method has the ability to identify potential risks with the prior knowledge in aviation field. Performing expert review on unknown risk events, we can identify the operational behavior that leads to the potential risk.

3   Methodology

The workflow of the QHSE discrimination model is described in Fig. 1: First we need to precisely define the flight phases to make the QHSE can be compared across different flights. In the second step, flight data of different phases will be transformed into a high dimensional space, where each vector is a snapshot of the selected flight parameters at the sampling time. The third step reduces the dimension of the matrix, and extract the optimal feature parameters to form the original feature vector (fv). The fourth step uses a random forest (RF) to discriminate the high-severity events and identify suspected potential risk patterns. Finally, the suspected flights will be reviewed to determine the abnormal operations.

3.1   Data Resampling

During the TKO and FNA-LNA, the aircrafts are in a high-density traffic condition around the airports, which make them the most risk-prone phase during the whole flight [8]. In order to analyze the flight data of the above two phases more efficiently, the data of the above two phases should be accurately extracted from the entire data frame. Further more the flights in the dataset always have different flight duration, which may lead the inconsistent lengths of data frame from different flights. In order to make the data can be compared directly, they must have the same dimensions and the corresponding flight parameters should be the same. According to the guidance of the experts, we identified the reference points for the two phases respectively.

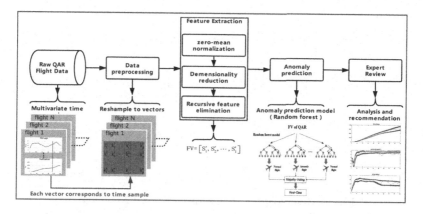

Fig. 1. Workflow of the high-severity events discrimination model.

For TKO phase, the point which pitch angle changes significantly is taken as the takeoff point. Then the data frame will be resampled at a fixed time interval from the takeoff point (Fig. 2), there are $t1$ sampling points in TKO phase.

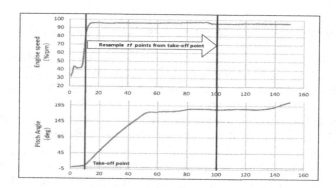

Fig. 2. Resample the time series in TKO phase

Correspondingly, for the FNA-LND phase, the point of speed brake application is taken as the touchdown point. Then the takeoff data is resampled at a same fixed time interval backtracking from touchdown point (Fig. 3), there are $t2$ sampling points in FNA-LND phase.

To ensure data of different flights have equal length, each sample of data from flight f at time t is represented by a vector as a form of:

$$k_f^t = \left[k_1^t, k_2^t, \ldots, k_p^t\right] \tag{1}$$

where k_p^t means the value of the pth flight parameter at time t.

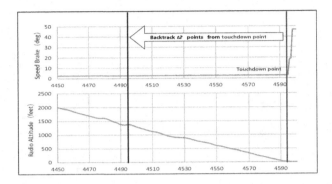

Fig. 3. Resample the time series in FNA-LND phase

Since there are a large number of different types and units of parameters in raw data, the vector k_f^t must be normalized to the vector $\hat{k}_f^t$. Then the original feature vectors (fv) will be constructed by $\hat{k}_f^t$, as a form of:

$$fv = \left[\hat{k}_f^1, \hat{k}_f^2, \ldots, \hat{k}_f^m\right] \tag{2}$$

3.2   Dimensionality Reduction and Feature Elimination

Extracting key features in high-dimensional data is an important step in anomaly detection, not only for the reduction of the computational complexity but also the improvement of accuracy of the model. A mass amount of previous experiments [9] show that training with all features often leads to high error rate while utilizing only the key features can achieve a better model performance. As a typical high-dimensional multivariate time series, the vector of QAR data from the resample stage usually have tens thousands of dimensions. Under the guidance of experts, in this paper, we set the sampling points $t1 = 90$ for the TKO phase and there are 118 flight parameters need to be focused on. Correspondingly, we set the sampling points $t2 = 100$ for the FNA-LND phase and we should pay attention to 127 flight parameters. The data dimensions of TKO phase are 10600 $(118 * 90)$ and for FNA-LND phase are 12700 $(127 * 100)$. It is necessary to introduce a feature selection method, which can effectively reduce the data dimension and select key feature in the fv.

We adopt an accelerated recursive feature elimination based on principal component analysis (PCA-RFE) to select the key features. When an principal component analysis (PCA) [10] is used to reduce the dimension of the original ultra-high dimensional data, thus greatly accelerate the calculation speed of the model. Given a data set of $N * D$ dimensions, the PCA aims to find the lower dimensional linear subspace D' of dimension D. Where D' is composed of a set of linear uncorrelated variables called principal components, which are arranged in descending order of variance in the original data. In this study, we choose to capture 90% of the variance in the data. In the case where the above criteria are used, after performing the PCA, the number of dimensions was reduced from 10620 to 107 for the TKO phase and from 12700 to 115 for the FNA-LND phase.

In a PCA-RFE method, recursive feature elimination (RFE) is used as an feature selector [11], which consists of three steps: (1) train the classifier; (2) compute the ranking criterion; (3) remove the features with smallest ranking scores. After the dimensionality reduction, we can select the feature subsets more efficiently by RFE and we used the Cross-validation method to make sure the optimal feature subset is determined by the ranking scores. A 10-fold cross-validation is accurate enough to be used here. After PCA-RFE, the number of dimensions was reduced from 107 to 36 for the TKO phase and from 115 to 48 for the FNA-LND phase.

3.3 Discrimination Method Based on RF

By analyzing the samples in the datasets, we found that there are only a small percentage of high-severity events, just about 5%, which means it is a typical imbalanced problem [12]. If the original data is used for training, the classification algorithm will be biased toward the category with a large number.

We use the synthetic minority oversampling technique (SMOTE) algorithm to reconstruct the training dataset, which is an oversampling method for unbalanced datasets. Traditional random oversampling methods always lead to overfitting, which can be greatly improved by synthetic minority oversampling. By inserting re-constructed new samples between the space of closely related minority samples. SMOTE can make the dataset achieve the predefined balance ratio.

Further more a machine learning method that are robust enough must be employed. In this paper the high-severity events discrimination model is based on the random forest (RF). RF has the advantages of computational efficiency. In RF, the decision tree is used as the basic classifier, then the classification results of multiple decision trees will be combined to form the final classification result. As is shown in Fig. 4, if more decision trees vote for normal flight, then the prediction result of the random forest is the majority-voting.

4 Experimental Results

We evaluate our method on a dataset that generated by a fleet of Boeing 737NG of a real world airline. When a flight is identified as an abnormal pattern but does not exist on the QHSE list, it will be reviewed by the experts to confirm whether they represent any safety risks, then the abnormal behavior must be investigated.

4.1 Evaluation Indicators

Table 1 is the confusion matrix for the 2-class (binary) classification problem. The labels for the majority classes and majority classes are positive and negative. TP and TN are the number of samples of the minority and majority classes that are

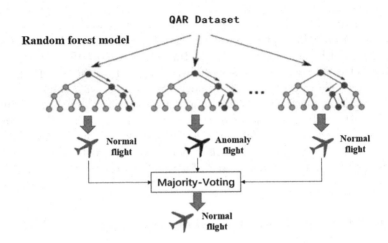

Fig. 4. Diagram of RF based high-severity events discrimination model

Table 1. Confusion matrix of binary classification model.

	Classified positive	Classified negative
Positive	TP	FN
Negative	FP	TN

correctly classified; FN and FP are the number of the minority and majority of the misclassification.

The evaluation indicators of the classifier are given by the following definitions.

Definition 1. Sensitivity (TPR)

$$TPR = \frac{TP}{TP + FN} \qquad (3)$$

The sensitivity of a test is also called the true positive rate (TPR), which means the true positive is the correct identification of anomalous data as such, e.g., classifying as "abnormal" data which is in fact abnormal.

Definition 2. Specificity (TNR)

$$TNR = \frac{TN}{TN + FP} \qquad (4)$$

The specificity of a test is also called the true negative rate (TNR), e.g., classifying as "normal" data which is in fact normal flight.

Definition 3. FPR

$$FPR = \frac{FP}{TN + FP} \qquad (5)$$

The false positive rate (FPR), which means the incorrect identification of normal data as such, i.e. classifying as "normal" data which is in fact abnormal.

In general, a combination of sensitivity and specificity can be used to evaluate the classification model's performance for a few classes in the data. When using the True Positive Rate (TPR) as the ordinate and the False Positive Rate (FPR) as the abscissa, we can plot the receiver operating characteristic (ROC) curve [8]. As shown in Fig. 1, the ROC curve of a good classification model should be as close as possible to the upper left corner of the graph.

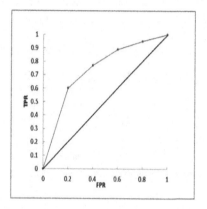

Fig. 5. Diagram of ROC curve

The classification model is usually evaluated by the area under the curve of ROC (AUC). If one classification model is better than the other, its AUC area is larger. Since the evaluation index AUC is not biased to any class, it reflects the general performance of the model. In the following experiment, we chose sensitivity, specificity and AUC as indicators to evaluate the performance of the algorithm.

4.2   QHSE Dataset

The safety of the aircraft is the result of a combination of external factors such as the pilot's basic driving skills, configuration of crew members, atmospheric conditions, aircraft maintenance status and so on. No matter how the influencing factors change, it will eventually reflected as a result of aircraft attitude and kinematic parameters.

According to the feature extraction method mentioned above, we extracted the FV from the raw QAR data. The data was generated by a large airline's Boeing 737NG fleet, of which 6107 flights are in the TKO phase and 4130 flights are in the FNA-LND phase. The aim of takeoff and landing is to let the aircraft takeoff and touchdown steady and smoothly, we focused on analyzing the most risk related events during both two phases. Finally, 22 high-severity events are selected and calculated as shown in Table 2 based on python programs.

Table 2. The most common Exceedance Events in TKO & FNA-LND.

Classification	Events	Snapshot in QAR	Unit
Kinematics related events	High rotate speed	Max vertical g	g
	High acceleration in flight	Longitudinal acceleration	g
	High liftoff pitch	Pitch angle	deg
	Early/Late approach pitch	Pitch angle	deg
	Low/High LND speed	Groundspeed	knots
	High touchdown acceleration	Max vertical g	g
	Localizer/Glideslope Dev	Localizer/Glideslope deviation	dots
Operational related events	Early rotation	Max vertical g	g
	Heading dev	Localizer deviation	dots
	Late /Early flare	Flare time	sec
	Long/Short touchdown	Touchdown distance	feet
	Bounced landing	Max vertical g	g
Configuration related events	Excessive takeoff weight	Gross weight	ton
	Early landing gear retraction	Compute air speed	knots
	Late thrust reduction	Throttle angle	degs
	Late landing flap setting	Flap position	degs
	Speedbrake not armed	Brake handle position	degs

In this paper, QHSE refers to two situations:

1. Flights with certain high-risk patterns, which has already included on the list of predefined events. 2. Flights may contain unknown high-risk patterns that do not exist on the list.

All the events in the above tables have been labeled as high-severity positive in the dataset and vice versa. The sample distribution of the QHSE dataset is shown in Table 3.

Table 3. Sample distribution of the QHSE dataset.

Dataset	Number of all samples	High-severity negative	High-severity positive	Sample ratio
Original TKO	6107	5819	288	1:20.2
Original FNA-LND	4130	3985	145	1:27.5

4.3   Results Overview

After performing up-sample to the unbalanced training dataset with a specified sample ratio, four sets of experiments were performed according to different balance ratios. The results are shown in Table 4.

Table 4. Evaluation results of improved imbalanced data.

Dataset	Sample ratio	Sensitivity	Specificity	AUC
Original TKO test	1:20.2	0.84	0.99	0.94
Up-sampled TKO test	1:4.3	0.95	0.96	0.95
Original FNA-LND test	1:27.5	0.89	0.99	0.95
Up-sampled FNA-LND test	1:5.6	0.94	0.95	0.96

Comparing the original imbalanced dataset with the up-sampled dataset that reduces numerical difference between the positive and negative samples, the classification performance of the model is improved, especially in the detection accuracy of the minority samples. Although the specificity has decreased slightly but the sensitivity of the model has increased significantly, and the AUC value has also been greatly improved.

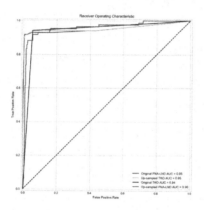

Fig. 6. ROC curve of classifiers

As is shown in Fig. 6 the discrimination model has obtained a good ROC curve,the AUC for TKO has reached 0.95 and the AUC for FNA-LND has reached 0.96, which means that our method can accurately classify the high-severity events in practical practice. After the above analysis, a small number of events in two phases are predicted as high-severity events but do not on the existing events list, which will be considered as a potential risk flight and will be further reviewed by the domain expert for examination.

4.4   Examples of Anomalous Flights

In this section we take two risky suspected flight as the example to explore the ability of the discrimination model.As is shown in Fig. 7, the red line indicates the anomaly flights. The dark blue band represents the 25th to 75th percentile of all flight data; the light blue band contains the 5th to 95th percentile. Respectively, the dark blue band contains 50% of the data, while the light blue band covers 90% of the flights.

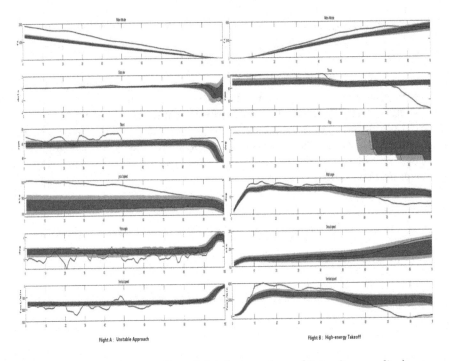

Fig. 7. Example anomalous flights in two phases (Color figure online)

The example of anomaly landing is shown in the left column, flight A is reviewed as operationally abnormal by domain experts. Within 50 s before landing, the aircraft's thrust is higher than most flights and continues to be unstable; the pitch angle is in a state of large fluctuations before landing. The vertical speed come to more than 1000 feet/min in 75 s before landing, which may lead to a hard landing event. Part of the domain experts believe that this landing does not meet the criteria of a stable approach and requires a go-around, which is a typical unstable approach mode.

The example of anomaly take off is shown in the right column, flight B is rated as operationally abnormal by domain experts. The flight has a similar flap configuration with other flights, it maintains a higher thrust and pitch angle from the take off point, which results a high take off energy during the TKO phase. Although the thrust has decreased dramatically since 45 s, the altitude profile is still higher than the majority until 75 s after takeoff. Experts state that the frequent operational changes in thrust and pitch make the vertical speed changes too fast, which greatly increases the acceleration in the cabin.

It is a typical high-energy take-off mode, which may further lead to potential serious consequences.

5   Conclusion

By using the traditional ED method to generate a QHSE dataset, a random forest based method is designed to discriminate the high-severity events using QAR data. Experiments show that the model can apply the prior knowledge to discriminate potential risk patterns using machine learning method. This method greatly improves the efficiency of experts in identifying flight anomalies and can locate the corresponding risk-prone characteristics, which are worthy of our attention. Preliminary evaluations show that the model is a promising approach for improve aviation safety, and as the amount of data increases in the QHSE dataset, the accuracy of the model will be further improved.

References

1. Guidance on the establishment of a Flight Data Analysis Program (FDAP) - Safety Management Systems (SMS)
2. Qi, M.: Astronautics: flight operations risk diagnosis method on quick-access-record exceedance. J. Beijing Univ. Aeronaut. Astronaut. **37**, 1207–1210 (2011)
3. Shao, X., Qi, M., Gao, M.: A risk analysis model of flight operations based on region partition. Kybernetes **41**, 1497–1508 (2012)
4. Sun, R.S., Han, W.L.: Technology: analysis on parameters characteristics of flight exceedance events based on distinction test. J. Saf. Sci. Technol. (2011)
5. Wang, L., Wu, C., Sun, R.: Pilot operating characteristics analysis of long landing based on flight QAR data. In: Harris, D. (ed.) EPCE 2013. LNCS, vol. 8020, pp. 157–166. Springer, Heidelberg (2013). https://doi.org/10.1007/978-3-642-39354-9_18
6. Li, L., Gariel, M., Hansman, R.J., Palacios, R.: Anomaly detection in onboard-recorded flight data using cluster analysis. In: Digital Avionics Systems Conference (2011)
7. Zheng, L., Hong, C., Shao, X.Y.: Pattern Recognition and Risk Analysis for Flight Operations (2017)
8. Janic, M.: An assessment of risk and safety in civil aviation. J. Air Transp. Manag. **6**, 43–50 (2000)
9. Granitto, P.M., Furlanello, C., Biasioli, F., Gasperi, F.: Recursive feature elimination with random forest for PTR-MS analysis of agroindustrial products. Chemom. Intell. Lab. Syst. **83**, 83–90 (2006)
10. Hotelling, H.: Analysis of a complex of statistical variables into principal components. J. Educ. Psychol. **24**, 417 (1933)
11. Kohavi, R., John, G.H.: Wrappers for feature subset selection. Artif. Intell. **97**, 273–324 (1997)
12. He, H., Garcia, E.A.: Learning from imbalanced data. Trans. Knowl. Data Eng. **21**, 1263–1284 (2009)

A New Method of Improving BERT for Text Classification

Shaomin Zheng and Meng Yang[✉]

School of Data and Computer Science, Sun yat-sen University,
Guangzhou, Guangdong, China
zhengshm8@mail2.sysu.edu.cn, yangm6@mail.sysu.edu.cn

Abstract. Text classification is a basic task in natural language processing. Recently, pre-training models such as BERT have achieved outstanding results compared with previous methods. However, BERT fails to take into account local information in the text such as a sentence and a phrase. In this paper, we present a BERT-CNN model for text classification. By adding CNN to the task-specific layers of BERT model, our model can get the information of important fragments in the text. In addition, we input the local representation along with the output of the BERT into the transformer encoder in order to take advantage of the self-attention mechanism and finally get the representation of the whole text through transformer layer. Extensive experiments demonstrate that our model obtains competitive performance against state-of-the-art baselines on four benchmark datasets.

Keywords: Text classification · Natural language processing · Deep neural network

1 Introduction

Text classification is one of the fundamental tasks of Natural Language Processing (NLP), with the goal of assigning text to different categories. The applications of text classification include sentiment analysis [1], question classification [2], and topic classification [3]. Today, deep learning-based approaches have become mainstream in text categorization, such as Convolutional Neural Networks (CNN) [4], Recurrent Neural Networks (RNN) [5] or some more complicated methods.

The method using deep learning for text classification requires inputting text into a deep network to obtain a representation of the text, and then inputting the text representation into the softmax function to obtain the probability of each category. The CNN-based models [4,6,7] can obtain representations of text with local information. RNN-based models [8,9] can obtain representations of text with long-term information. Therefore, some methods are intended to be modeled by combining the advantages of CNN and RNN such as C-LSTM [10], CNN-LSTM [11] and DRNN [12].

© Springer Nature Switzerland AG 2019
Z. Cui et al. (Eds.): IScIDE 2019, LNCS 11936, pp. 442–452, 2019.
https://doi.org/10.1007/978-3-030-36204-1_37

On this basis, some models use attention mechanisms to enable the model to focus on key information in the text. For example, the HAN [13] model adopts a hierarchical attention mechanism to divide the text into two levels of sentences and words, and uses the bidirectional RNN as encoder. DCCNN [14] first uses a multi-layer CNN to capture representations of different n-gram feature, and then uses the attention mechanism to obtain representations by selecting the more important feature. MEAN [15] tries alleviate the problem by integrating three kinds of sentiment linguistic knowledge into the deep neural network via attention mechanisms. DiSAN [16] is a novel attention mechanism in which the attention between elements from input sequences is directional and multi-dimensional. There are also some models that use the attention mechanism as the main means. For example, Bi-BloSAN [17] proposes the Block Self-Attention mechanism as the encoder for text, and also uses gate network to extract feature. These models all use the attention mechanism to select more important feature, which is more in line with people's observation mode than the traditional max-pooling and mean-pooling.

In addition, language model pre-training has shown to be effective for learning universal language representations by leveraging large amounts of unlabeled data. Some of the most outstanding examples are ELMo [18], GPT [19], ULMFiT [20] and BERT [21]. These are neural network language models trained on text data using unsupervised objectives. For example, BERT is based on a multi-layer bidirectional Transformer, and is trained on plain text for masked word prediction and next sentence prediction tasks. In order to apply a pre-trained model to specific tasks, we need to fine-tune them using task-specific training data and design additional task-specific layers after pre-training module. For example, to perform text categorization tasks, BERT adds a simple software layer after the pre-trained model and can be fine-tuned this way to create state-of-the-art models for text classification tasks on some datasets.

The BERT model performs well in text classification tasks due to its language comprehension capabilities. However, it does not focus on the information of some fragments or phrases in the text. To solve the problem, in this paper, we propose a BERT-CNN model for text classification. Our model have two advantages. One is that our model can obtain local information in text more effectively by adopting CNN in task-specific layers of BERT. Through CNN, our model can obtain new text representation, which can focus on local information in text. Another is that we input the local representation along with the output of the BERT into the transformer encoder. This allows the use of the self-attention mechanism to make the final text representation focus on important segments of the text. The two characteristic enable our model to obtain competitive results on four text classification datasets. Especially on the SST-1 and SST-2 datasets, our model achieved state-of-the-art result. In addition, we also analyzed the combination of our model and BERT model.

2   Related Work

2.1   Deep Neural Networks

Recently, Deep Neural Networks have achieved good results in natural language processing. Recurrent Neural Networks (RNN), including Long Short-Term Memory (LSTM) and Gated Recurrent Units (GRU), is very suitable for processing words sequences. Several variants are also proposed, such as Tree-LSTM [8] and TG-LSTM [9]. CNN is also one of the popular Deep Neural Networks. VDCNN [6] tries to construct a deeper CNN for text classification. [7] adopted multiple filters with different window sizes to extract multi-scale convolutional features for text classification. DCNN [22] uses a dynamic k-max pooling mechanism. DPCNN [23] aims to deepen CNNs without increasing much computational cost. [24] presents a novel weight initialization method to improve the CNNs for text classification. LK-MTL [25] is a multi-task convolutional neural network with the Leaky Unit, which has memory and forgetting mechanism to filter the feature flows between tasks. Unlike the above methods, char-CNN [4] is characters-level model which encodes characters of text as input.

Naturally, some methods try to combine the CNN and RNN models in order to get the advantages of both. C-LSTM [10] first uses CNN to capture local information of the text, and then uses the LSTM network to encode every output of the convolution kernel to capture global information. CNN-RNN [11] also uses a similar structure, but the difference between the two models is that the CNN layer is connected to the RNN layer differently. DRNN [12] uses RNN unit to replace the convolution kernel. In fact, it uses the structure of CNN and the encoding of RNN.

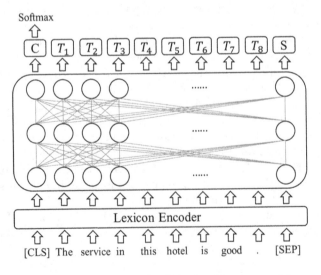

Fig. 1. The structure of BERT classification model.

2.2   Pre-training Model

Recently, similar to computer vision research, the pre-training model has achieved very good results in multiple tasks of natural language processing. They usually learn universal language representations by leveraging large amounts of unlabeled data and adopt additional task-specific layers after pre-training module for different tasks. ELMo [18] is devoted to extract context-sensitive features from a language model. It advances the state-of-the-art for several major NLP benchmarks including question answering [26], sentiment analysis [1], and named entity recognition [27]. GPT [19] and ULMFiT [20] pre-train some model architecture on a LM objective before fine-tuning that same model for a supervised downstream task such as text classification. BERT [21] is based on a multi-layer bidirectional Transformer, and is trained on plain text for masked word prediction and next sentence prediction tasks.

Figure 1 is the structure of BERT when it performs text categorization tasks. [CLS] and [SEP] in the input are beginning and ending marks of a sentence. The lexicon encoder will output the sum of the token embeddings, the segmentation embeddings and the position embeddings. Through multi-layer self-attention mechanism (transformer encoder) in the box, for each input token, there will be corresponding output value. Among them, C is the representation of the whole text because it obtains the information of all words. Finally, we input C to the softmax layer to get the classification results. In fact, C pays attention to the importance of every word in the text, and every word is equal and independent. However, it does not pay attention to the information of some fragments or phrases in the text.

3   BERT-CNN Model

Our proposed BERT-CNN model has two characteristics. One is to use the CNN to transform the task-specific layer of the BERT so that our model can obtain the local representation of the text. The other is that after the CNN layer, we input the local representation along with the output C of the BERT into the transformer encoder. This allows the use of the self-attention mechanism to make the final text representation focus on important segments of the text.

The architecture of the BERT-CNN model is shown in Fig. 2. Unlike BERT, which directly uses C as the representation of text, our model uses CNN and a layer of transformer encoder at task-specific layers. In what follows, we elaborate on the model in detail.

3.1   Lexicon Encoder

The input of ours model is a sequence of l tokens, $X = \{x_1, x_2, x_3, ..., x_l\}$. Since we consider text as a sentence, the first token is [CLS] and the last token is [SEP]. Following BERT [21], the input representation is constructed by summing the corresponding token, segment and position embeddings.

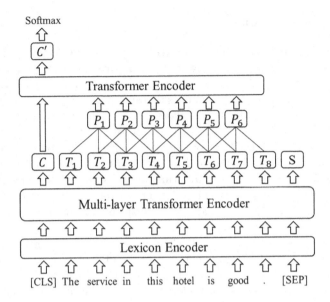

Fig. 2. The structure of BERT-CNN.

3.2   Multi-layer Transformer Encoder

We use a multi-layer bidirectional Transformer encoder [28] to map the input representation into a sequence of contextual embedding vectors $C = \{c, T, s\}, C \in \mathbb{R}^{d \times l}$. c and s are the are contextual representations corresponding to [CLS] and [SEP], respectively. $T = \{T_1, T_2, T_3, ..., T_m\}$ is the are contextual representations corresponding to real meaning tokens and $m = l - 2$.

3.3   Local CNN Encoder

In order to make the representation of text focus on local information in the text, such as a short sentence and phrase. We use a convolution filter to extract features from T. Assume that the size of the convolution window is $1 \times k$, the output of the local CNN encoder is:

$$P = CNN_k(T), P \in \mathbb{R}^{d \times (m-k+1)} \tag{1}$$

P is $\{P_1, P_2, P_3, ..., P_n\}$ and $n = m - k + 1$. Through convolution operation, we can obtain the representation of phrases in windows with size K. For example, P_n is the representation of $\{T_{n-1}, T_n, T_{n+1}\}$.

3.4   Transformer Encoder

Similar to multi-layer transformer encoder, to integrate information in P, we adopt transformer encoder to map the local representation P into the representation of whole text. The input of the transformer encoder is $\{c, p_1, p_2, ..., p_n\}$

and we use c' which corresponds to c as the representation of the whole text. Obviously, through the transformer encoder, c' gets independent token information from c and local information of text from P.

3.5 Output Layer

The output of the model is represented as follows:

$$\widehat{y} = softmax(tanh(Wc' + b)) \tag{2}$$

We use cross entropy as the loss function:

$$L = \sum_i y_i \log \widehat{y}_i \tag{3}$$

The y_i and $\widehat{y}_i$ are the labels and outputs of the ith text respectively.

4 Experiments

4.1 Data Sets

In order to evaluate the effectiveness of our model, we used four data sets of text categories for experimentation. These data sets include sentiment analysis and topic classification. The detailed statistics of all the datasets are listed in Table 1.

- **AG's news** [4]: This is a news data set for topic classification, which includes 4 news categories, *world, sports, business, sci/tech.*
- **SST-1** [29]: Stanford sentiment treebank a data set for sentiment analysis consisting of about 10K movie reviews with 5 classes i.e., *very negative, negative, neutral, positive, very positive.*
- **SST-2** [29]: This is another version of the Stanford sentiment treebank, which removes the reviews with the neutral category from SST-1, and then regards *very negative* and *negative* as *negative, very positive* and *positive* as *positive.* So it has two categories.
- **Yelp Review Full** [4]: This is a user review data set for sentiment classification. There are five levels of ratings from 1 to 5 (higher is better).

4.2 Experiments Setup

Our implementation of BERT-CNN is based on the TensorFlow implementation of BERT (https://github.com/google-research/bert). Due to the limited computational performance, we choose the *"BERT-Base, Uncased: 12-layer, 768-hidden, 12-heads, 110M parameters"* pre-training model as the BERT part of ours model.

Table 1. Statistics of the four datasets for text classification.

Datasets	Target	Train size	Test size	Classes	Avg. length
AG's news	Topic	120K	7.6K	4	45
SST-1	Sentiment	8544	2210	5	19
SST-2	Sentiment	6920	1821	2	19
Yelp-F	Sentiment	650K	50K	5	158

We set the batch size to 32 and the max length of sequence to 128 for AG's news, SST-1 and SST-2 dataset. Because the text length of Yelp-F dataset is large, we set the batch size to 8 and the max length of sequence to 256 due to GPU memory limitations. We set the size of CNN window k is 7 for AG's news, SST-1 and SST-2 dataset and set the size is 15 for Yelp-F. We use a dropout operation [30] after the multi-layer transformer encoder and the local transformer encoder. Its keep prob is set to 0.9. We used Adamax [31] as our optimizer with a learning rate of 1e−5. The maximum number of epochs was set to 5.

4.3   Baselines and Result

We compared our model with several genres of popular models: **char-CNN** [4], **fastText** [32], **VDCNN** [6], **DRNN** [12], **CNN-non-static+UNI** [24], **MEAN** [15], **Capsule Networks** [33] and **LK-MTL** [25]. As always, we use accuracy as an evaluation indicator.

Table 2. Experimental results for different model. The results of the baselines marked with * are re-printed from the references.

Model	AG's news	SST-1	SST-2	Yelp-F
char-CNN* [4]	87.2	–	–	62.0
fastText* [32]	92.5	–	–	63.9
VDCNN* [6]	91.3	–	–	64.7
DRNN [12]	93.6	47.3	86.4	65.3
CNN-non-static+UNI* [24]	–	50.8	89.4	–
MEAN* [15]	–	51.4	–	–
Capsule Networks* [33]	92.6	–	86.8	–
LK-MTL* [25]	–	49.7	88.5	–
BERT [21]	94.5	50.1	89.3	65.78
Our model	**94.9**	**51.7**	**90.1**	**66.13**

The experimental results are shown in Table 2. The experimental results of BERT model are obtained by using *"BERT-Base, Uncased: 12-layer, 768-hidden, 12-heads, 110M parameters"* pre-training model.

From the experimental results, we can find that our model performs well on short text datasets such as SST-1 and SST-2. This shows that through simple convolution operation, our model can obtain key local information in short text. But in long text, our model does not have a good advantage over ULMFiT model. This may be due to the large amount of interference information in the long text, our model can not extract the key local information well.

Compared with BERT model, our model performs better on four data sets, which proves that it is effective to use CNN to improve task-specific layers of BERT. One obvious benefit is that this approach achieves better results without sacrificing too much computational cost.

4.4   Further Analysis

Since our model is improved based on the BERT model, a natural idea is to combine the classification results of the two model. We suppose the model output of the BERT is $\widehat{y}_{bert}$, and the output of BERT-CNN is $\widehat{y}_{ours}$. Then we can get weighted summation of them:

$$\widehat{y} = w\widehat{y}_{ours} + (1 - w)\widehat{y}_{bert} \tag{4}$$

$w \in [0, 1]$ is the weight of $\widehat{y}_{ours}$. We use $\widehat{y}$ to test in SST-2 data sets. With the change of w, accuracy is fluctuating and shown in Fig. 3.

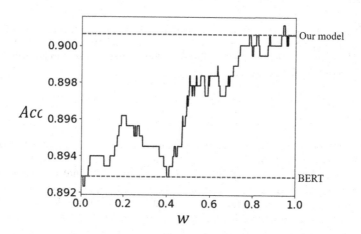

Fig. 3. The result of combining our model with BERT on the SST-2 dataset.

We can see that when $W = 0$, this is the result of BERT and when $W = 1$ this is the result of BERT-CNN. When w is $0.942 - 0.949$, we found that this combination method is more accurate than the BERT-CNN model and the accuracy is 0.9012. So this simple combination method is effective.

As shown in Table 3, we applied this approach to four datasets and achieved better results. When the accuracy is the highest, the value of W is in an interval. So we use the average of its interval.

Table 3. Experimental results for combination of BERT and BERT-CNN.

Model	AG's news	SST-1	SST-2	Yelp-F
BERT [21]	94.5	50.1	89.3	65.78
BERT-CNN	94.7	51.6	90.1	66.12
Combination	**95.0** $(W = 0.792)$	**51.8** $(W = 0.844)$	**90.3** $(W = 0.946)$	**66.31** $(W = 0.776)$

5   Conclusion

In this paper, we propose a new model called BERT-CNN. In order to get the information of important fragments in the text, we adopt CNN in task-specific layers of BERT. In addition, we use a transformer encoder layer to enable the final text representation to focus on key information in different fragments. Our model achieves good results on four data sets and the comparison with BERT model proves the availability of our idea.

Acknowledgement. This work is partially supported by the National Natural Science Foundation of China (Grant no. 61772568), the Guangzhou Science and Technology Program (Grant no. 201804010288), and the Fundamental Research Funds for the Central Universities (Grant no. 18lgzd15).

References

1. Maas, A.L., et al.: Learning word vectors for sentiment analysis. In: Proceedings of the 49th Annual Meeting of the Association for Computational Linguistics: Human Language Technologies, vol. 1. Association for Computational Linguistics (2011)
2. Zhang, D., Lee, W.S.: Question classification using support vector machines. In: Proceedings of the 26th Annual International ACM SIGIR Conference on Research and Development in Informaion Retrieval. ACM (2003)
3. Wang, S., Manning, C.D.: Baselines and bigrams: simple, good sentiment and topic classification. In: Proceedings of the 50th Annual Meeting of the Association for Computational Linguistics: Short Papers, vol. 2. Association for Computational Linguistics (2012)
4. Zhang, X., Zhao, J., LeCun, Y.: Character-level convolutional networks for text classification. In: Advances in Neural Information Processing Systems (2015)

5. Chung, J., et al.: Empirical evaluation of gated recurrent neural networks on sequence modeling. arXiv preprint arXiv:1412.3555 (2014)
6. Conneau, A., et al.: Very deep convolutional networks for text classification. arXiv preprint arXiv:1606.01781 (2016)
7. Kim, Y.: Convolutional neural networks for sentence classification. arXiv preprint arXiv:1408.5882 (2014)
8. Tai, K.S., Socher, R., Manning, C.D.: Improved semantic representations from tree-structured long short-term memory networks. arXiv preprint arXiv:1503.00075 (2015)
9. Huang, M., Qian, Q., Zhu, X.: Encoding syntactic knowledge in neural networks for sentiment classification. ACM Trans. Inf. Syst. (TOIS) **35**(3), 26 (2017)
10. Zhou, C., et al.: A C-LSTM neural network for text classification. arXiv preprint arXiv:1511.08630 (2015)
11. Xiao, Y., Cho, K.: Efficient character-level document classification by combining convolution and recurrent layers. arXiv preprint arXiv:1602.00367 (2016)
12. Wang, B.: Disconnected recurrent neural networks for text categorization. In: Proceedings of the 56th Annual Meeting of the Association for Computational Linguistics, Long Papers, vol. 1 (2018)
13. Yang, Z., et al.: Hierarchical attention networks for document classification. In: Proceedings of the 2016 Conference of the North American Chapter of the Association for Computational Linguistics: Human Language Technologies (2016)
14. Wang, S., Huang, M., Deng, Z.: Densely connected CNN with multi-scale feature attention for text classification. In: IJCAI (2018)
15. Lei, Z., et al.: A multi-sentiment-resource enhanced attention network for sentiment classification. arXiv preprint arXiv:1807.04990 (2018)
16. Shen, T., et al.: DiSAN: directional self-attention network for RNN/CNN-free language understanding. In: Thirty-Second AAAI Conference on Artificial Intelligence (2018)
17. Shen, T., et al.: Bi-directional block self-attention for fast and memory-efficient sequence modeling. arXiv preprint arXiv:1804.00857 (2018)
18. Peters, M.E., et al.: Deep contextualized word representations. arXiv preprint arXiv:1802.05365 (2018)
19. Radford, A., et al.: Improving language understanding by generative pre-training (2018). https://s3-us-west-2.amazonaws.com/openai-assets/research-covers/langu ageunsupervised/languageunderstandingpaper.pdf
20. Howard, J., Ruder, S.: Universal language model fine-tuning for text classification. arXiv preprint arXiv:1801.06146 (2018)
21. Devlin, J., et al.: Bert: pre-training of deep bidirectional transformers for language understanding. arXiv preprint arXiv:1810.04805 (2018)
22. Kalchbrenner, N., Grefenstette, E., Blunsom, P.: A convolutional neural network for modelling sentences. arXiv preprint arXiv:1404.2188 (2014)
23. Johnson, R., Zhang, T.: Deep pyramid convolutional neural networks for text categorization. In: Proceedings of the 55th Annual Meeting of the Association for Computational Linguistics, Long Papers, vol. 1 (2017)
24. Li, S., et al.: Initializing convolutional filters with semantic features for text classification. In: Proceedings of the 2017 Conference on Empirical Methods in Natural Language Processing (2017)
25. Xiao, L., et al.: Learning what to share: leaky multi-task network for text classification. In: Proceedings of the 27th International Conference on Computational Linguistics (2018)

26. Rajpurkar, P., et al.: SQuAD: 100,000+ questions for machine comprehension of text. arXiv preprint arXiv:1606.05250 (2016)
27. Sang, E.F., De Meulder, F.: Introduction to the CoNLL-2003 shared task: language-independent named entity recognition. arXiv preprint cs/0306050 (2003)
28. Vaswani, A., et al.: Attention is all you need. In: Advances in Neural Information Processing Systems (2017)
29. Socher, R., et al.: Recursive deep models for semantic compositionality over a sentiment treebank. In: Proceedings of the 2013 Conference on Empirical Methods in Natural Language Processing (2013)
30. Srivastava, N., et al.: Dropout: a simple way to prevent neural networks from overfitting. J. Mach. Learn. Res. **15**(1), 1929–1958 (2014)
31. Kingma, D.P., Ba, J.: Adam: a method for stochastic optimization. arXiv preprint arXiv:1412.6980 (2014)
32. Joulin, A., et al.: Bag of tricks for efficient text classification. arXiv preprint arXiv:1607.01759 (2016)
33. Zhao, W., et al.: Investigating capsule networks with dynamic routing for text classification. arXiv preprint arXiv:1804.00538 (2018)

Author Index

Printed in the United States
By Bookmasters